Advanced Engineering Dynamics

Advanced Engineering Dynamics

H. R. Harrison

Formerly Department of Mechanical Engineering & Aeronautics
City University
London

T. Nettleton

Formerly Department of Mechanical Engineering & Aeronautics
City University
London

ARNOLD

A member of the Hodder Headline Group
LONDON • SYDNEY • AUCKLAND

Copublished in North, Central and South America by
John Wiley & Sons Inc., New York • Toronto

First Published in Great Britain in 1997 by Arnold,
a member of the Hodder Headline Group,
338 Euston Road, London NW1 3BH

Copublished in North, Central and South America by
John Wiley & Sons, Inc., 605 Third Avenue,
New York, NY 10158–0012

British Library Cataloguing in Publication Data
A catalogue record for this book is available from the British Library

Library of Congress Cataloging-in-Publication Data
A catalog record for this book is available from the Library of Congress

ISBN 0 340 64571 7
ISBN 0 470 23592 6 (Wiley)

Typeset in 10/12pt Times by
J&L Composition Ltd, Filey, North Yorkshire
Printed and bound in Great Britain by
J W Arrowsmith Ltd, Bristol

This book is dedicated to Trevor Nettleton

Contents

Preface

The subject referred to as dynamics is usually taken to mean the study of the kinematics and kinetics of particles, rigid bodies and deformable solids. When applied to fluids it is referred to as fluid dynamics or hydrodynamics or aerodynamics and is not covered in this book.

The object of this book is to form a bridge between elementary dynamics and advanced specialist applications in engineering. Our aim is to incorporate the terminology and notation used in various disciplines such as road vehicle stability, aircraft stability and robotics. Any one of these topics is worthy of a complete textbook but we shall concentrate on the fundamental principles so that engineering dynamics can be seen as a whole.

Chapter 1 is a reappraisal of Newtonian principles to ensure that definitions and symbols are all carefully defined. Chapters 2 and 3 expand into so-called analytical dynamics typified by the methods of Lagrange and by Hamilton's principle.

Chapter 4 deals with rigid body dynamics to include gyroscopic phenomena and the stability of spinning bodies.

Chapter 5 discusses four types of vehicle: satellites, rockets, aircraft and cars. Each of these highlights different aspects of dynamics.

Chapter 6 covers the fundamentals of the dynamics of one-dimensional continuous media. We restrict our discussion to wave propagation in homogeneous, isentropic, linearly elastic solids as this is adequate to show the differences in technique when compared with rigid body dynamics. The methods are best suited to the study of impact and other transient phenomena. The chapter ends with a treatment of strain wave propagation in helical springs. Much of this material has hitherto not been published.

Chapter 7 extends the study into three dimensions and discusses the types of wave that can exist within the medium and on its surface. Reflection and refraction are also covered. Exact solutions only exist for a limited number of cases. The majority of engineering problems are best solved by the use of finite element and finite difference methods; these are outside the terms of reference of this book.

Chapter 8 forges a link between conventional dynamics and the highly specialized and distinctive approach used in robotics. The Denavit–Hartenberg system is studied as an extension to the kinematics already encountered.

Chapter 9 is a brief excursion into the special theory of relativity mainly to define the boundaries of Newtonian dynamics and also to reappraise the fundamental definitions. A practical application of the theory is found in the use of the Doppler effect in light propagation. This forms the basis of velocity measuring equipment which is in regular use.

There are three appendices. The first is a summary of tensor and matrix algebra. The second concerns analytical dynamics and is included to embrace some methods which are less well known than the classical Lagrangian dynamics and Hamilton's principle. One such approach is that known as the Gibbs–Appell method. The third demonstrates the use of curvilinear co-ordinates with particular reference to vector analysis and second-order tensors.

As we have already mentioned, almost every topic covered could well be expanded into a complete text. Many such texts exist and a few of them are listed in the Bibliography which, in turn, leads to a more comprehensive list of references.

The important subject of vibration is not dealt with specifically but methods by which the equations of motion can be set up are demonstrated. The fundamentals of vibration and control are covered in our earlier book *The Principles of Engineering Mechanics*, 2nd edn, published by Edward Arnold in 1994.

The author and publisher would like to thank Brüel and Kjaer for information on the Laser Velocity Transducer and SP Tyres UK Limited for data on tyre cornering forces.

It is with much personal sadness that I have to inform the reader that my co-author, friend and colleague, Trevor Nettleton, became seriously ill during the early stages of the preparation of this book. He died prematurely of a brain tumour some nine months later. Clearly his involvement in this book is far less than it would have been; I have tried to minimize this loss.

Ron Harrison
January 1997

1
Newtonian Mechanics

1.1 Introduction

The purpose of this chapter is to review briefly the assumptions and principles underlying Newtonian mechanics in a form that is generally accepted today. Much of the material to be presented is covered in more elementary texts (Harrison and Nettleton 1994) but in view of the importance of having clear definitions of the terms used in dynamics all such terms will be reviewed.

Many of the terms used in mechanics are used in everyday speech so that misconceptions can easily arise. The concept of force is one that causes misunderstanding even among those with some knowledge of mechanics. The question as to whether force is the servant or the master of mechanics often lies at the root of any difficulties. We shall consider force to be a useful servant employed to provide communication between the various aspects of physics. The newer ideas of relativity and quantum mechanics demand that all definitions are reappraised; however, our definitions in Newtonian mechanics must be precise so that any modification required will be apparent. Any new theory must give the same results, to within experimental accuracy, as the Newtonian theory when dealing with macroscopic bodies moving at speeds which are slow relative to that of light. This is because the degree of confidence in Newtonian mechanics is of a very high order based on centuries of experiment.

1.2 Fundamentals

The earliest recorded writings on the subject of mechanics are those of Aristotle and Archimedes some two thousand years ago. Although some knowledge of the principles of levers was known then there was no clear concept of dynamics. The main problem was that it was firmly held that the natural state of a body was that of rest and therefore any motion required the intervention of some agency at all times. It was not until the sixteenth century that it was suggested that straight line steady motion might be a natural state as well as rest. The accurate measurement of the motion of the planets by Tycho Brahe led Kepler to enunciate his three laws of planetary motion in the early part of the seventeenth century. Galileo added another important contribution to the development of dynamics by describing the motion of projectiles, correctly defining acceleration. Galileo was also responsible for the specification of inertia, which is a body's natural resistance to a change velocity and is associated with its mass.

Newton acknowledged the contributions of Kepler and Galileo and added two more axioms before stating the laws of motion. One was to propose that earthly objects obeyed the same laws as did the Moon and the planets and, consequently, accepted the notion of action at a distance without the need to specify a medium or the manner in which the force was transmitted.

The first law states

a body shall continue in a state of rest or of uniform motion in a straight line unless impressed upon by a force.

This repeats Galileo's idea of the natural state of a body and defines the nature of force. The question of the frame of reference is now raised. To clarify the situation we shall regard force to be the action of one body upon another. Thus an isolated body will move in a straight line at constant speed relative to an inertial frame of reference. This statement could be regarded as defining an inertial frame; more discussion occurs later.

The second law is

the rate of change of momentum is proportional to the impressed force and takes place in the same direction as the force.

This defines the magnitude of a force in terms of the time rate of change of the product of mass and velocity. We need to assume that mass is some measure of the amount of matter in a body and is to be regarded as constant.

The first two laws are more in the form of definitions but the third law which states that

to every action (force) there is an equal and opposite reaction (force)

is a law which can be tested experimentally.

Newton's law of gravity states that

the gravitational force of attraction between two bodies, one of mass m_1 *and one of mass* m_2*, separated by a distance* d*, is proportional to* $m_1 m_2/d^2$ *and lies along the line joining the two centres.*

This assumes that action at a distance is instantaneous and independent of any motion.

Newton showed that by choosing a frame of reference centred on the Sun and not rotating with respect to the distant stars his laws correlated to a high degree of accuracy with the observations of Tycho Brahe and to the laws deduced by Kepler. This set of axes can be regarded as an inertial set. According to Galileo any frame moving at a constant speed relative to an inertial set of axes with no relative rotation is itself an inertial set.

1.3 Space and time

Space and time in Newtonian mechanics are independent of each other. Space is three dimensional and Euclidean so that relative positions have unique descriptions which are independent of the position and motion of the observer. Although the actual numbers describing the location of a point will depend on the observer, the separation between two points and the angle between two lines will not. Since time is regarded as absolute the time between two events will not be affected by the position or motion of the observer. This last assumption is challenged by Einstein's special theory of relativity.

The unit of length in SI units is the metre and is currently defined in terms of the wavelength of radiation of the krypton-86 atom. An earlier definition was the distance between two marks on a standard bar.

The unit of time is the second and this is defined in terms of the frequency of radiation of the caesium-133 atom. The alternative definition is as a given fraction of the tropical year 1900, known as ephemeris time, and is based on a solar day of 24 hours.

1.4 Mass

The unit of mass is the kilogram and is defined by comparison with the international prototype of the kilogram. We need to look closer at the ways of comparing masses, and we also need to look at the possibility of there being three types of mass.

From Newton's second law we have that force is proportional to the product of mass and acceleration; this form of mass is known as inertial mass. From Newton's law of gravitation we have that force on body A due to the gravitational attraction of body B is proportional to the mass of A times the mass of B and inversely proportional to the square of their separation. The gravitational field is being produced by B so the mass of B can be regarded as an *active mass* whereas body A is reacting to the field and its mass can be regarded as *passive*. By Newton's third law the force that B exerts on A is equal and opposite to the force that A exerts on B, and therefore from the symmetry the active mass of A must equal the passive mass of A.

Let inertial mass be denoted by m and gravitational mass by μ. Then the force on mass A due to B is

$$F_{A/B} = \frac{\mu_A G \mu_B}{d^2} \tag{1.1}$$

where G is the *universal gravitational* constant and d is the separation. By Newton's second law

$$F_{A/B} = \frac{d}{dt}(m_A v_A) = m_A a_A \tag{1.2}$$

where v is velocity and a is acceleration.

Equating the expressions for force in equations (1.1) and (1.2) gives the acceleration of A

$$a_A = \frac{\mu_A}{m_A}\left(\frac{G\mu_B}{d^2}\right) = \frac{\mu_A}{m_A} g \tag{1.3}$$

where $g = G\mu_B/d^2$ is the *gravitational field strength* due to B. If the mass of B is assumed to be large compared with that of body A and also of a third body C, as seen in Fig. 1.1, we can write

$$a_C = \frac{\mu_C}{m_C}\left(\frac{G\mu_B}{d^2}\right) = \frac{\mu_C}{m_C} g \tag{1.4}$$

on the assumption that, even though A is close to C, the mutual attraction between A and C in negligible compared with the effect of B.

If body A is made of a different material than body C and if the measured free fall acceleration of body A is found to be the same as that of body C it follows that $\mu_A/m_A = \mu_C/m_C$.

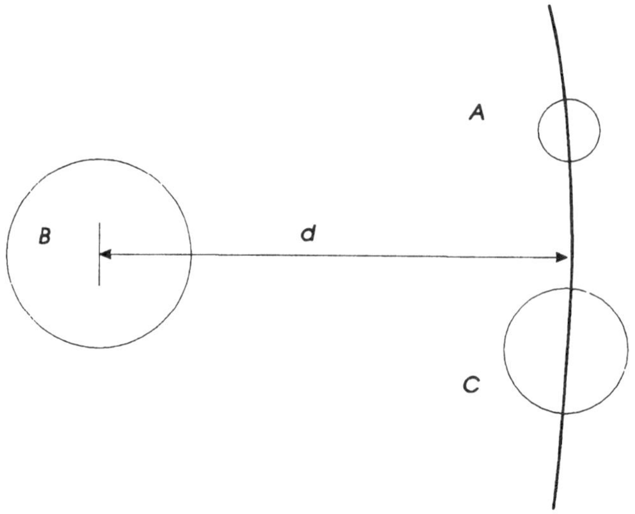

Fig. 1.1

More sophisticated experiments have been devised to detect any change in the ratio of inertial to gravitational mass but to date no measurable variation has been found. It is now assumed that this ratio is constant, so by suitable choice of units the inertial mass can be made equal to the gravitational mass.

The mass of a body can be evaluated by comparison with the standard mass. This can be done either by comparing their weights in a sensibly constant gravitational field or, in principle, by the results of a collision experiment. If two bodies, as shown in Fig. 1.2, are in colinear impact then, owing to Newton's third law, the momentum gained by one body is equal to that lost by the other. Consider two bodies A and B having masses m_A and m_B initially moving at speeds u_A and u_B, $u_A > u_B$. After collision their speeds are v_A and v_B. Therefore, equating the loss of momentum of A to the gain in momentum of B we obtain

$$m_A (u_A - v_A) = m_B (v_B - u_B) \tag{1.5}$$

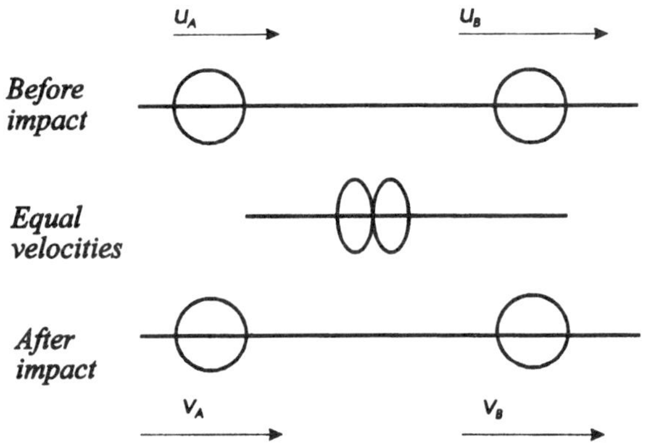

Fig. 1.2

so that

$$\frac{m_B}{m_A} = \frac{u_A - v_A}{v_B - u_B} \tag{1.6}$$

Thus if the mass of A is known then the mass of B can be calculated.

1.5 Force

We shall formally define force to be

the action of one body upon another which, if acting alone, would cause an acceleration measured in an inertial frame of reference.

This definition excludes terms such as inertia force which are to be regarded as fictitious forces. When non-inertial axes are used (discussed in later chapters) then it is convenient to introduce fictitious forces such as Coriolis force and centrifugal force to maintain thereby a Newtonian form to the equations of motion.

If experiments are conducted in a lift cage which has a constant acceleration it is, for a small region of space, practically impossible to tell whether the lift is accelerating or the local value of the strength of the gravity field has changed. This argument led Einstein to postulate the principle of equivalence which states that

all local, freely falling, non-rotating laboratories are fully equivalent for the performance of all physical experiments.

This forms the basis of the general theory of relativity but in Newtonian mechanics freely falling frames will be considered to be accelerating frames and therefore non-inertial.

1.6 Work and power

We have now accepted space, time and mass as the fundamental quantities and defined force in terms of these three. We also tacitly assumed the definitions of velocity and acceleration. That is,

velocity is the time rate of change of position and acceleration is the time rate of change of velocity.

Since position is a vector quantity and time is a scalar it follows that velocity and acceleration are also vectors. By the definition of force it also is a vector.

Work is formally defined as

the product of a constant force and the distance moved, in the direction of the force, by the particle on which the force acts.

If $\boldsymbol{F}$ is a variable force and $\mathrm{d}\boldsymbol{s}$ is the displacement of the particle then the work done is the integral of the scalar product as below

$$W = \int \boldsymbol{F} \cdot \mathrm{d}\boldsymbol{s} \tag{1.7}$$

Typical misuse of the definition of work is the case of a wheel rolling without slip. The tangential force at the contact point of the rim of the wheel and the ground does not do any work because the particle on the wheel at the contact point does not move in the direction of the force but normal to it. As the wheel rolls the point of application of the force moves along the ground but no work is done. If sliding takes place the work definition cannot be applied because the particle motion at the contact point is complex, a stick/slip situation occurring between the two surfaces. Also the heat which is generated may be passing in either direction.

Power is simply the rate of doing work.

1.7 Kinematics of a point

The position of a point relative to the origin is represented by the *free vector* $\boldsymbol{r}$. This is represented by the product of the *scalar magnitude*, r, and a *unit vector* $\boldsymbol{e}$

$$\boldsymbol{r} = r\boldsymbol{e} \tag{1.8}$$

Velocity is by definition

$$\boldsymbol{v} = \frac{\mathrm{d}\boldsymbol{r}}{\mathrm{d}t} = \frac{\mathrm{d}r}{\mathrm{d}t}\boldsymbol{e} + r\frac{\mathrm{d}\boldsymbol{e}}{\mathrm{d}t} \tag{1.9}$$

The change in a unit vector is due only to a change in direction since by definition its magnitude is a constant unity. From Fig. 1.3 it is seen that the magnitude of d$\boldsymbol{e}$ is $1 \cdot \mathrm{d}\theta$ and is in a direction normal to $\boldsymbol{e}$. The angle $\mathrm{d}\theta$ can be represented by a vector normal to both $\boldsymbol{e}$ and d$\boldsymbol{e}$.

It is important to know that finite angles are not vector quantities since they do not obey the parallelogram law of vector addition. This is easily demonstrated by rotating a box about orthogonal axes and then altering the order of rotation.

Non-vectorial addition takes place if the axes are fixed or if the axes are attached to the box. A full discussion of this point is to be found in the chapter on robot dynamics.

The change in the unit vector can be expressed by a vector product thus

$$\mathrm{d}\boldsymbol{e} = \mathrm{d}\boldsymbol{\theta} \times \boldsymbol{e} \tag{1.10}$$

Dividing by the time increment $\mathrm{d}t$

$$\frac{\mathrm{d}\boldsymbol{e}}{\mathrm{d}t} = \boldsymbol{\omega} \times \boldsymbol{e} \tag{1.11}$$

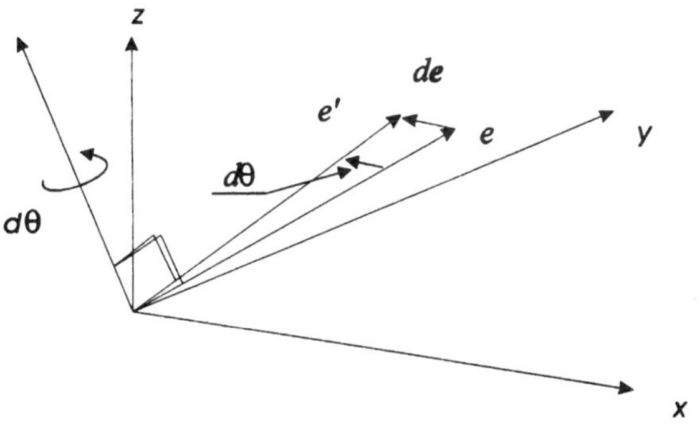

Fig. 1.3

where $\boldsymbol{\omega} = \mathrm{d}\theta/\mathrm{d}t$ is the angular velocity of the unit vector $\boldsymbol{e}$. Thus we may write $\boldsymbol{v} = \dot{\boldsymbol{r}} = \dot{r}\boldsymbol{e} + r(\boldsymbol{\omega} \times \boldsymbol{e}) = r\boldsymbol{e} + \boldsymbol{\omega} \times \boldsymbol{r}$. It is convenient to write this equation as

$$\boldsymbol{v} = \frac{\mathrm{d}\boldsymbol{r}}{\mathrm{d}t} = \frac{\partial \boldsymbol{r}}{\partial t} + \boldsymbol{\omega} \times \boldsymbol{r} \tag{1.12}$$

where the partial differentiation is the rate of change of $\boldsymbol{r}$ as seen from the moving axes. The form of equation (1.12) is applicable to any vector $\boldsymbol{V}$ expressed in terms of moving co-ordinates, so

$$\frac{\mathrm{d}\boldsymbol{V}}{\mathrm{d}t} = \frac{\partial \boldsymbol{V}}{\partial t} + \boldsymbol{\omega} \times \boldsymbol{V} \tag{1.13}$$

Acceleration is by definition

$$\boldsymbol{a} = \frac{\mathrm{d}\boldsymbol{v}}{\mathrm{d}t} = \dot{\boldsymbol{v}}$$

and by using equation (1.13)

$$\boldsymbol{a} = \frac{\partial \boldsymbol{v}}{\partial t} + \boldsymbol{\omega} \times \boldsymbol{v} \tag{1.14}$$

Using equation (1.12)

$$\boldsymbol{a} = \frac{\partial \boldsymbol{v}}{\partial t} + \omega \times \frac{\partial \boldsymbol{r}}{\partial t} + \omega \times (\omega \times \boldsymbol{r}) \tag{1.14a}$$

In Cartesian co-ordinates

Here the unit vectors, *see* Fig. 1.4, are fixed in direction so differentiation is simple

$$\boldsymbol{r} = x\boldsymbol{i} + y\boldsymbol{j} + z\boldsymbol{k} \tag{1.15}$$

$$\boldsymbol{v} = x\boldsymbol{i} + y\boldsymbol{j} + z\boldsymbol{k} \tag{1.16}$$

and

$$\boldsymbol{a} = \ddot{x}\boldsymbol{i} + \ddot{y}\boldsymbol{j} + \ddot{z}\boldsymbol{k} \tag{1.17}$$

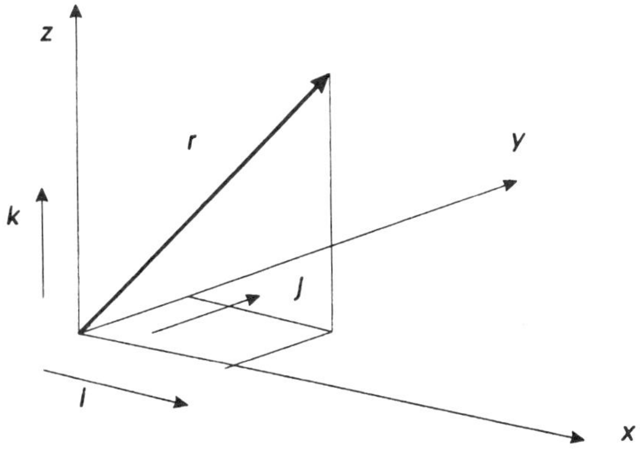

Fig. 1.4 Cartesian co-ordinates

In cylindrical co-ordinates

From Fig. 1.5 we see that

$$\boldsymbol{r} = R\boldsymbol{e}_R + z\boldsymbol{k} \tag{1.18}$$

and

$$\boldsymbol{\omega} = \dot{\theta}\boldsymbol{k}$$

so, using equation (1.12),

$$\begin{aligned}\boldsymbol{v} &= (\dot{R}\boldsymbol{e}_R + \dot{z}\boldsymbol{k}) + \boldsymbol{\omega} \times (R\boldsymbol{e}_R + z\boldsymbol{k}) \\ &= \dot{R}\boldsymbol{e}_R + \dot{z}\boldsymbol{k} + R\dot{\theta}\boldsymbol{e}_\theta \\ &= \dot{R}\boldsymbol{e}_R + R\dot{\theta}\boldsymbol{e}_\theta + \dot{z}\boldsymbol{k}\end{aligned} \tag{1.19}$$

Differentiating once again

$$\begin{aligned}\boldsymbol{a} &= \ddot{R}\boldsymbol{e}_R + \dot{R}\dot{\theta}\boldsymbol{e}_\theta + R\ddot{\theta}\boldsymbol{e}_\theta + \ddot{z}\boldsymbol{k} + \boldsymbol{\omega} \times \boldsymbol{v} \\ &= \ddot{R}\boldsymbol{e}_R + \dot{R}\dot{\theta}\boldsymbol{e}_\theta + R\ddot{\theta}\boldsymbol{e}_\theta + \ddot{z}\boldsymbol{k} + \dot{\theta}(\dot{R}\boldsymbol{e}_\theta - R\dot{\theta}\boldsymbol{e}_R) \\ &= (\ddot{R} - R\dot{\theta}^2)\boldsymbol{e}_R + (R\ddot{\theta} + 2\dot{R}\dot{\theta})\boldsymbol{e}_\theta + \ddot{z}\boldsymbol{k}\end{aligned} \tag{1.20}$$

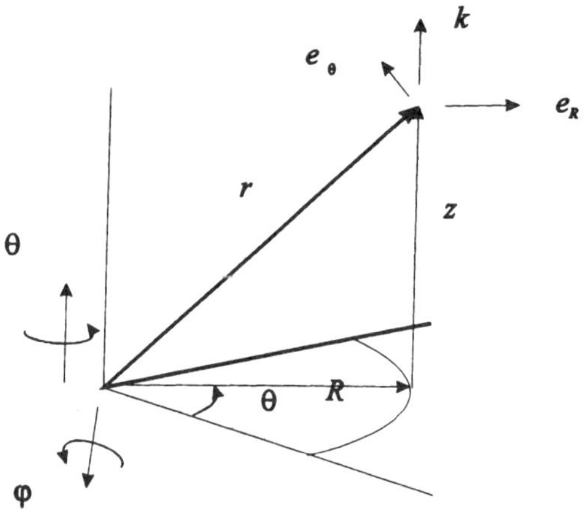

Fig. 1.5 Cylindrical co-ordinates

In spherical co-ordinates

From Fig. 1.6 we see that ω has three components

$$\boldsymbol{\omega} = \dot{\theta}\sin\theta\, \boldsymbol{e}_r - \dot{\phi}\boldsymbol{e}_\theta + \dot{\theta}\cos\phi\, \boldsymbol{e}_\phi$$

Now

$$\boldsymbol{r} = r\boldsymbol{e}_r \tag{1.21}$$

Therefore

$$\begin{aligned}\boldsymbol{v} &= \dot{r}\boldsymbol{e}_r + \boldsymbol{\omega} \times \boldsymbol{r} \\ &= \dot{r}\boldsymbol{e}_r + \dot{\phi}r\boldsymbol{e}_\phi + \dot{\theta}\cos\phi\, r\boldsymbol{e}_\theta \\ &= \dot{r}\boldsymbol{e}_r + r\dot{\theta}\cos\phi\, \boldsymbol{e}_\theta + r\dot{\phi}\, \boldsymbol{e}_\phi\end{aligned} \tag{1.22}$$

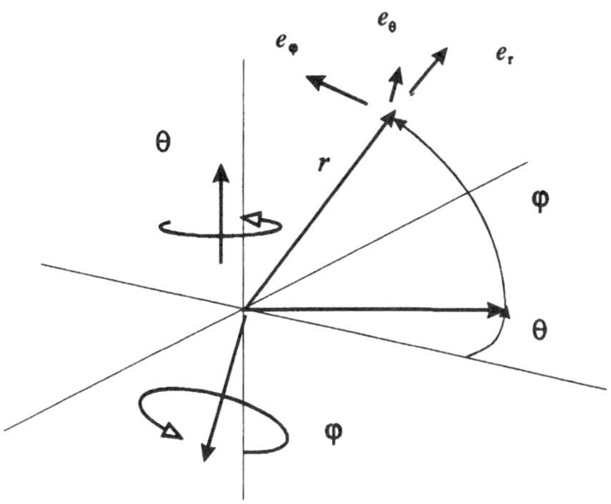

Fig. 1.6 Spherical co-ordinates

Differentiating again

$$
\begin{aligned}
\boldsymbol{a} &= \frac{\partial \boldsymbol{v}}{\partial t} + \boldsymbol{\omega} \times \boldsymbol{v} \\
&= \ddot{r}\boldsymbol{e}_r + (\dot{r}\dot{\theta}\cos\o + r\ddot{\theta}\cos\o - r\dot{\theta}\sin\o\,\dot{\o})\boldsymbol{e}_\theta + (\dot{r}\dot{\o} + r\ddot{\o})\boldsymbol{e}_\o \\
&\quad + \begin{vmatrix} \boldsymbol{e}_r & \boldsymbol{e}_\theta & \boldsymbol{e}_\o \\ \dot{\theta}\sin\o & -\dot{\o} & \dot{\theta}\cos\o \\ \dot{r} & r\dot{\theta}\cos\o & r\dot{\o} \end{vmatrix} \\
&= (\ddot{r} - r\dot{\o}^2 - r\dot{\theta}^2\cos\o)\boldsymbol{e}_r \\
&\quad + (\dot{r}\dot{\theta}\cos\o + r\ddot{\theta}\cos\o - r\dot{\theta}\sin\o\,\dot{\o} - r\dot{\o}\dot{\theta}\sin\o + \dot{r}\dot{\theta}\cos\o)\boldsymbol{e}_\theta \\
&\quad + (\dot{r}\dot{\o} + r\ddot{\o} + r\dot{\theta}^2\sin\o\cos\o + \dot{r}\dot{\o})\boldsymbol{e}_\o \\
&= (\ddot{r} - r\dot{\o}^2 - r\dot{\theta}^2\cos\o)\boldsymbol{e}_r \\
&\quad + (2\dot{r}\dot{\theta}\cos\o + r\ddot{\theta}\cos\o - 2r\dot{\theta}\dot{\o}\sin\o)\boldsymbol{e}_\theta \\
&\quad + (2\dot{r}\dot{\o} + r\ddot{\o} + r\dot{\theta}^2\sin\o\cos\o)\boldsymbol{e}_\o
\end{aligned}
\tag{1.23}
$$

In Path co-ordinates

Figure 1.7 shows a general three-dimensional curve in space. The distance measured along the curve from some arbitrary origin is denoted by s. At point P the unit vector $\boldsymbol{t}$ is tangent to the curve in the direction of increasing s. The unit vector $\boldsymbol{n}$ is normal to the curve and points towards the centre of curvature of the osculating circle which has a local radius ρ. The unit vector $\boldsymbol{b}$ is the bi-normal and completes the right-handed triad.

The position vector is not usually quoted but is

$$\boldsymbol{r} = \boldsymbol{r}_0 + \int \boldsymbol{t}\,\mathrm{d}s$$

The velocity is

$$\boldsymbol{v} = \dot{s}\boldsymbol{t} \tag{1.24}$$

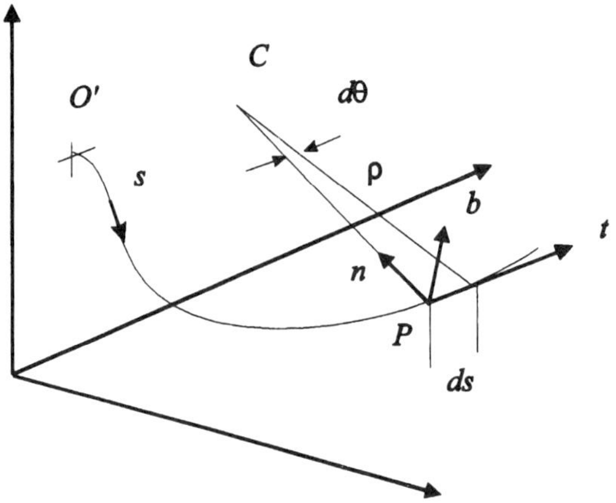

Fig. 1.7 Path co-ordinates

From Fig. 1.8 it is seen that the angular velocity of the unit vector triad is $\boldsymbol{\omega} = \dot{\theta}\boldsymbol{b} + \dot{\gamma}\boldsymbol{t}$. There cannot be a component in the $\boldsymbol{n}$ direction since by definition there is no curvature when the curve is viewed in the direction of arrow A.

The acceleration is therefore

$$\begin{aligned} \boldsymbol{a} &= \ddot{s}\boldsymbol{t} + \boldsymbol{\omega} \times \boldsymbol{v} = \ddot{s}\boldsymbol{t} + \dot{\theta}\boldsymbol{b} \times \dot{s}\boldsymbol{t} \\ &= \ddot{s}\boldsymbol{t} + \dot{s}\dot{\theta}\boldsymbol{n} \end{aligned} \tag{1.25}$$

It is also seen that $\dot{s} = \rho\dot{\theta}$; hence the magnitude of the centripetal acceleration is

$$\dot{s}\dot{\theta} = \rho\dot{\theta}^2 = \dot{s}^2/\rho \tag{1.26}$$

From Fig. 1.7 we see that $\mathrm{d}s = \rho\ \mathrm{d}\theta$ and also that the change in the tangential unit vector is

$$\mathrm{d}\boldsymbol{t} = \mathrm{d}\theta\ \boldsymbol{n}$$

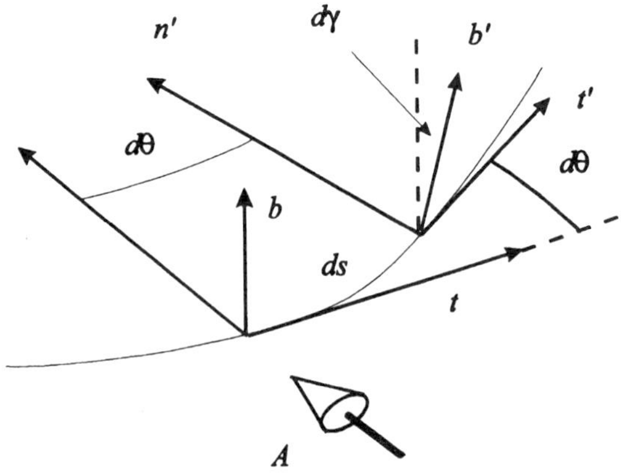

Fig. 1.8 Details of path co-ordinates

Dividing by ds gives

$$\frac{\mathrm{d}\boldsymbol{t}}{\mathrm{d}s} = \frac{\mathrm{d}\theta}{\mathrm{d}s}\boldsymbol{n} = \frac{1}{\rho}\boldsymbol{n} \tag{1.27}$$

The reciprocal of the radius of curvature is known as the curvature κ. Note that curvature is always positive and is directed towards the centre of curvature. So

$$\frac{\mathrm{d}\boldsymbol{t}}{\mathrm{d}s} = \kappa\boldsymbol{n} \tag{1.28}$$

The rate at which the bi-normal, $\boldsymbol{b}$, rotates about the tangent with distance along the curve is known as the torsion or tortuosity of the curve τ.
By definition

$$\frac{\mathrm{d}\boldsymbol{b}}{\mathrm{d}s} = -\tau\boldsymbol{n} \tag{1.29}$$

The negative sign is chosen so that the torsion of a right-handed helix is positive.
Now $\boldsymbol{n} = \boldsymbol{b} \times \boldsymbol{t}$ so

$$\frac{\mathrm{d}\boldsymbol{n}}{\mathrm{d}s} = \frac{\mathrm{d}\boldsymbol{b}}{\mathrm{d}s} \times \boldsymbol{t} + \boldsymbol{b} \times \frac{\mathrm{d}\boldsymbol{t}}{\mathrm{d}s}$$

Substituting from equations (1.29) and (1.28) we have

$$\frac{\mathrm{d}\boldsymbol{n}}{\mathrm{d}s} = -\tau\boldsymbol{n} \times \boldsymbol{t} + \boldsymbol{b} \times \kappa\boldsymbol{n}$$

$$= \tau\boldsymbol{b} - \chi\boldsymbol{t} \tag{1.30}$$

Equations (1.28) to (1.30) are known as the *Serret–Frenet formulae*. From equation (1.27) we see that $\dot{\theta} = \kappa\dot{s}$ and from Fig. 1.8 we have $\dot{\gamma} = \tau\dot{s}$.

1.8 Kinetics of a particle

In the previous sections we considered the kinematics of a point; here we are dealing with a particle. A particle could be a point mass or it could be a body in circumstances where its size and shape are of no consequence, its motion being represented by that of some specific point on the body.

A body of mass m moving at a velocity $\boldsymbol{v}$ has, by definition, a *momentum*

$$\boldsymbol{p} = m\boldsymbol{v}$$

By Newton's second law the force $\boldsymbol{F}$ is given by

$$\boldsymbol{F} = \frac{\mathrm{d}\boldsymbol{p}}{\mathrm{d}t} = m\frac{\mathrm{d}\boldsymbol{v}}{\mathrm{d}t} = m\boldsymbol{a} \tag{1.31}$$

It is convenient to define a quantity known as the *moment of a force*. This takes note of the line of action of a force $\boldsymbol{F}$ passing through the point P, as shown in Fig. 1.9. The *moment of a force* about some chosen reference point is defined to have a magnitude equal to the magnitude of the force times the shortest distance from that line of action to that point. The direction of the moment vector is taken to be normal to the plane containing $\boldsymbol{F}$ and $\boldsymbol{r}$ and the sense is that given by the right hand screw rule. The moment of the force $\boldsymbol{F}$ about O is

$$\boldsymbol{M} = |\boldsymbol{F}|\,\mathrm{d}\boldsymbol{e} = |\boldsymbol{F}||\boldsymbol{r}|\sin\alpha\,\boldsymbol{e}$$

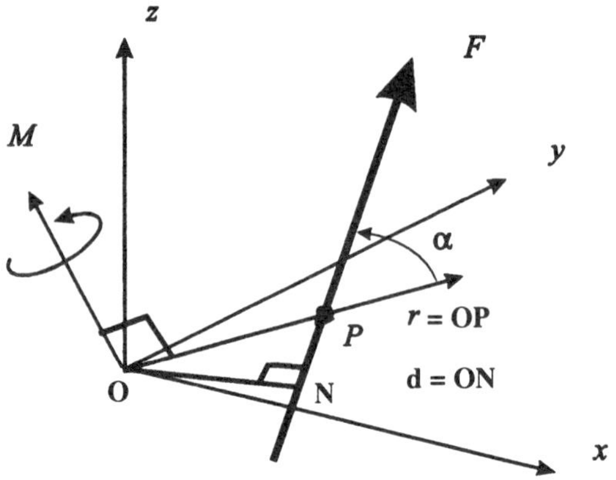

Fig. 1.9 Moment of a force

and, from the definition of the vector product of two vectors,

$$\boldsymbol{M} = \boldsymbol{r} \times \boldsymbol{F}$$

So, from equation (1.31) we have

$$\boldsymbol{r} \times \boldsymbol{F} = \boldsymbol{r} \times \frac{\mathrm{d}\boldsymbol{p}}{\mathrm{d}t} = \frac{\mathrm{d}}{\mathrm{d}t}(\boldsymbol{r} \times \boldsymbol{p}) \tag{1.32}$$

The last equality is true because $\dot{\boldsymbol{r}} = \boldsymbol{v}$ which is parallel to $\boldsymbol{p}$. We can therefore state that

the moment of force about a given point is equal to the rate of change of moment of momentum about that same point.

Here we prefer to use 'moment of momentum' rather than 'angular momentum', which we reserve for rigid body rotation.

1.9 Impulse

Integrating equation (1.31) with respect to time we have

$$\int_1^2 \boldsymbol{F}\,\mathrm{d}t = \Delta\boldsymbol{p} = \boldsymbol{p}_2 - \boldsymbol{p}_1 \tag{1.33}$$

The integral is known as the *impulse*, so in words

impulse equals the change in momentum

From equation 1.32 we have

$$\boldsymbol{M} = \frac{\mathrm{d}}{\mathrm{d}t}(\boldsymbol{r} \times \boldsymbol{p})$$

so integrating both sides with respect to time we have

$$\int_1^2 \boldsymbol{M}\,\mathrm{d}t = \Delta(\boldsymbol{r} \times \boldsymbol{p}) = (\boldsymbol{r} \times \boldsymbol{p})_2 - (\boldsymbol{r} \times \boldsymbol{p})_1 \tag{1.34}$$

The integral is known as the *moment of the impulse*, so in words

moment of the impulse equals the change in the moment of momentum

1.10 Kinetic energy

Again from equation (1.31)

$$\boldsymbol{F} = m\frac{\mathrm{d}\boldsymbol{v}}{\mathrm{d}t}$$

so integrating with respect to displacement we have

$$\int \boldsymbol{F}\cdot \mathrm{d}\boldsymbol{s} = \int m\frac{\mathrm{d}\boldsymbol{v}}{\mathrm{d}t}\cdot \mathrm{d}\boldsymbol{s} = \int m\frac{\mathrm{d}\boldsymbol{s}}{\mathrm{d}t}\cdot \mathrm{d}\boldsymbol{v} = \int m\boldsymbol{v}\cdot \mathrm{d}\boldsymbol{v}$$

$$= \frac{m}{2}\boldsymbol{v}\cdot\boldsymbol{v} + \text{constant} = \frac{m}{2}v^2 + \text{constant}$$

The term $mv^2/2$ is called the *kinetic energy of the particle*. Integrating between limits 1 and 2

$$\int_1^2 \boldsymbol{F}\cdot \mathrm{d}\boldsymbol{s} = \frac{m}{2}v_2^2 - \frac{m}{2}v_1^2 \tag{1.35}$$

or, in words,

the work done is equal to the change in kinetic energy

1.11 Potential energy

If the work done by a force depends only on the end conditions and is independent of the path taken then the force is said to be *conservative*. It follows from this definition that if the path is a closed loop then the work done by a conservative force is zero. That is

$$\oint \boldsymbol{F}\cdot \mathrm{d}\boldsymbol{s} = 0 \tag{1.36}$$

Consider a conservative force acting on a particle between positions 1 and 2. Then

$$\int_1^2 \boldsymbol{F}\cdot \mathrm{d}\boldsymbol{s} = W_2 - W_1 = \frac{m}{2}v_2^2 - \frac{m}{2}v_1^2 \tag{1.37}$$

Here W is called the *work function* and its value depends only on the positions of points 1 and 2 and not on the path taken.

The potential energy is defined to be the negative of the work function and is, here, given the symbol ϕ. Equation (1.37) may now be written

$$0 = \left(\phi_2 + \frac{m}{2}v_2^2\right) - \left(\phi_1 + \frac{m}{2}v_1^2\right) \tag{1.38}$$

Potential energy may be measured from any convenient datum because it is only the difference in potential energy which is important.

1.12 Coriolis's theorem

It is often advantageous to use reference axes which are moving with respect to inertial axes. In Fig. 1.10 the $x'y'z'$ axes are translating and rotating, with an angular velocity $\boldsymbol{\omega}$, with respect to the xyz axes.

The position vector, OP, is

$$\boldsymbol{r} = \boldsymbol{R} + \boldsymbol{r}' = \mathrm{OO}' + \mathrm{O}'\mathrm{P} \tag{1.39}$$

Differentiating equation (1.39), using equation (1.13), gives the velocity

$$\dot{\boldsymbol{r}} = \dot{\boldsymbol{R}} + \boldsymbol{v}' + \boldsymbol{\omega} \times \boldsymbol{r}' \tag{1.40}$$

where $\boldsymbol{v}'$ is the velocity as seen from the moving axes.

Differentiating again

$$\begin{aligned}\ddot{\boldsymbol{r}} &= \ddot{\boldsymbol{R}} + \boldsymbol{a}' + \dot{\boldsymbol{\omega}} \times \boldsymbol{r}' + \boldsymbol{\omega} \times \boldsymbol{v}' + \boldsymbol{\omega} \times (\boldsymbol{v}' + \boldsymbol{\omega} \times \boldsymbol{r}') \\ &= \ddot{\boldsymbol{R}} + \boldsymbol{a}' + \dot{\boldsymbol{\omega}} \times \boldsymbol{r}' + 2\boldsymbol{\omega} \times \boldsymbol{v}' + \boldsymbol{\omega} \times (\boldsymbol{\omega} \times \boldsymbol{r}')\end{aligned} \tag{1.41}$$

where $\boldsymbol{a}'$ is the acceleration as seen from the moving axes.

Using Newton's second law

$$\boldsymbol{F} = m\ddot{\boldsymbol{r}} = m[\ddot{\boldsymbol{R}} + \boldsymbol{a}' + \dot{\boldsymbol{\omega}} \times \boldsymbol{r}' + 2\boldsymbol{\omega} \times \boldsymbol{v}' + \boldsymbol{\omega} \times (\boldsymbol{\omega} \times \boldsymbol{r}')] \tag{1.42}$$

Expanding the triple vector product and rearranging gives

$$\boldsymbol{F} - m\ddot{\boldsymbol{R}} - m\dot{\boldsymbol{\omega}} \times \boldsymbol{r}' - 2m\boldsymbol{\omega} \times \boldsymbol{v}' - m[\boldsymbol{\omega}(\boldsymbol{\omega} \cdot \boldsymbol{r}') - \omega^2 \boldsymbol{r}'] = m\boldsymbol{a}' \tag{1.43}$$

This is known as *Coriolis's theorem.*

The terms on the left hand side of equation (1.43) comprise one real force, $\boldsymbol{F}$, and four fictitious forces. The second term is the inertia force due to the acceleration of the origin O′, the third is due to the angular acceleration of the axes, the fourth is known as the Coriolis force and the last term is the centrifugal force. The centrifugal force through P is normal to and directed away from the ω axis, as can be verified by forming the scalar product with $\boldsymbol{\omega}$. The Coriolis force is normal to both the relative velocity vector, $\boldsymbol{v}'$, and to $\boldsymbol{\omega}$.

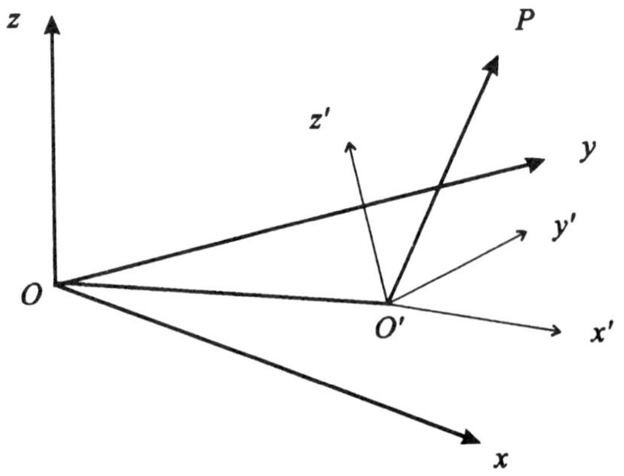

Fig. 1.10

1.13 Newton's laws for a group of particles

Consider a group of n particles, three of which are shown in Fig. 1.11, where the ith particle has a mass m_i and is at a position defined by $\boldsymbol{r}_i$ relative to an inertial frame of reference. The force on the particle is the vector sum of the forces due to each other particle in the group and the resultant of the external forces. If $\boldsymbol{f}_{ij}$ is the force on particle i due to particle j and $\boldsymbol{F}_i$ is the resultant force due to bodies external to the group then summing over all particles, except for $j = i$, we have for the ith particle

$$\sum_j \boldsymbol{f}_{ij} + \boldsymbol{F}_i = m_i \ddot{\boldsymbol{r}}_i \tag{1.44}$$

We now form the sum over all particles in the group

$$\sum_i \sum_j \boldsymbol{f}_{ij} + \sum_i \boldsymbol{F}_i = \sum_i m_i \ddot{\boldsymbol{r}}_i \tag{1.45}$$

The first term sums to zero because, by Newton's third law, $\boldsymbol{f}_{ij} = -\boldsymbol{f}_{ji}$. Thus

$$\sum_i \boldsymbol{F}_i = \sum_i m_i \ddot{\boldsymbol{r}}_i \tag{1.46}$$

The position vector of the *centre of mass* is defined by

$$\sum_i m_i \boldsymbol{r}_i = \left(\sum_i m_i\right) \boldsymbol{r}_{\mathrm{G}} = m\boldsymbol{r}_{\mathrm{G}} \tag{1.47}$$

where m is the total mass and $\boldsymbol{r}_{\mathrm{G}}$ is the location of the centre of mass. It follows that

$$\sum_i m_i \dot{\boldsymbol{r}}_i = m\dot{\boldsymbol{r}}_{\mathrm{G}} \tag{1.48}$$

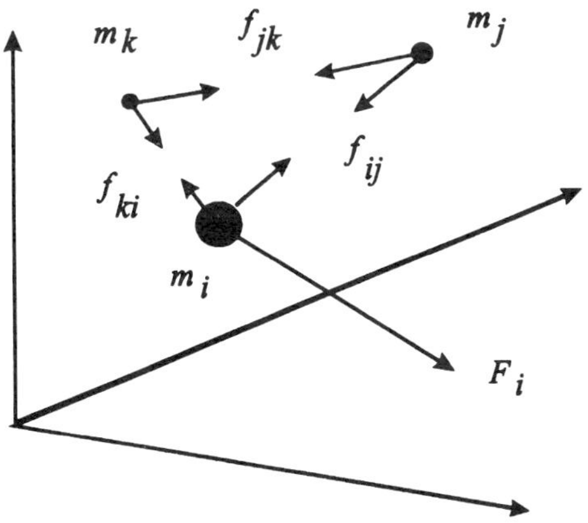

Fig. 1.11

and

$$\sum_i m_i \ddot{\boldsymbol{r}}_i = m\ddot{\boldsymbol{r}}_G \tag{1.49}$$

Therefore equation (1.46) can be written

$$\sum_i \boldsymbol{F}_i = m\ddot{\boldsymbol{r}}_G = \frac{\mathrm{d}}{\mathrm{d}t}(m\dot{\boldsymbol{r}}_G) \tag{1.50}$$

This may be summarized by stating

the vector sum of the external forces is equal to the total mass times the acceleration of the centre of mass or to the time rate of change of momentum.

A moment of momentum expression for the ith particle can be obtained by forming the vector product with $\boldsymbol{r}_i$ of both sides of equation (1.44)

$$\sum_j \boldsymbol{r}_i \times \boldsymbol{f}_{ij} + \boldsymbol{r}_i \times \boldsymbol{F}_i = \boldsymbol{r}_i \times m_i \ddot{\boldsymbol{r}}_i \tag{1.51}$$

Summing equation (1.51) over n particles

$$\sum_i \boldsymbol{r}_i \times \boldsymbol{F}_i + \sum_i \boldsymbol{r}_i \times \sum_j \boldsymbol{f}_{ij} = \sum_i \boldsymbol{r}_i \times m_i \ddot{\boldsymbol{r}}_i = \frac{\mathrm{d}}{\mathrm{d}t} \sum_i \boldsymbol{r}_i \times m_i \dot{\boldsymbol{r}}_i \tag{1.52}$$

The double summation will vanish if Newton's third law is in its strong form, that is $\boldsymbol{f}_{ij} = -\boldsymbol{f}_{ji}$ and also they are colinear. There are cases in electromagnetic theory where the equal but opposite forces are not colinear. This, however, is a consequence of the special theory of relativity.

Equation (1.52) now reads

$$\sum_i \boldsymbol{r}_i \times \boldsymbol{F}_i = \frac{\mathrm{d}}{\mathrm{d}t} \sum_i \boldsymbol{r}_i \times m_i \dot{\boldsymbol{r}}_i \tag{1.53}$$

and using $\boldsymbol{M}$ to denote moment of force and $\boldsymbol{L}$ the moment of momentum

$$\boldsymbol{M}_O = \frac{\mathrm{d}}{\mathrm{d}t} \boldsymbol{L}_O$$

Thus,

the moment of the external forces about some arbitrary point is equal to the time rate of change of the moment of momentum (or the moment of the rate of change of momentum) about that point.

The position vector for particle i may be expressed as the sum of the position vector of the centre of mass and the position vector of the particle relative to the centre of mass, or

$$\boldsymbol{r}_i = \boldsymbol{r}_G + \boldsymbol{\rho}_i$$

Thus equation (1.53) can be written

$$\sum r_i \times F_i = \frac{d}{dt}\sum (r_G + \rho) \times m_i(\dot{r}_G + \dot{\rho}_i)$$

$$= \frac{d}{dt}\left[r_G \times (m\dot{r}_G) + r_G \sum m_i\dot{\rho}_i + \left(\sum m_i\rho_i\right) \times \dot{r}_G \right.$$

$$\left. + \sum \rho_i \times (m_i\dot{\rho}_i) \right]$$

$$= \frac{d}{dt}\left[r_G \times (m\dot{r}_G) + \sum \rho_i \times (m_i\dot{\rho}_i) \right] \tag{1.53a}$$

1.14 Conservation of momentum

Integrating equation (1.46) with respect to time gives

$$\sum_i \int F_i \, dt = \Delta \sum_i m_i\dot{r}_i \tag{1.54}$$

That is,

the sum of the external impulses equals the change in momentum of the system.

It follows that if the external forces are zero then the momentum is conserved.

Similarly from equation (1.53) we have that

the moment of the external impulses about a given point equals the change in moment of momentum about the same point.

$$\sum_i \int r_i \times F_i \, dt = \Delta \sum_i r_i \times m_i\dot{r}_i$$

From which it follows that if the moment of the external forces is zero the moment of momentum is conserved.

1.15 Energy for a group of particles

Integrating equation (1.45) with respect to displacement yields

$$\sum_i \int \sum_j f_{ij}\cdot dr_i + \sum_i \int F_i\cdot dr_i = \sum_i \int m_i\ddot{r}_i\cdot dr_i$$

$$= \sum_i \int m_i \frac{d\dot{r}_i}{dt}\cdot dr_i = \sum_i \int m_i\dot{r}_i\cdot d\dot{r}_i = \sum_i \frac{m_i}{2}\dot{r}_i^2 \tag{1.55}$$

The first term on the left hand side of the equation is simply the work done by the external forces. The second term does not vanish despite $f_{ij} = -f_{ji}$ because the displacement of the ith particle, resolved along the line joining the two particles, is only equal to that of the jth particle in the case of a rigid body. In the case of a deformable body energy is either stored or dissipated.

If the stored energy is recoverable, that is the process is reversible, then the energy stored is a form of potential energy which, for a solid, is called *strain energy*.

The energy equation may be generalized to

$$\text{work done by external forces} = \Delta V + \Delta T + \text{losses} \tag{1.56}$$

where ΔV is the change in any form of potential energy and ΔT is the change in kinetic energy. The losses account for any energy forms not already included.

The kinetic energy can be expressed in terms of the motion of the centre of mass and motion relative to the centre of mass. Here ρ is the position of a particle relative to the centre of mass, as shown in Fig. 1.12.

$$T = \frac{1}{2}\sum_i m_i \dot{\boldsymbol{r}}_i \cdot \dot{\boldsymbol{r}}_i = \frac{1}{2}\sum_i m_i (\dot{\boldsymbol{r}}_G + \dot{\boldsymbol{\rho}}_i) \cdot (\dot{\boldsymbol{r}}_G + \dot{\boldsymbol{\rho}}_i)$$

$$= \frac{1}{2} m \dot{r}_G^2 + \frac{1}{2}\sum_i m_i \dot{\rho}_{ii}^2 \qquad m = \sum_i m_i \tag{1.57}$$

The other two terms of the expansion are zero by virtue of the definition of the centre of mass. From this expression we see that the kinetic energy can be written as that of a point mass, equal to the total mass, at the centre of mass plus that due to motion relative to the centre of mass.

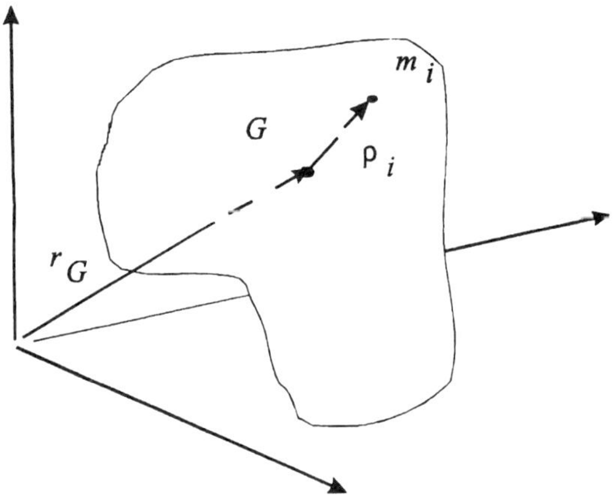

Fig. 1.12

1.16 The principle of virtual work

The concept of virtual work evolved gradually, as some evidence of the idea is inherent in the ancient treatment of the principle of levers. Here the weight or force at one end of a lever times the distance moved was said to be the same as that for the other end of the lever. This notion was used in the discussion of equilibrium of a lever or balance in the static case. The motion was one which *could* take place rather than any actual motion.

The formal definition of *virtual displacement*, $\delta \boldsymbol{r}$, is any displacement which could take place subject to any constraints. For a system having many degrees of freedom all displacements save one may be held fixed leaving just one degree of freedom.

From this definition *virtual work* is defined as $\boldsymbol{F}\cdot\delta\boldsymbol{r}$ where $\boldsymbol{F}$ is the force acting on the particle at the original position and at a specific time. That is, the force is constant during the virtual displacement. For equilibrium

$$\sum_i \boldsymbol{F}_i\cdot\mathrm{d}\boldsymbol{r} = 0 = \delta W \tag{1.58}$$

Since there is a choice of which co-ordinates are fixed and which one is free it means that for a system with n degrees of freedom n independent equations are possible.

If the force is conservative then $\boldsymbol{F}\cdot\delta\boldsymbol{r} = \delta W$, the variation of the work function. By definition the potential energy is the negative of the work function; therefore $\boldsymbol{F}\cdot\delta\boldsymbol{r} = -\delta V$.

In general if both conservative and non-conservative forces are present

$$\sum_i (\boldsymbol{F}_{i,\,\text{non-con}} + \boldsymbol{F}_{i,\,\text{con}})\cdot\delta\boldsymbol{r}_i = 0$$

or

$$\sum_i (\boldsymbol{F}_{i,\,\text{non-con}})\cdot\delta\boldsymbol{r}_i = \delta V \tag{1.59}$$

That is,

the virtual work done by the non-conservative forces = δV

1.17 D'Alembert's principle

In 1743 D'Alembert extended the principle of virtual work into the field of dynamics by postulating that the work done by the active forces less the 'inertia forces' is zero. If $\boldsymbol{F}$ is a real force not already included in any potential energy term then the principle of virtual work becomes

$$\sum_i (\boldsymbol{F}_i - m_i\ddot{\boldsymbol{r}}_i)\cdot\delta\boldsymbol{r}_i = \delta V \tag{1.60}$$

This is seen to be in agreement with Newton's laws by considering the simple case of a particle moving in a gravitational field as shown in Fig. 1.13. The potential energy $V = mgy$ so D'Alembert's principle gives

$$[(F_x - m\ddot{x})\boldsymbol{i} + (F_y - m\ddot{y})\boldsymbol{j}]\cdot(\delta x\boldsymbol{i} + \delta y\boldsymbol{j}) = \delta V = \frac{\partial V}{\partial x}\delta x + \frac{\partial V}{\partial y}\delta y$$

$$(F_x - m\ddot{x})\,\delta x + (F_y - m\ddot{y})\,\delta y = mg\,\delta y \tag{1.61}$$

Because δx and δy are independent we have

$$(F_x - m\ddot{x})\,\delta x = 0$$

or

$$F_x = m\ddot{x} \tag{1.62}$$

and

$$(F_y - m\ddot{y})\,\delta y = mg\,\delta y$$

or

$$F_y - mg = m\ddot{y} \tag{1.63}$$

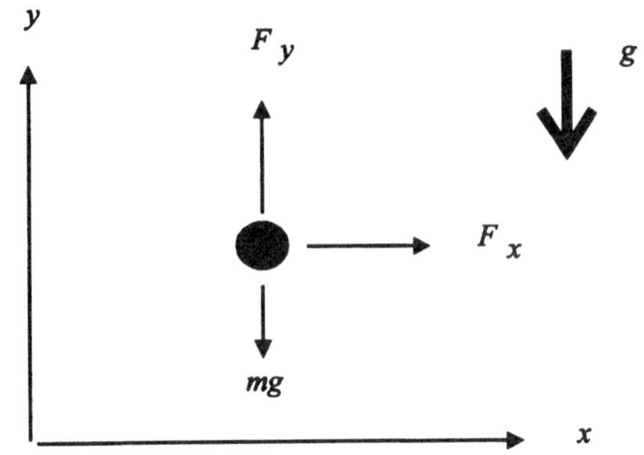

Fig. 1.13

As with the principle of virtual work and D'Alembert's principle the forces associated with workless constraints are not included in the equations. This reduces the number of equations required but of course does not furnish any information about these forces.

2

Lagrange's Equations

2.1 Introduction

The dynamical equations of J.L. Lagrange were published in the eighteenth century some one hundred years after Newton's *Principia*. They represent a powerful alternative to the Newton–Euler equations and are particularly useful for systems having many degrees of freedom and are even more advantageous when most of the forces are derivable from potential functions.

The equations are

$$\frac{\mathrm{d}}{\mathrm{d}t}\left(\frac{\partial \mathfrak{L}}{\partial \dot{q}_i}\right) - \left(\frac{\partial \mathfrak{L}}{\partial q_i}\right) = Q_i \quad 1 \le i \le n \tag{2.1}$$

where

$\mathfrak{L}$ is the Lagrangian defined to be $T - V$,
T is the kinetic energy (relative to inertial axes),
V is the potential energy,
n is the number of degrees of freedom,
q_1 to q_n are the generalized co-ordinates,
Q_1 to Q_n are the generalized forces

and d/dt means differentiation of the scalar terms with respect to time. *Generalized co-ordinates* and *generalized forces* are described below.

Partial differentiation with respect to $\dot{q}_i$ is carried out assuming that all the other $\dot{q}$, all the q and time are held fixed. Similarly for differentiation with respect to q_i all the other q, all $\dot{q}$ and time are held fixed.

We shall proceed to prove the above equations, starting from Newton's laws and D'Alembert's principle, during which the exact meaning of the definitions and statements will be illuminated. But prior to this a simple application will show the ease of use.

EXAMPLE

A mass is suspended from a point by a spring of natural length *a* and stiffness *k*, as shown in Fig. 2.1. The mass is constrained to move in a vertical plane in which the gravitational field strength is *g*. Determine the equations of motion in terms of the distance *r* from the support to the mass and the angle θ which is the angle the spring makes with the vertical through the support point.

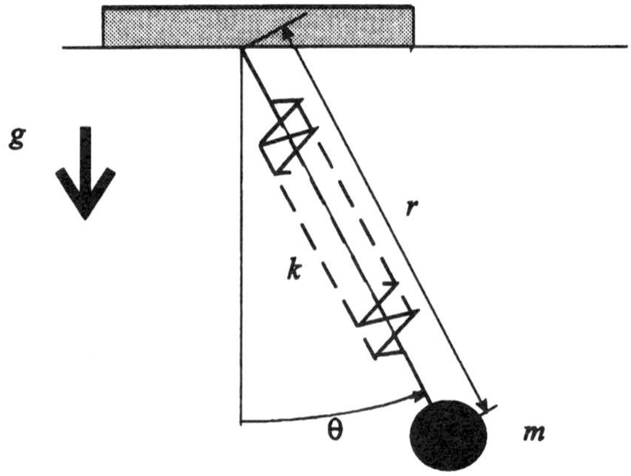

Fig. 2.1

The system has two degrees of freedom and r and θ, which are independent, can serve as generalized co-ordinates. The expression for kinetic energy is

$$T = \frac{m}{2}\left[\dot{r}^2 + (r\dot{\theta})^2\right]$$

and for potential energy, taking the horizontal through the support as the datum for gravitational potential energy,

$$V = -mgr\cos\theta + \frac{k}{2}(r-a)^2$$

so

$$\mathfrak{L} = T - V = \frac{m}{2}\left[\dot{r}^2 + (r\dot{\theta})^2\right] + mgr\cos\theta - \frac{k}{2}(r-a)^2$$

Applying Lagrange's equation with $q_1 = r$ we have

$$\frac{\partial \mathfrak{L}}{\partial \dot{r}} = m\dot{r}$$

so

$$\frac{\mathrm{d}}{\mathrm{d}t}\left(\frac{\partial \mathfrak{L}}{\partial \dot{r}}\right) = m\ddot{r}$$

and

$$\frac{\partial \mathfrak{L}}{\partial r} = mr\dot{\theta}^2 + mg\cos\theta - k(r-a)$$

From equation (2.1)

$$\frac{\mathrm{d}}{\mathrm{d}t}\left(\frac{\partial \mathfrak{L}}{\partial \dot{r}}\right) - \left(\frac{\partial \mathfrak{L}}{\partial r}\right) = Q_r$$

$$m\ddot{r} - mr\dot{\theta}^2 - mg\cos\theta + k(r-a) = 0 \tag{i}$$

The generalized force $Q_r = 0$ because there is no externally applied radial force that is not included in V.

Taking θ as the next generalized co-ordinate

$$\frac{\partial \mathfrak{L}}{\partial \dot{\theta}} = mr^2\dot{\theta}$$

so

$$\frac{\mathrm{d}}{\mathrm{d}t}\left(\frac{\partial \mathfrak{L}}{\partial \dot{\theta}}\right) = 2mr\dot{r}\dot{\theta} + mr^2\ddot{\theta}$$

and

$$\frac{\partial \mathfrak{L}}{\partial \theta} = mgr\sin\theta$$

Thus the equation of motion in θ is

$$\frac{\mathrm{d}}{\mathrm{d}t}\left(\frac{\partial \mathfrak{L}}{\partial \dot{\theta}}\right) - \left(\frac{\partial \mathfrak{L}}{\partial \theta}\right) = Q_\theta = 0$$

$$2mr\dot{r}\dot{\theta} + mr^2\ddot{\theta} - mgr\sin\theta = 0 \qquad \text{(ii)}$$

The generalized force in this case would be a torque because the corresponding generalized co-ordinate is an angle. Generalized forces will be discussed later in more detail.

Dividing equation (ii) by r gives

$$2m\dot{r}\dot{\theta} + mr\ddot{\theta} - mg\sin\theta = 0 \qquad \text{(iia)}$$

and rearranging equations (i) and (ii) leads to

$$mg\cos\theta - k(r - a) = m(\ddot{r} - r\dot{\theta}^2) \qquad \text{(ia)}$$

and

$$-mg\sin\theta = m(2\dot{r}\dot{\theta} + r\ddot{\theta}) \qquad \text{(iib)}$$

which are the equations obtained directly from Newton's laws plus a knowledge of the components of acceleration in polar co-ordinates.

In this example there is not much saving of labour except that there is no requirement to know the components of acceleration, only the components of velocity.

2.2 Generalized co-ordinates

A set of generalized co-ordinates is one in which each co-ordinate is independent and the number of co-ordinates is just sufficient to specify completely the configuration of the system. A system of N particles, each free to move in a three-dimensional space, will require $3N$ co-ordinates to specify the configuration. If Cartesian co-ordinates are used then the set could be

$$\{x_1\ y_1\ z_1\ x_2\ y_2\ z_2 \ldots x_N\ y_N\ z_N\}$$

or

$$\{x_1\ x_2\ x_3\ x_4\ x_5\ x_6 \ldots x_{n-2}\ x_{n-1}\ x_n\}$$

where $n = 3N$.

This is an example of a set of generalized co-ordinates but other sets may be devised involving different displacements or angles. It is conventional to designate these co-ordinates as

$$\{q_1\ q_2\ q_3\ q_4\ q_5\ q_6\ \ldots\ q_{n-2}\ q_{n-1}\ q_n\}$$

If there are constraints between the co-ordinates then the number of independent co-ordinates will be reduced. In general if there are r equations of constraint then the number of degrees of freedom n will be $3N - r$. For a particle constrained to move in the xy plane the equation of constraint is $z = 0$. If two particles are rigidly connected then the equation of constraint will be

$$(x_2 - x_1)^2 + (y_2 - y_1)^2 + (z_2 - z_1)^2 = L^2$$

That is, if one point is known then the other point must lie on the surface of a sphere of radius L. If $x_1 = y_1 = z_1 = 0$ then the constraint equation simplifies to

$$x_2^2 + y_2^2 + z_2^2 = L^2$$

Differentiating we obtain

$$2x_2\,\mathrm{d}x_2 + 2y_2\,\mathrm{d}y_2 + 2z_2\,\mathrm{d}z_2 = 0$$

This is a perfect differential equation and can obviously be integrated to form the constraint equation. In some circumstances there exist constraints which appear in differential form and cannot be integrated; one such example of a rolling wheel will be considered later. A system for which all the constraint equations can be written in the form $f(q_1 \ldots q_n)$ = constant or a known function of time is referred to as *holonomic* and for those which cannot it is called *non-holonomic*.

If the constraints are moving or the reference axes are moving then time will appear explicitly in the equations for thc Lagrangian. Such systems are called *rheonomous* and those where time does not appear explicitly are called *scleronomous*.

Initially we will consider a holonomic system (rheonomous or scleronomous) so that the Cartesian co-ordinates can be expressed in the form

$$x_i = x_i\,(q_1\ q_2 \ldots q_n\ t) \tag{2.2}$$

By the rules for partial differentiation the differential of equation (2.2) with respect to time is

$$v_i = \frac{\mathrm{d}x_i}{\mathrm{d}t} = \sum_j \frac{\partial x_i}{\partial q_j}\dot{q}_j + \frac{\partial x_i}{\partial t} \tag{2.3}$$

so

$$v_i = v_i\,(q_1\ q_2 \ldots q_n\ \dot{q}_1\ \dot{q}_2 \ldots \dot{q}_n t) \tag{2.4}$$

thus

$$\frac{\mathrm{d}v_i}{\mathrm{d}t} = \sum_j \frac{\partial v_i}{\partial q_j}\dot{q}_j + \sum_j \frac{\partial v_i}{\partial \dot{q}_j}\ddot{q}_j + \frac{\partial^2 x_i}{\partial t^2} \tag{2.5}$$

Differentiating equation (2.3) directly gives

$$\frac{\mathrm{d}v_i}{\mathrm{d}t} = \sum_j \frac{\partial \dot{x}_i}{\partial q_j}\dot{q}_j + \sum_j \frac{\partial x_i}{\partial q_j}\ddot{q}_j + \frac{\partial^2 x_i}{\partial t^2} \tag{2.6}$$

and comparing equation (2.5) with equation (2.6), noting that $v_i = \dot{x}_i$, we see that

$$\frac{\partial \dot{x}_i}{\partial \dot{q}_j} = \frac{\partial x_i}{\partial q_j} \tag{2.7}$$

a process sometimes referred to as the cancellation of the dots.

From equation (2.2) we may write

$$\mathrm{d}x_i = \sum_j \frac{\partial x_i}{\partial q_j}\,\mathrm{d}q_j + \frac{\partial x_i}{\partial t}\,\mathrm{d}t \tag{2.8}$$

Since, by definition, virtual displacements are made with time constant

$$\delta x_i = \sum_j \frac{\partial x_i}{\partial q_j}\,\delta q_j \tag{2.9}$$

These relationships will be used in the proof of Lagrange's equations.

2.3 Proof of Lagrange's equations

The proof starts with D'Alembert's principle which, it will be remembered, is an extension of the principle of virtual work to dynamic systems. D'Alembert's equation for a system of N particles is

$$\sum_i (\boldsymbol{F}_i - m_i\ddot{\boldsymbol{r}}_i)\cdot\delta\boldsymbol{r}_i = 0 \quad 1 \le i \le N \tag{2.10}$$

where $\delta\boldsymbol{r}_i$ is any virtual displacement, consistent with the constraints, made with time fixed.

Writing $\boldsymbol{r}_1 = x_1\boldsymbol{i} + x_2\boldsymbol{j} + x_3\boldsymbol{k}$ etc. equation (2.10) may be written in the form

$$\sum_i (F_i - m_i\ddot{x}_i)\delta x_i = 0 \quad 1 \le i \le n = 3N \tag{2.11}$$

Using equation (2.9) and changing the order of summation, the first summation in equation (2.11) becomes

$$\sum_i F_i\delta x_i = \sum_i F_i\left(\sum_j \frac{\partial x_i}{\partial q_j}\delta q_j\right) = \sum_j\left(\sum_i F_i\frac{\partial x_i}{\partial q_j}\right)\delta q_j = \delta W \tag{2.12}$$

the virtual work done by the forces. Now $W = W(q_j)$ so

$$\delta W = \sum_j \frac{\partial W}{\partial q_j}\,\delta q_j \tag{2.13}$$

and by comparison of the coefficients of δq in equations (2.12) and (2.13) we see that

$$\frac{\partial W}{\partial q_j} = \sum_i F_i\,\frac{\partial x_i}{\partial q_j} \tag{2.14}$$

This term is designated Q_j and is known as a generalized force. The dimensions of this quantity need not be those of force but the product of the generalized force and the associated generalized co-ordinate must be that of work. In most cases this reduces to force and displacement or torque and angle. Thus we may write

$$\delta W = \sum_j Q_j\,\delta q_j \tag{2.15}$$

In a large number of problems the force can be derived from a position-dependent potential V, in which case

$$Q_j = -\frac{\partial V}{\partial q_j} \tag{2.16}$$

Equation (2.13) may now be written

$$\sum_i F_i \delta x_i = \sum_j \left(-\frac{\partial V}{\partial q_j} + Q_j \right) \delta q_j \tag{2.17}$$

where Q_j now only applies to forces not derived from a potential.

Now the second summation term in equation (2.11) is

$$\sum_i m_i \ddot{x}_i \, \delta x_i = \sum_i m_i \ddot{x}_i \left(\sum_j \frac{\partial x_i}{\partial q_j} \delta q_j \right)$$

or, changing the order of summation,

$$\sum_i m_i \ddot{x}_i \, \delta x_i = \sum_j \left(\sum_i m_i \ddot{x}_i \frac{\partial x_i}{\partial q_j} \right) \delta q_j \tag{2.18}$$

We now seek a form for the right hand side of equation (2.18) involving the kinetic energy of the system in terms of the generalized co-ordinates.

The kinetic energy of the system of N particles is

$$T = \sum_{i=1}^{i=N} \frac{m_i}{2} \dot{r}_i \cdot \dot{r}_i = \sum_{i=1}^{i=3N} \frac{m_i}{2} \dot{x}_i \dot{x}_i$$

Thus

$$\frac{\partial T}{\partial \dot{q}_j} = \sum_i m_i \dot{x}_i \frac{\partial \dot{x}_i}{\partial \dot{q}_j} = \sum_i m_i \dot{x}_i \frac{\partial x_i}{\partial q_j}$$

because the dots may be cancelled, *see* equation (2.7). Differentiating with respect to time gives

$$\frac{\mathrm{d}}{\mathrm{d}t}\left(\frac{\partial T}{\partial \dot{q}_j} \right) = \sum_i m_i \ddot{x}_i \frac{\partial x_i}{\partial q_j} + \sum_i m_i \dot{x}_i \frac{\partial \dot{x}_i}{\partial q_j}$$

but

$$\frac{\partial T}{\partial q_j} = \sum_i m_i \dot{x}_i \frac{\partial \dot{x}_i}{\partial q_j}$$

so

$$\frac{\mathrm{d}}{\mathrm{d}t}\left(\frac{\partial T}{\partial \dot{q}_j} \right) - \frac{\partial T}{\partial q_j} = \sum_i m_i \ddot{x}_i \frac{\partial x_i}{\partial q_j} \tag{2.19}$$

Substitution of equation (2.19) into equation (2.18) gives

$$\sum_i m_i \ddot{x}_i \, \delta x_i = \sum_j \left[\frac{\mathrm{d}}{\mathrm{d}t}\left(\frac{\partial T}{\partial \dot{q}_j} \right) - \frac{\partial T}{\partial q_j} \right] \delta q_j \tag{2.20}$$

Substituting from equations (2.17) and (2.20) into equation 2.11 leads to

$$\sum_j \left[-\frac{\partial V}{\partial q_j} + Q_j - \frac{\mathrm{d}}{\mathrm{d}t}\left(\frac{\partial T}{\partial \dot{q}_j}\right) + \frac{\partial T}{\partial q_j} \right] \delta q_j = 0$$

Because the q are independent we can choose δq_j to be non-zero whilst all the other δq are zero. So

$$\frac{\mathrm{d}}{\mathrm{d}t}\left(\frac{\partial T}{\partial \dot{q}_j}\right) - \frac{\partial T}{\partial q_j} + \frac{\partial V}{\partial q_j} = Q_j$$

Alternatively since V is taken not to be a function of the generalized velocities we can write the above equation in terms of the Lagrangian $\mathfrak{L} = T - V$

$$\frac{\mathrm{d}}{\mathrm{d}t}\left(\frac{\partial \mathfrak{L}}{\partial \dot{q}_j}\right) - \left(\frac{\partial \mathfrak{L}}{\partial q_j}\right) = Q_j \quad 1 \leq j \leq n \tag{2.21}$$

In the above analysis we have taken n to be $3N$ but if we have r holonomic equations of constraint then $n = 3N - r$. In practice it is usual to write expressions for T and V directly in terms of the reduced number of generalized co-ordinates. Further, the forces associated with workless constraints need not be included in the analysis.

For example, if a rigid body is constrained to move in a vertical plane with the y axis vertically upwards then

$$T = \frac{m}{2}(\dot{x}_G^2 + \dot{y}_G^2) + \frac{I_G}{2}\dot{\theta}^2 \text{ and } V = mgy_G$$

The constraint equations are fully covered by the use of total mass and moment of inertia and the suppression of the z_G co-ordinate.

2.4 The dissipation function

If there are forces of a viscous nature that depend linearly on velocity then the force is given by

$$F_i = -\sum_j c_{ij}\dot{x}_j$$

where c_{ij} are constants.

The power dissipated is

$$P = \sum_i F_i \dot{x}_i$$

In terms of generalized forces

$$P = \sum_i Q_i \dot{q}_i$$

and

$$Q_i = -\sum_j C_{ij}\dot{q}_j$$

where C_{ij} are related to c_{ij} (the exact relationship does not concern us at this point).

The power dissipated is

$$P = -\sum_j Q_j \dot{q}_j$$

By differentiation

$$\frac{\partial P}{\partial \dot{q}_i} = -Q_i - \sum_j \frac{\partial Q_j}{\partial \dot{q}_i}\dot{q}_j = -Q_i + \sum_j C_{ij}\dot{q}_j = -2Q_i$$

If we now define $\mathfrak{F} = P/2$ then

$$\frac{\partial \mathfrak{F}}{\partial \dot{q}_i} = -Q_i \tag{2.22}$$

The term $\mathfrak{F}$ is known as *Rayleigh's dissipative function* and is half the rate at which power is being dissipated.

Lagrange's equations are now

$$\frac{\mathrm{d}}{\mathrm{d}t}\left(\frac{\partial \mathfrak{L}}{\partial \dot{q}_j}\right) - \left(\frac{\partial \mathfrak{L}}{\partial q_j}\right) + \frac{\partial \mathfrak{F}}{\partial \dot{q}_j} = Q_j \tag{2.23}$$

where Q_j is the generalized force not obtained from a position-dependent potential or a dissipative function.

EXAMPLE

For the system shown in Fig. 2.2 the scalar functions are

$$T = \frac{m_1}{2}\dot{x}_1^2 + \frac{m_2}{2}\dot{x}_2^2$$

$$V = \frac{k_1}{2}x_1^2 + \frac{k_2}{2}(x_2 - x_1)^2$$

$$\mathfrak{F} = \frac{c_i}{2}\dot{x}_1^2 + \frac{c_2}{2}(\dot{x}_2 - \dot{x}_1)^2$$

The virtual work done by the external forces is

$$\delta W = F_1\,\delta x_1 + F_2\,\delta x_2$$

For the generalized co-ordinate x_1 application of Lagrange's equation leads to

$$m_1\ddot{x}_1 + k_1 x_1 - k_2(x_2 - x_1) + c_1\dot{x}_1 - c_2(\dot{x}_2 - \dot{x}_1) = F_1$$

and for x_2

$$m_2\ddot{x}_2 + k_2(x_2 - x_1) + c_2(\dot{x}_2 - \dot{x}_1) = F_2$$

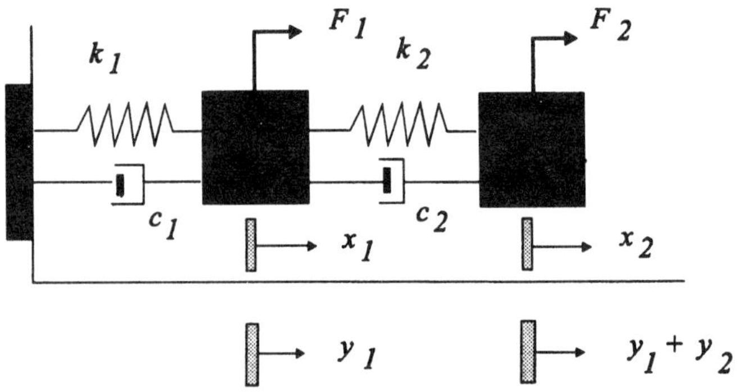

Fig. 2.2

Alternatively we could have used co-ordinates y_1 and y_2 in which case the appropriate functions are

$$T = \frac{m_1}{2}\dot{y}_1^2 + \frac{m_2}{2}(\dot{y}_1 + \dot{y}_2)^2$$

$$V = \frac{k_1}{2}y_1^2 + \frac{k_2}{2}y_2^2$$

$$\mathfrak{F} = \frac{c_1}{2}\dot{y}_1^2 + \frac{c_2}{2}\dot{y}_2^2$$

and the virtual work is

$$\delta W = F_1\delta y_1 + F_2\delta(y_1 + y_2) = (F_1 + F_2)\delta y_1 + F_2\delta y_2$$

Application of Lagrange's equation leads this time to

$$m_1\ddot{y}_1 + m_2(\ddot{y}_1 + \ddot{y}_2) + k_1 y_1 + c_1\dot{y}_1 = F_1 + F_2$$

$$m_2(\ddot{y}_1 + \ddot{y}_2) + k_2 y_2 + c_2\dot{y}_2 = F_2$$

Note that in the first case the kinetic energy has no term which involves products like $\dot{q}_i\dot{q}_j$ whereas in the second case it does. The reverse is true for the potential energy regarding terms like q_iq_j. Therefore the coupling of co-ordinates depends on the choice of co-ordinates and de-coupling in the kinetic energy does not imply that de-coupling occurs in the potential energy. It can be proved, however, that there exists a set of co-ordinates which leads to uncoupled co-ordinates in both the kinetic energy and the potential energy; these are known as principal co-ordinates.

2.5 Kinetic energy

The kinetic energy of a system is

$$T = \frac{1}{2}\Sigma\, m_i\dot{x}_i^2 = \frac{1}{2}\,(\dot{x})^{\mathrm{T}}[m](\dot{x})$$

where

$$(x) = (x_1\, x_2 \ldots x_{3N})^{\mathrm{T}}$$

and

$$[m] = \begin{bmatrix} m_1 & & \\ & \ddots & \\ & & m_{3N} \end{bmatrix}_{\text{diagonal}}$$

Now

$$x_i = x_i(q_1\, q_2 \ldots q_n t)$$

so

$$\dot{x}_i = \sum_j \frac{\partial x_i}{\partial q_j}\dot{q}_j + \frac{\partial x_i}{\partial t}$$

We shall, in this section, use the notation ( ) to mean a column matrix and [ ] to indicate a square matrix. Thus with

$$(\dot{x}) = (\dot{x}_1 \dot{x}_2 \ldots \dot{x}_n)^{\mathrm{T}}$$

then we may write

$$(\dot{x}) = [A](\dot{q}) + (b)$$

where

$$[A] = \begin{bmatrix} \dfrac{\partial x_1}{\partial q_1} \cdots \dfrac{\partial x_1}{\partial q_n} \\ \dfrac{\partial x_n}{\partial q_1} \cdots \dfrac{\partial x_n}{\partial q_n} \end{bmatrix} = f(q_1\, q_2 \ldots q_n)$$

$$(\mathrm{b}) = \left[\frac{\partial x_1}{\partial t} \cdots \frac{\partial x_n}{\partial t}\right]^{\mathrm{T}}$$

and

$$(\dot{q}) = (\dot{q}_1\, \dot{q}_2 \ldots \dot{q}_n)^{\mathrm{T}}$$

Hence we may write

$$T = \frac{1}{2}\left((b)^{\mathrm{T}} + (\dot{q})^{\mathrm{T}}\, [A]^{\mathrm{T}}\right)[m]\left([A](\dot{q}) + (b)\right)$$

$$= \frac{1}{2}(\dot{q})^{\mathrm{T}}\, [A]^{\mathrm{T}}\, [m]\, [A]\, (\dot{q}) + (b)^{\mathrm{T}}[m]\, [A]\, (\dot{q}) + (b)^{\mathrm{T}}\, [m]\, (b) \tag{2.24}$$

Note that use has been made of the fact that $[m]$ is symmetrical. This fact also means that $[A]^{\mathrm{T}}[m][A]$ is symmetrical.

Let us write the kinetic energy as

$$T = T_2 + T_1 + T_0$$

where T_2, the first term of equation (2.24), is a quadratic in $\dot{q}$ and does not contain time explicitly. T_1 is linear in $\dot{q}$ and the coefficients contain time explicitly. T_0 contains time but is independent of $\dot{q}$. If the system is scleronomic with no moving constraints or moving axes then $T_1 = 0$ and $T_0 = 0$.

T_2 has the form

$$T_2 = \frac{1}{2}\sum_{ij} a_{ij}\, \dot{q}_i \dot{q}_j, \quad a_{ij} = a_{ji} = f(q)$$

and in some cases terms like $\dot{q}_i \dot{q}_j$ are absent and T_2 reduces to

$$T_2 = \frac{1}{2}\sum_i a_i \dot{q}_i^2$$

Here the co-ordinates are said to be orthogonal with respect to the kinetic energy.

2.6 Conservation laws

We shall now consider systems for which the forces are only those derivable from a position-dependent potential so that Lagrange's equations are of the form

$$\frac{\mathrm{d}}{\mathrm{d}t}\left(\frac{\partial \mathfrak{L}}{\partial \dot{q}_i}\right) - \left(\frac{\partial \mathfrak{L}}{\partial q_i}\right) = 0$$

If a co-ordinate does not appear explicitly in the Lagrangian but only occurs as its time derivative then

$$\frac{\mathrm{d}}{\mathrm{d}t}\left(\frac{\partial \mathfrak{L}}{\partial \dot{q}_i}\right) = 0$$

Therefore

$$\frac{\partial \mathfrak{L}}{\partial \dot{q}_i} = \text{constant}$$

In this case q_i is said to be a *cyclic* or *ignorable co-ordinate.*

Consider now a group of particles such that the forces depend only on the relative positions and motion between the particles. If we choose Cartesian co-ordinates relative to an arbitrary set of axes which are drifting in the x direction relative to an inertial set of axes as seen in Fig. 2.3, the Lagrangian is

$$\mathfrak{L} = \sum_{i=1}^{i=N} \frac{1}{2} m_i \left[(\dot{X} + \dot{x}_i)^2 + \dot{y}_i^2 + \dot{z}_i^2\right] - V(x_i\, y_i\, z_i)$$

Because X does not appear explicitly and is therefore ignorable

$$\frac{\partial \mathfrak{L}}{\partial \dot{X}} = \sum_{i=1}^{i=N} m_i(\dot{X} + \dot{x}_i) = \text{constant}$$

If $\dot{X} \rightarrow 0$ then

$$\sum_{i=1}^{i=N} m_i \dot{x}_i = \text{constant} \tag{2.25}$$

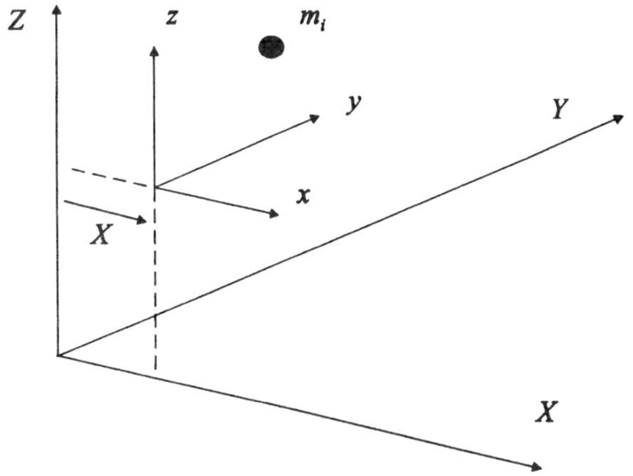

Fig. 2.3

This may be interpreted as consistent with the Lagrangian being independent of the position in space of the axes and this also leads to the linear momentum in the arbitrary x direction being constant or conserved.

Consider now the same system but this time referred to an arbitrary set of cylindrical co-ordinates. This time we shall superimpose a rotational drift of $\dot{\gamma}$ of the axes about the z axis, *see* Fig. 2.4. Now the Lagrangian is

$$\mathfrak{L} = \sum_i \frac{m_i}{2}\left[r_i^2(\dot{\theta}_i + \dot{\gamma})^2 + \dot{r}_i^2 + \dot{z}_i^2\right] - V(r_i, \theta_i, z_i)$$

Because γ is a cyclic co-ordinate

$$\frac{\partial \mathfrak{L}}{\partial \dot{\gamma}} = \sum_i m_i r_i^2(\dot{\theta}_i + \dot{\gamma}) = \text{constant}$$

If we now consider $\dot{\gamma}$ to tend to zero then

$$\frac{\partial \mathfrak{L}}{\partial \dot{\gamma}} = \sum_i m_i r_i^2 \dot{\theta}_i = \text{constant} \tag{2.26}$$

This implies that the conservation of the moment of momentum about the z axis is associated with the independence of $\partial \mathfrak{L}/\partial \dot{\gamma}$ to a change in angular position of the axes.

Both the above show that $\partial \mathfrak{L}/\partial \dot{q}$ is related to a momentum or moment of momentum. We now define $\partial \mathfrak{L}/\partial \dot{q}_i = p_i$ to be the *generalized momentum*, the dimensions of which will depend on the choice of generalized co-ordinate.

Consider the total time differential of the Lagrangian

$$\frac{d\mathfrak{L}}{dt} = \sum_j \frac{\partial \mathfrak{L}}{\partial q_j}\dot{q}_j + \sum_j \frac{\partial \mathfrak{L}}{\partial \dot{q}_j}\ddot{q}_j + \frac{\partial \mathfrak{L}}{\partial t} \tag{2.27}$$

If all the generalized forces, Q_j, are zero then Lagrange's equation is

$$\frac{d}{dt}\left(\frac{\partial \mathfrak{L}}{\partial \dot{q}_j}\right) - \left(\frac{\partial \mathfrak{L}}{\partial q_j}\right) = 0$$

Substitution into equation (2.27) gives

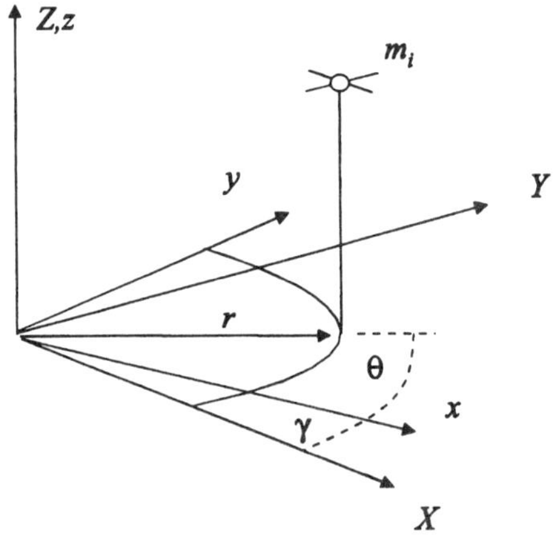

Fig. 2.4

$$\frac{d\mathfrak{L}}{dt} = \sum_j \frac{d}{dt}\left(\frac{\partial \mathfrak{L}}{\partial \dot{q}_j}\right)\dot{q}_j + \sum_j \frac{\partial \mathfrak{L}}{\partial \dot{q}_j}\ddot{q}_j + \frac{\partial \mathfrak{L}}{\partial t}$$

Thus

$$\frac{d}{dt}\left(\sum_j \frac{\partial \mathfrak{L}}{\partial \dot{q}_j}\dot{q}_j - \mathfrak{L}\right) = -\frac{\partial \mathfrak{L}}{\partial t} \tag{2.28}$$

and if the Lagrangian does not depend explicitly on time then

$$\sum_j \frac{\partial \mathfrak{L}}{\partial \dot{q}_j}\dot{q}_j - \mathfrak{L} = \text{constant} \tag{2.29}$$

Under these conditions $\mathfrak{L} = T - V = T_2(q_j\dot{q}_j) - V(q_j)$. Now

$$T_2 = \frac{1}{2}\sum_j\sum_i a_{ij}\,\dot{q}_i\dot{q}_j, \qquad a_{ij} = a_{ji}$$

so

$$\frac{\partial T_2}{\partial \dot{q}_j} = \frac{1}{2}\sum_i a_{ij}\dot{q}_i + \frac{1}{2}\sum_j a_{ij}\dot{q}_j = \sum_i a_{ij}\dot{q}_i$$

because $a_{ij} = a_{ji}$.

We can now write

$$\sum_j \frac{\partial \mathfrak{L}}{\partial \dot{q}_j}\dot{q}_j = \sum_j \frac{\partial T_2}{\partial \dot{q}_j}\dot{q}_j = \sum_j\left(\dot{q}_j\sum_i a_{ij}\dot{q}_i\right) = 2T_2$$

so that

$$\begin{aligned}\sum_j \frac{\partial \mathfrak{L}}{\partial \dot{q}_j}\dot{q}_j - \mathfrak{L} &= 2T_2 - (T_2 - V)\\ &= T_2 + V = T + V\\ &= E\end{aligned}$$

the total energy.

From equation (2.29) we see that the quantity conserved when there are (a) no generalized forces and (b) the Lagrangian does not contain time explicitly is the total energy. Thus conservation of energy is a direct consequence of the Lagrangian being independent of time. This is often referred to as symmetry in time because time could in fact be reversed without affecting the equations. Similarly we have seen that symmetry with respect to displacement in space yields the conservation of momentum theorems.

2.7 Hamilton's equations

The quantity between the parentheses in equation (2.28) is known as the *Hamiltonian H*

$$H = \sum_j \frac{\partial \mathfrak{L}}{\partial \dot{q}_j}\dot{q}_j - \mathfrak{L}(q_j\dot{q}_jt) \tag{2.30}$$

or in terms of momenta

$$H = \sum_j p_j\dot{q}_j - \mathfrak{L}(q_j\dot{q}_jt) \tag{2.31}$$

Since $\dot{q}$ can be expressed in terms of p the Hamiltonian may be considered to be a function of generalized momenta, co-ordinates and time, that is $H = H(q_j p_j t)$. The differential of H is

$$\mathrm{d}H = \sum_j \frac{\partial H}{\partial q_j}\mathrm{d}q_j + \sum_j \frac{\partial H}{\partial p_j}\mathrm{d}p_j + \frac{\partial H}{\partial t}\mathrm{d}t \tag{2.32}$$

From equation (2.32)

$$\mathrm{d}H = \sum_j (p_j d\dot{q}_j + \dot{q}_j dp_j) - \sum_j \frac{\partial \mathfrak{L}}{\partial q_j}\mathrm{d}q_j - \sum_j \frac{\partial \mathfrak{L}}{\partial \dot{q}_j}\mathrm{d}\dot{q}_j - \frac{\partial \mathfrak{L}}{\partial t}\mathrm{d}t \tag{2.33}$$

By definition $\partial \mathfrak{L}/\partial \dot{q}_j = p_j$ and from Lagrange's equations we have

$$\frac{\partial \mathfrak{L}}{\partial q_j} = \frac{\mathrm{d}}{\mathrm{d}t}\left(\frac{\partial \mathfrak{L}}{\partial \dot{q}_j}\right) = \dot{p}_j$$

Therefore, substituting into equation (2.33) the first and fourth terms cancel leaving

$$\mathrm{d}H = \sum_j \dot{q}_j \mathrm{d}p_j - \sum_j p_j \mathrm{d}q_j - \frac{\partial \mathfrak{L}}{\partial t}\mathrm{d}t \tag{2.34}$$

Comparing the coefficients of the differentials in equations (2.32) and (2.34) we have

$$\frac{\partial H}{\partial q_j} = -\dot{p}_j \quad \frac{\partial H}{\partial p_j} = \dot{q}_j \tag{2.35}$$

and

$$\frac{\partial H}{\partial t} = -\frac{\partial \mathfrak{L}}{\partial t}$$

Equations (2.35) are called *Hamilton's canonical equations*. They constitute a set of $2n$ first-order equations in place of a set of n second-order equations defined by Lagrange's equations.

It is instructive to consider a system with a single degree of freedom with a moving foundation as shown in Fig. 2.5. First we shall use the absolute motion of the mass as the generalized co-ordinate.

$$\mathfrak{L} = \frac{m\dot{x}^2}{2} - \frac{k}{2}(x - x_0)^2$$

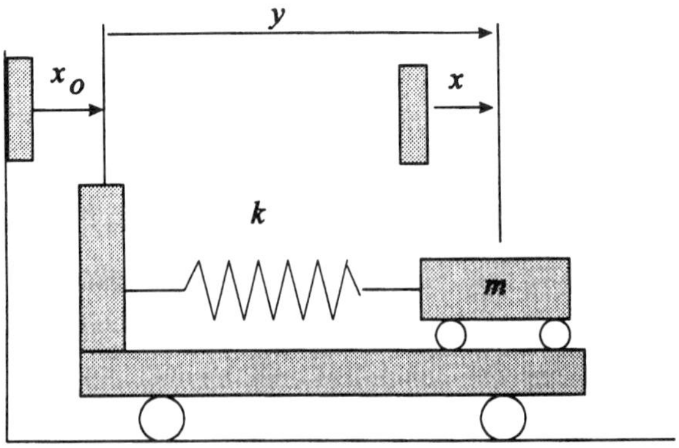

Fig. 2.5

$$\frac{\partial \mathfrak{L}}{\partial \dot{x}} = m\dot{x} = p$$

Therefore $\dot{x} = p/m$. From equation (2.32)

$$H = p(p/m) - \left[\frac{p^2}{2m} - \frac{k}{2}(x-x_0)^2\right]$$

$$= \frac{p^2}{2m} + \frac{k}{2}(x-x_0)^2 \qquad (2.36)$$

In this case it is easy to see that H is the total energy but it is not conserved because x_0 is a function of time and hence so is H. Energy is being fed in and out of the system by whatever forces are driving the foundation.

Using y as the generalized co-ordinate we obtain

$$\mathfrak{L} = \frac{m}{2}(\dot{y} + \dot{x}_0)^2 - \frac{k}{2}y^2$$

$$\frac{\partial \mathfrak{L}}{\partial \dot{y}} = m(\dot{y} + \dot{x}_2) = p$$

Therefore $\dot{y} = (p/m) - \dot{x}_0$ and

$$H = p\left[\frac{p}{m} - \dot{x}_0\right] - \left[\frac{p^2}{2m} - \frac{ky^2}{2}\right]$$

$$= \frac{p^2}{2m} - p\dot{x}_0 + \frac{ky^2}{2} \qquad (2.37)$$

Taking specific values for x_0 and x (and hence y) it is readily shown that the numerical value of the Lagrangian is the same in both cases whereas the value of the Hamiltonian is different, in this example by the amount $p\dot{x}_0$.

If we choose $\dot{x}_0$ to be constant then time does not appear explicitly in the second case; therefore H is conserved but it is not the total energy. Rewriting equation (2.37) in terms of y and x_0 we get

$$H = \left(\frac{1}{2}m\dot{y}^2 + \frac{1}{2}ky^2\right) - \frac{1}{2}m\dot{x}_0^2 \qquad (2.38)$$

where the term in parentheses is the total energy as seen from the moving foundation and the last term is a constant providing, of course, that $\dot{x}_0$ is a constant.

We have seen that choosing different co-ordinates changes the value of the Hamiltonian and also affects conservation properties, but the value of the Lagrangian remains unaltered. However, the equations of motion are identical whichever form of $\mathfrak{L}$ or H is used.

2.8 Rotating frame of reference and velocity-dependent potentials

In all the applications of Lagrange's equations given so far the kinetic energy has always been written strictly relative to an inertial set of axes. Before dealing with moving axes in general we shall consider the case of axes rotating at a constant speed relative to a fixed axis.

Assume that in Fig. 2.6 the *XYZ* axes are inertial and the *xyz* axes are rotating at a constant speed Ω about the *Z* axis. The position vector relative to the inertial axes is ***r*** and relative to the rotating axes is ***ρ***.

Now

$$\boldsymbol{r} = \boldsymbol{\rho}$$

and

$$\dot{\boldsymbol{r}} = \frac{\partial \boldsymbol{\rho}}{\partial t} + \boldsymbol{\Omega}\times\boldsymbol{\rho}$$

The kinetic energy for a particle is

$$T = \frac{1}{2} m\dot{\boldsymbol{r}}\cdot\dot{\boldsymbol{r}}$$

or

$$T = \frac{m}{2}\left(\frac{\partial \boldsymbol{\rho}}{\partial t}\cdot\frac{\partial \boldsymbol{\rho}}{\partial t} + (\boldsymbol{\Omega}\times\boldsymbol{\rho})\cdot(\boldsymbol{\Omega}\times\boldsymbol{\rho}) + 2\frac{\partial \boldsymbol{\rho}}{\partial t}\cdot(\boldsymbol{\Omega}\times\boldsymbol{\rho})\right) \tag{2.39}$$

Let $\boldsymbol{\Omega}\times\boldsymbol{\rho} = \boldsymbol{A}$, a vector function of position, so the kinetic energy may be written

$$T = \frac{m}{2}\left(\frac{\partial \boldsymbol{\rho}}{\partial t}\right)^2 + \frac{m}{2}A^2 + m\frac{\partial \boldsymbol{\rho}}{\partial t}\cdot\boldsymbol{A}$$

and the Lagrangian is

$$\mathfrak{L} = \frac{m}{2}\left(\frac{\partial \boldsymbol{\rho}}{\partial t}\right)^2 - \left(-\frac{m}{2}A^2 - m\frac{\partial \boldsymbol{\rho}}{\partial t}\cdot\boldsymbol{A}\right) - V \tag{2.39a}$$

The first term is the kinetic energy as seen from the rotating axes. The second term relates to a position-dependent potential function $ø = -A^2/2$. The third term is the negative of a velocity-dependent potential energy *U*. *V* is the conventional potential energy assumed to depend only on the relative positions of the masses and therefore unaffected by the choice of reference axes

$$\mathfrak{L} = \frac{m}{2}\left(\frac{\partial \boldsymbol{\rho}}{\partial t}\right)^2 - \left(mø + U\right) - V \tag{2.39b}$$

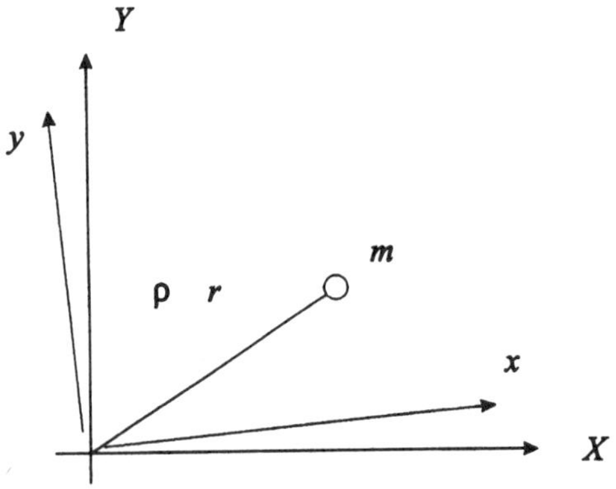

Fig. 2.6

It is interesting to note that for a charged particle, of mass m and charge $\bar{q}$, moving in a magnetic field $\boldsymbol{B} = \nabla \times \boldsymbol{A}$, where $\boldsymbol{A}$ is the magnetic vector potential, and an electric field $\boldsymbol{E} = -\nabla ø - \frac{\partial \boldsymbol{A}}{\partial t}$, where ø is a scalar potential, the Lagrangian can be shown to be

$$\mathfrak{L} = -\frac{m}{2}\left(\frac{\partial \boldsymbol{\rho}}{\partial t}\right)^2 - \left(\bar{q}ø - \bar{q}\frac{\partial \boldsymbol{\rho}}{\partial t}\cdot\boldsymbol{A}\right) - V \tag{2.40}$$

This has a similar form to equation (2.39b).

From equation (2.40) the generalized momentum is

$$p_x = m\dot{x} + \bar{q}A_x$$

From equation (2.40b) the generalized momentum is

$$p_x = m\dot{x} + mA_x = m\dot{x} + m(\omega_y z - \omega_z y)$$

In neither of these expressions for generalized momentum is the momentum that as seen from the reference frame. In the electromagnetic situation the extra momentum is often attributed to the momentum of the field. In the purely mechanical problem the momentum is the same as that referenced to a coincident inertial frame. However, it must be noted that the xyz frame is rotating so the time rate of change of momentum will be different to that in the inertial frame.

EXAMPLE

An important example of a rotating co-ordinate frame is when the axes are attached to the Earth. Let us consider a special case for axes with origin at the centre of the Earth, as shown in Fig. 2.7 The z axis is inclined by an angle α to the rotational axis and the x axis initially intersects the equator. Also we will consider only small movements about the point where the z axis intersects the surface. The general form for the Lagrangian of a particle is

$$\begin{aligned}\mathfrak{L} &= \frac{m}{2}\frac{\partial \boldsymbol{\rho}}{\partial t}\cdot\frac{\partial \boldsymbol{\rho}}{\partial t} + \frac{m}{2}(\boldsymbol{\Omega}\times\boldsymbol{\rho})\cdot(\boldsymbol{\Omega}\times\boldsymbol{\rho}) + m\frac{\partial \boldsymbol{\rho}}{\partial t}\cdot(\boldsymbol{\Omega}\times\boldsymbol{\rho}) - V\\ &= T' - U_1 - U_2 - V\end{aligned}$$

with

$$\boldsymbol{\Omega} = \omega_x\boldsymbol{i} + \omega_y\boldsymbol{j} + \omega_z\boldsymbol{k} \quad \text{and} \quad \boldsymbol{\rho} = x\boldsymbol{i} + y\boldsymbol{j} + z\boldsymbol{k}$$

$$\boldsymbol{A} = \boldsymbol{\Omega}\times\boldsymbol{\rho} = \boldsymbol{i}(\omega_y z - \omega_z y) + \boldsymbol{j}(\omega_z x - \omega_x z) + \boldsymbol{k}(\omega_x y - \omega_y x)$$

$$\begin{aligned}\frac{m}{2}\boldsymbol{A}\cdot\boldsymbol{A} &= \frac{m}{2}[(\omega_y z - \omega_z y)^2 + (\omega_z x - \omega_x z)^2 + (\omega_x y - \omega_y x)^2]\\ &= -U_1\end{aligned}$$

and

$$\begin{aligned}m\frac{\partial \boldsymbol{\rho}}{\partial t}\cdot\boldsymbol{A} &= m\dot{x}(\omega_y z - \omega_z y) + m\dot{y}(\omega_z x - \omega_x z) + m\dot{z}(\omega_x y - \omega_y x)\\ &= -U_2\end{aligned}$$

where $\dot{x} = \frac{\partial x}{\partial t}$, etc. the velocities as seen from the moving axes.

When Lagrange's equations are applied to these functions U_1 gives rise to position-dependent fictitious forces and U_2 to velocity and position-dependent

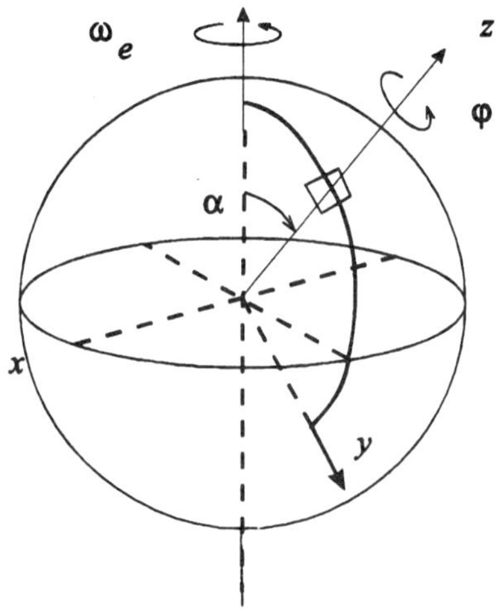

Fig. 2.7

fictitious forces. Writing $U = U_1 + U_2$ we can evaluate the x component of the fictitious force from

$$\frac{\mathrm{d}}{\mathrm{d}t}\left(\frac{\partial U}{\partial \dot{x}}\right) - \left(\frac{\partial U}{\partial x}\right) = -Q_{\mathrm{f}x}$$

$$m(\omega_y \dot{z} - \omega_z \dot{y}) - m(\omega_z x - \omega_x z)\omega_z - m(\omega_x y - \omega_y x)(-\omega_y) - m(\dot{y}\omega_z - \dot{z}\omega_y) = -Q_{\mathrm{f}x}$$

or

$$-Q_{\mathrm{f}x} = m[(\omega_z^2 + \omega_y^2)x - \omega_x\omega_y y - \omega_x\omega_z z] + 2m(\omega_y \dot{z} - \omega_z \dot{y})$$

Similarly

$$-Q_{\mathrm{f}y} = m[(\omega_x^2 + \omega_z^2)y - \omega_y\omega_z z - \omega_y\omega_x x] + 2m(\omega_z \dot{x} - \omega_x \dot{z})$$

$$-Q_{\mathrm{f}z} = m[(\omega_y^2 + \omega_x^2)z - \omega_z\omega_x x - \omega_z\omega_y y] + 2m(\omega_x \dot{y} - \omega_y \dot{x})$$

For small motion in a tangent plane parallel to the xy plane we have $\dot{z} = 0$ and $z = R$, since $x \ll z$ and $y \ll z$, thus

$$-Q_{\mathrm{f}x} = m[-\omega_x\omega_z R] - 2m\omega_z \dot{y} \tag{i}$$

$$-Q_{\mathrm{f}y} = m[-\omega_y\omega_z R] + 2m\omega_z \dot{x} \tag{ii}$$

$$-Q_{\mathrm{f}z} = m(\omega_y^2 + \omega_x^2)R - 2m(\omega_x \dot{y} - \omega_y \dot{x}) \tag{iii}$$

We shall consider two cases:

Case 1, where the xyz axes remain fixed to the Earth:

$$\omega_x = 0 \quad \omega_y = -\omega_\mathrm{e}\sin\alpha \quad \text{and} \quad \omega_z = \omega_\mathrm{e}\cos\alpha$$

Equations (i) to (iii) are now

$$-Q_{\mathrm{f}x} = -2m\omega_\mathrm{e}\cos\alpha\,\dot{y}$$

$$-Q_{\mathrm{f}y} = m(\omega_\mathrm{e}^2\sin\alpha\ \cos\alpha\, R) + 2m\omega_\mathrm{e}\cos\alpha\,\dot{x}$$

$$-Q_{\mathrm{f}z} = m(\omega_\mathrm{e}^2\sin^2\alpha)R - 2m\omega_\mathrm{e}\sin\alpha\,\dot{x}$$

from which we see that there are fictitious Coriolis forces related to $\dot{x}$ and $\dot{y}$ and also some position-dependent fictitious centrifugal forces. The latter are usually absorbed in the modified gravitational field strength. In practical terms the value of g is reduced by some 0.3% and a plumb line is displaced by about 0.1°.

Case 2, where the xyz axes rotate about the z axis by angle ϕ:

$$\omega_x = \omega_e \sin\alpha \sin\phi, \quad \omega_y = -\omega_e \sin\alpha \cos\phi \quad \text{and } \omega_z = \omega_e \cos\alpha + \dot{\phi}$$

We see that if $\dot{\phi} = -\omega_e \cos\alpha$ then $\omega_z = 0$, so the Coriolis terms in equations (i) and (ii) disappear. Motion in the tangent plane is now the same as that in a plane fixed to a non-rotating Earth.

2.9 Moving co-ordinates

In this section we shall consider the situation in which the co-ordinate system moves with a group of particles. These axes will be translating and rotating relative to an inertial set of axes. The absolute position vector will be the sum of the position vector of a reference point to the origin plus the position vector relative to the moving axes. Thus, referring to Fig. 2.8, $\boldsymbol{r}_j = \boldsymbol{R} + \boldsymbol{\rho}_j$ so the kinetic energy will be

$$T = \sum_j \frac{m_j}{2} \dot{\boldsymbol{r}}_j \cdot \dot{\boldsymbol{r}}_j = \sum_j \frac{m_j}{2} (\dot{\boldsymbol{R}} \cdot \dot{\boldsymbol{R}} + \dot{\boldsymbol{\rho}}_j \cdot \dot{\boldsymbol{\rho}}_j + 2\dot{\boldsymbol{R}} \cdot \dot{\boldsymbol{\rho}}_j)$$

Denoting $\sum_j m_j = m$, the total mass,

$$T = \frac{m}{2} \dot{\boldsymbol{R}} \cdot \dot{\boldsymbol{R}} + \sum_j \frac{m_j}{2} \dot{\boldsymbol{\rho}}_j \cdot \dot{\boldsymbol{\rho}}_j = \dot{\boldsymbol{R}} \cdot \sum_j m_j \cdot \dot{\boldsymbol{\rho}}_j \tag{2.41}$$

Here the dot above the variables signifies differentiation with respect to time as seen from the inertial set of axes. In the following arguments the dot will refer to scalar differentiation.

If we choose the reference point to be the centre of mass then the third term will vanish. The first term on the right hand side of equation (2.41) will be termed T_0 and is the kinetic energy of a single particle of mass m located at the centre of mass. The second term will be

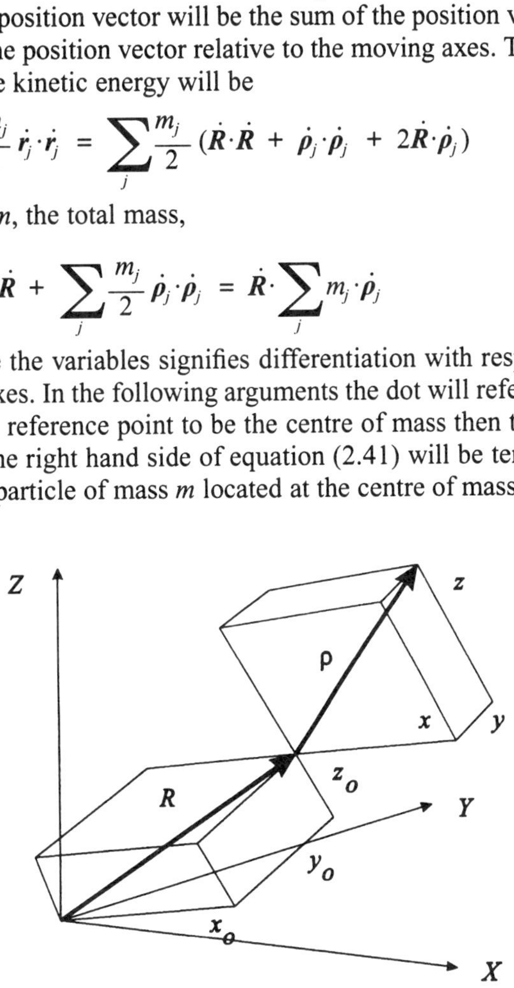

Fig. 2.8

denoted by T_G and is the kinetic energy due to motion relative to the centre of mass, but still as seen from the inertial axes.

The position vector $\boldsymbol{R}$ can be expressed in the moving co-ordinate system xyz, the specific components being x_0, y_0 and z_0,

$$\boldsymbol{R} = x_0\boldsymbol{i} + y_0\boldsymbol{j} + z_0\boldsymbol{k}$$

By the rules for differentiation with respect to rotating axes

$$\begin{aligned}\frac{\mathrm{d}\boldsymbol{R}}{\mathrm{d}t_{xyz}} &= \frac{\partial \boldsymbol{R}}{\partial t_{xyz}} + \omega \times \boldsymbol{R} \\ &= \dot{x}_0\boldsymbol{i} + \dot{y}_0\boldsymbol{j} + \dot{x}_0\boldsymbol{k} + (\omega_y z_0 - \omega_z y_0)\boldsymbol{i} \\ &\qquad + (\omega_z x_0 - \omega_x z_0)\boldsymbol{j} + (\omega_x y_0 - \omega_y x_0)\boldsymbol{k}\end{aligned}$$

so

$$T_0 = \frac{m}{2}\Big[\dot{x}_0\boldsymbol{i} + \dot{y}_0\boldsymbol{j} + \dot{x}_0\boldsymbol{k} + (\omega_y z_0 - \omega_z y_0)\boldsymbol{i} + (\omega_z x_0 - \omega_x z_0)\boldsymbol{j} + (\omega_x y_0 - \omega_y x_0)\boldsymbol{k}\Big]^2 \tag{2.42}$$

Similarly with $\boldsymbol{\rho}_j = x_j\boldsymbol{i} + y_j\boldsymbol{j} + z_j\boldsymbol{k}$ we have

$$T_G = \sum_j \frac{m_j}{2}\Big[\dot{x}_j\boldsymbol{i} + \dot{y}_j\boldsymbol{j} + \dot{x}_j\boldsymbol{k} + (\omega_y z_j - \omega_z y_j)\boldsymbol{i} + (\omega_z x_j - \omega_x z_j)\boldsymbol{j} + (\omega_x y_j - \omega_y x_j)\boldsymbol{k}\Big]^2 \tag{2.43}$$

The Lagrangian is

$$\mathfrak{L} = T_0(x_0\, y_0\, z_0\, \dot{x}_0\, \dot{y}_0\, \dot{z}_0) + T_G(x_j\, y_j\, z_j\, \dot{x}_j\, \dot{y}_j\, \dot{z}_j) - V \tag{2.44}$$

Let the linear momentum of the system be $\boldsymbol{p}$. Then the resultant force $\boldsymbol{F}$ acting on the system is

$$\boldsymbol{F} = \frac{\mathrm{d}}{\mathrm{d}t_{XYZ}}\boldsymbol{p} = \frac{\mathrm{d}}{\mathrm{d}t_{xyz}}\boldsymbol{p} + \omega \times \boldsymbol{p}$$

and the component in the x direction is

$$F_x = \frac{\mathrm{d}}{\mathrm{d}t_{XYZ}}p_x = \frac{\mathrm{d}}{\mathrm{d}t_{xyz}}p_x + (\omega_y p_z - \omega_z p_y)$$

In this case the momenta are generalized momenta so we may write

$$F_x = Q_x = \frac{\mathrm{d}}{\mathrm{d}t_{xyz}}\left(\frac{\partial \mathfrak{L}}{\partial \dot{x}_0}\right) - \left(\omega_z \frac{\partial \mathfrak{L}}{\partial \dot{y}_0} - \omega_y \frac{\partial \mathfrak{L}}{\partial \dot{z}_0}\right) \tag{2.45}$$

If Lagrange's equations are applied to the Lagrangian, equation (2.44), exactly the same equations are formed, so it follows that in this case the contents of the last term are equivalent to $\partial \mathfrak{L}/\partial x_0$.

If the system is a rigid body with the xyz axes aligned with the principal axes then the kinetic energy of the body for motion relative to the centre of mass T_G is

$$T_G = \frac{1}{2}I_x\omega_x^2 + \frac{1}{2}I_y\omega_y^2 + \frac{1}{2}I_z\omega_z^2$$, see section 4.5

The modified form of Lagrange's equation for angular motion

$$Q_{\omega x} = \frac{\mathrm{d}}{\mathrm{d}t_{xyz}}\left(\frac{\partial \mathfrak{L}}{\partial \dot{\omega}_x}\right) - \left(\omega_z \frac{\partial \mathfrak{L}}{\partial \dot{\omega}_y} - \omega_y \frac{\partial \mathfrak{L}}{\partial \dot{\omega}_z}\right) \tag{2.46}$$

yields

$$Q_{\omega x} = I_x \dot{\omega}_x - \left(\omega_z I_y \omega_y - \omega_y I_z \omega_z\right) \tag{2.47}$$

In this equation ω_x is treated as a generalized velocity but there is not an equivalent generalized co-ordinate. This, and the two similar ones in $Q_{\omega y}$ and $Q_{\omega z}$, form the well-known Euler's equations for the rotation of rigid bodies in space.

For flexible bodies T_G is treated in the usual way, noting that it is not a function of x_0, $\dot{x}_0$ etc., but still involves ω.

2.10 Non-holonomic systems

In the preceding part of this chapter we have always assumed that the constraints are holonomic. This usually means that it is possible to write down the Lagrangian such that the number of generalized co-ordinates is equal to the number of degrees of freedom. There are situations where a constraint can only be written in terms of velocities or differentials.

One often-quoted case is the problem of a wheel rolling without slip on an inclined plane (*see* Fig. 2.9).

Assuming that the wheel remains normal to the plane we can write the Lagrangian as

$$\mathfrak{L} = \frac{1}{2}m(\dot{x}^2 + \dot{y}^2) + \frac{1}{2}I_1\dot{\varnothing}^2 + \frac{1}{2}I_2\dot{\psi}^2 - mg(\sin\alpha\, y + \cos\alpha\, r)$$

The equation of constraint may be written

$$\mathrm{d}s = r\,\mathrm{d}\varnothing$$

or as

$$\begin{aligned} \mathrm{d}x &= \mathrm{d}s \sin\psi = r \sin\psi\, \mathrm{d}\varnothing \\ \mathrm{d}y &= \mathrm{d}s \cos\psi = r \cos\psi\, \mathrm{d}\varnothing \end{aligned}$$

We now introduce the concept of the *Lagrange undetermined multipliers* λ. Notice that each of the constraint equations may be written in the form $\Sigma a_{jk}\, dq_j = 0$; this is similar in form to the expression for virtual work. Multiplication by λ_k does not affect the equality but the dimensions of λ_k are such that each term has the dimensions of work. A modified virtual work expression can be formed by adding all such sums to the existing expression for virtual work. So $\delta W' = \delta W + \Sigma(\lambda_k \Sigma a_{jk} \mathrm{d}q_j)$; this means that extra generalized forces will be formed and thus included in the resulting Lagrange equations.

Applying this scheme to the above constraint equations gives

$$\begin{aligned} \lambda_1 \mathrm{d}x - \lambda_1(\mathrm{r}\sin\psi)\mathrm{d}\varnothing &= 0 \\ \lambda_2 \mathrm{d}y - \lambda_2(\mathrm{r}\cos\psi)\mathrm{d}\varnothing &= 0 \end{aligned}$$

The only term in the virtual work expression is that due to the couple C applied to the shaft, so $\delta W = C\,\delta\varnothing$. Adding the constraint equation gives

$$\delta W' = C\,\delta\varnothing + \lambda_1 \mathrm{d}x + \lambda_2 \mathrm{d}y - [\lambda_1(r\sin\psi) + \lambda_2(r\cos\psi)]\mathrm{d}\varnothing$$

Applying Lagrange's equations to $\mathfrak{L}$ for $q = x, y, \varnothing$ and ψ in turn yields

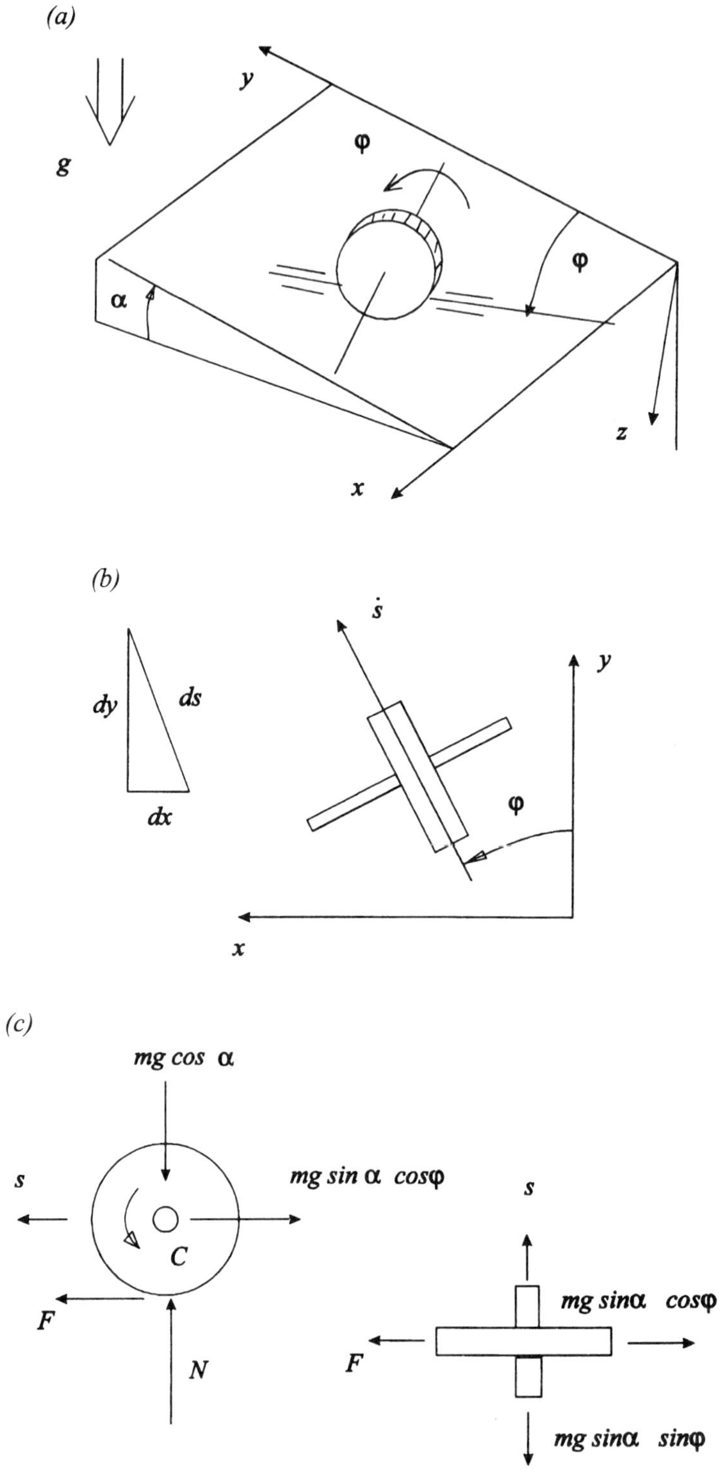

Fig. 2.9 (a), (b) and (c)

$$
\begin{aligned}
m\ddot{x} &= \lambda_1 \\
m\ddot{y} + mg\sin\alpha &= \lambda_2 \\
I_1\ddot{\phi} &= C - [\lambda_1(\text{r}\sin\psi) + \lambda_2(r\cos\psi)] \\
I_2\ddot{\psi} &= 0
\end{aligned}
$$

In addition we still have the constraint equations

$$
\begin{aligned}
\dot{x} &= r\sin\psi\,\dot{\phi} \\
\dot{y} &= r\cos\psi\,\dot{\phi}
\end{aligned}
$$

Simple substitution will eliminate λ_1 and λ_2 from the equations.

From a free-body diagram approach it is easy to see that

$$
\begin{aligned}
\lambda_1 &= F\sin\psi \\
\lambda_2 &= F\cos\psi
\end{aligned}
$$

and

$$[\lambda_1(r\sin\psi) + \lambda_2(r\cos\psi)] = -Fr$$

The use of Lagrange multipliers is not restricted to non-holonomic constraints, they may be used with holonomic constraints; if the force of constraint is required. For example, in this case we could have included $\lambda_3 \mathrm{d}z = 0$ to the virtual work expression as a result of the motion being confined to the xy plane. (It is assumed that gravity is sufficient to maintain this condition.) The equation of motion in the z direction is

$$-mg\cos\alpha = \lambda_3$$

It is seen here that $-\lambda_3$ corresponds to the normal force between the wheel and the plane.

However, non-holonomic systems are in most cases best treated by free-body diagram methods and therefore we shall not pursue this topic any further. (*See* Appendix 2 for methods suitable for non-holonomic systems.)

2.11 Lagrange's equations for impulsive forces

The force is said to be impulsive when the duration of the force is so short that the change in the position co-ordinates is negligible during the application of the force. The variation in any body forces can be neglected but contact forces, whether elastic or not, are regarded as external. The Lagrangian will thus be represented by the kinetic energy only and by the definition of short duration $\partial \mathrm{T}/\partial q$ will also be negligible. So we write

$$\frac{\mathrm{d}}{\mathrm{d}t}\left(\frac{\partial T}{\partial \dot{q}_j}\right) = Q_j$$

Integrating over the time of the impulse τ gives

$$\Delta\left(\frac{\partial T}{\partial \dot{q}_j}\right) = \int_0^t Q_j\,\mathrm{d}t \tag{2.48}$$

or

$$\Delta[\text{generalized momentum}] = \text{generalized impulse}$$

$$\Delta p_j = \mathrm{J}_j$$

EXAMPLE

The two uniform equal rods shown in Fig. 2.10 are pinned at B and are moving to the right at a speed *V*. End A strikes a rigid stop. Determine the motion of the two bodies immediately after the impact. Assume that there are no friction losses, no residual vibration and that the impact process is elastic.

The kinetic energy is given by

$$T = \frac{m}{2}\dot{x}_1^2 + \frac{m}{2}\dot{x}_2^2 + \frac{I}{2}\dot{\theta}_1^2 + \frac{I}{2}\dot{\theta}_2^2$$

The virtual work done by the impact force at A is

$$\delta W = F(-\mathrm{d}x_1 + a\,\mathrm{d}\theta_1)$$

and the constraint equation for the velocity of point B is

$$\dot{x}_1 + a\dot{\theta}_1 = \dot{x}_2 - a\dot{\theta}_2 \qquad \text{(ia)}$$

or, in differential form,

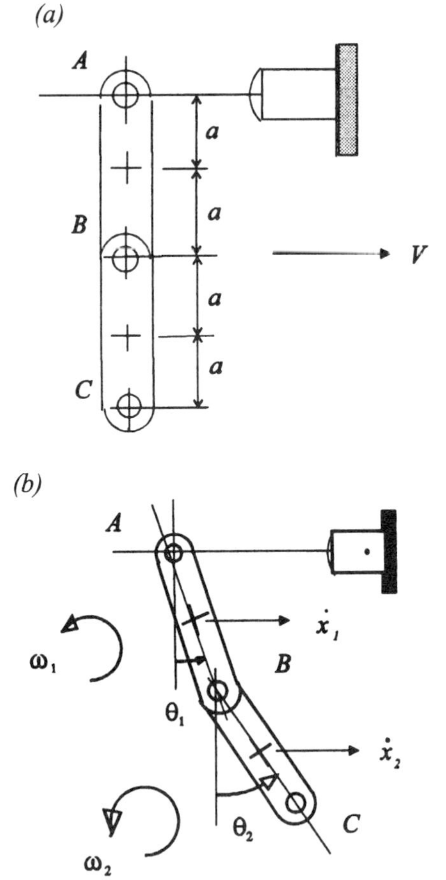

Fig. 2.10 (a) and (b)

$$dx_1 - dx_2 + a d\theta_1 + a d\theta_2 = 0 \qquad \text{(ib)}$$

There are two ways of using the constraint equation: one is to use it to eliminate one of the variables in T and the other is to make use of Lagrange multipliers. Neither has any great advantage over the other; we shall choose the latter. Thus the extra terms to be added to the virtual work expression are

$$\lambda[dx_1 - dx_2 + a d\theta_1 + a d\theta_2]$$

Thus the effective virtual work expression is

$$\delta W' = F(-dx_1 + a d\theta_1) + \lambda[dx_1 - dx_2 + a d\theta_1 + a d\theta_2]$$

Applying the Lagrange equations for impulsive forces

$$m(\dot{x}_1 - V) = -\int F dt + \int \lambda dt \qquad \text{(ii)}$$
$$m(\dot{x}_2 - V) = -\int \lambda\, dt \qquad \text{(iii)}$$
$$I\dot{\theta}_1 = \int aF\, dt + \int a\lambda dt \qquad \text{(iv)}$$
$$I\dot{\theta}_2 = \int a\lambda dt \qquad \text{(v)}$$

There are six unknowns but only five equations (including the equation of constraint, equation (i)). We still need to include the fact that the impact is elastic. This means that at the impact point the displacement–time curve must be symmetrical about its centre, in this case about the time when point A is momentarily at rest. The implication of this is that, at the point of contact, the speed of approach is equal to the speed of recession. It is also consistent with the notion of reversibility or time symmetry.

Our final equation is then

$$V = a\dot{\theta}_1 - \dot{x}_1 \qquad \text{(vi)}$$

Alternatively we may use conservation of energy. Equating the kinetic energies before and after the impact and multiplying through by 2 gives

$$mV^2 = m\dot{x}_1^2 + m\dot{x}_2^2 + I\dot{\theta}_1^2 + I\dot{\theta}_2^2 \qquad \text{(vi a)}$$

It can be demonstrated that using this equation in place of equation (vi) gives the same result. From a free-body diagram approach it can be seen that λ is the impulsive force at B.

We can eliminate the impulses from equations (ii) to (v). One way is to add equation (iii) times 'a' to equation (v) to give

$$m(\dot{x}_2 - V)a + I\dot{\theta}_2 = 0 \qquad \text{(vii)}$$

Also by adding 3 times equation (iii) to the sum of equations (ii), (iv) and (v) we obtain

$$m(\dot{x}_1 - V)a + 3m(\dot{x}_2 - V)a + I\dot{\theta}_1 + I\dot{\theta}_2 = 0 \qquad \text{(viii)}$$

This equation may be obtained by using conservation of moment of momentum for the whole system about the impact point and equation (vii) by the conservation of momentum for the lower link about the hinge B.

Equations (ia), (vi), (vii) and (viii) form a set of four linear simultaneous equations in the unknown velocities $\dot{x}_1$, $\dot{x}_2$, $\dot{\theta}_1$ and $\dot{\theta}_2$. These may be solved by any of the standard methods.

3

Hamilton's Principle

3.1 Introduction

In the previous chapters the equations of motion have been presented as differential equations. In this chapter we shall express the equations in the form of stationary values of a time integral. The idea of zero variation of a quantity was seen in the method of virtual work and extended to dynamics by means of D'Alembert's principle. It has long been considered that nature works so as to minimize some quantity often called action. One of the first statements was made by Maupertuis in 1744. The most commonly used form is that devised by Sir William Rowan Hamilton around 1834.

Hamilton's principle could be considered to be a basic statement of mechanics, especially as it has wide applications in other areas of physics, but we shall develop the principle directly from Newtonian laws. For the case with conservative forces the principle states that the time integral of the Lagrangian is stationary with respect to variations in the 'path' in configuration space. That is, the correct displacement–time relationships give a minimum (or maximum) value of the integral.

In the usual notation

$$\delta \int_{t_1}^{t_2} \mathfrak{L}\, \mathrm{d}t = 0 \qquad (3.1)$$

or

$$\delta I = 0$$

where

$$I = \int_{t_1}^{t_2} \mathfrak{L}\, \mathrm{d}t = 0 \qquad (3.2)$$

This integral is sometimes referred to as the action integral. There are several different integrals which are also known as action integrals.

The calculus of variations has an interesting history with many applications but we shall develop only the techniques necessary for the problem in hand.

3.2 Derivation of Hamilton's principle

Consider a single particle acted upon by non-conservative forces F_i, F_j, F_k and conservative forces f_i, f_j, f_k which are derivable from a position-dependent potential function. Referring to Fig. 3.1 we see that, with p designating momentum, in the x direction

$$F_i + f_i = \frac{\mathrm{d}}{\mathrm{d}t}(p_i)$$

with similar expressions for the y and z directions.

For a system having N particles D'Alembert's principle gives

$$\sum_i \left(F_i + f_i - \frac{\mathrm{d}}{\mathrm{d}t}(p_i) \right) \delta x_i = 0, \; 1 \le i \le 3N$$

We may now integrate this expression over the time interval t_1 to t_2

$$\int_{t_1}^{t_2} \sum_i \left(F_i + f_i - \frac{\mathrm{d}}{\mathrm{d}t}(p_i) \right) \delta x_i \,\mathrm{d}t = 0$$

Now $f_i = -\dfrac{\partial V}{\partial x_i}$ and the third term can be integrated by parts. So interchanging the order of summation and integration and then integrating the third term we obtain

$$\sum_i \left(\int_{t_1}^{t_2} F_i \delta x_i \,\mathrm{d}t - \int_{t_1}^{t_2} \frac{\partial V}{\partial x_i} \delta x_i \mathrm{d}t - [p_i \delta x_i]_{t_1}^{t_2} + \int_{t_1}^{t_2} (p_i) \frac{\mathrm{d}}{\mathrm{d}t}(\delta x_i) \,\mathrm{d}t \right) = 0 \qquad (3.3)$$

We now impose a restriction on the variation such that it is zero at the extreme points t_1 and t_2; therefore the third term in the above equation vanishes. Reversing the order of summation and integration again, equation (3.3) becomes

$$\int_{t_1}^{t_2} \left(\sum_i F_i \,\delta x_i - \delta V + \sum_i p_i \,\delta \dot{x}_i \right) \mathrm{d}t = 0 \qquad (3.4)$$

Let us assume that the momentum is a function of velocity but not necessarily a linear one. With reference to Fig. 3.2 if P is the resultant force acting on a particle then by definition

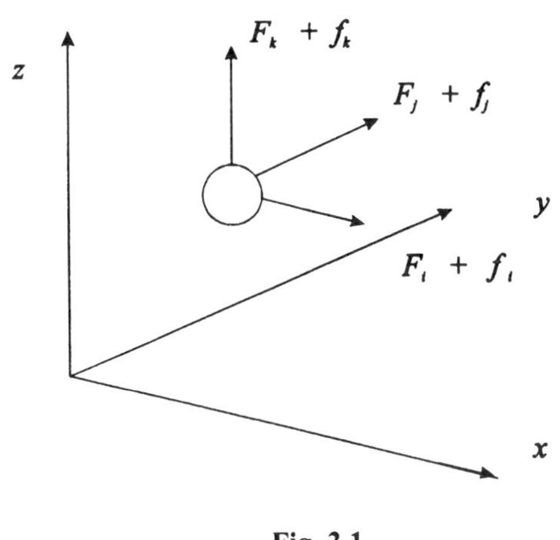

Fig. 3.1

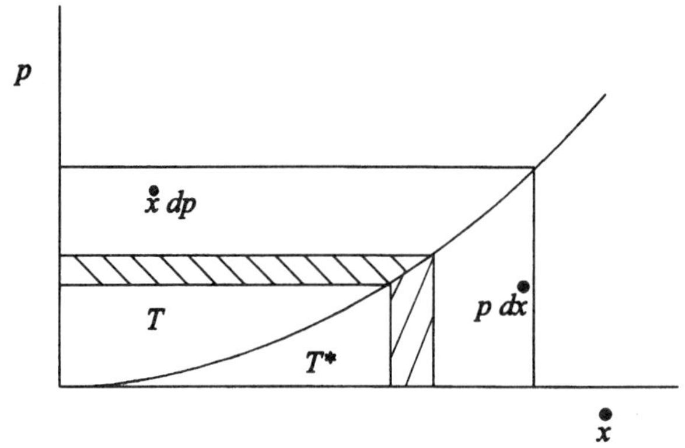

Fig. 3.2

$$P_i = \frac{\mathrm{d}p_i}{\mathrm{d}t}$$

so the work done over an elemental displacement is

$$P_i \mathrm{d}x_i = \frac{\mathrm{d}p_i}{\mathrm{d}t}\, dx_i = \dot{x}_i \mathrm{d}p_i$$

The kinetic energy of the particle is equal to the work done, so

$$T = \int \dot{x}_i \mathrm{d}p_i$$

Let the *complementary kinetic energy*, or *co-kinetic energy*, be defined by

$$T^* = \int p_i \mathrm{d}\dot{x}_i$$

It follows that $\delta T^* = p_i \delta \dot{x}_i$ so substitution into equation (3.4) leads to

$$\int_{t_1}^{t_2} \left(\delta(T^* - V) + \sum_i F_i \delta x_i \right) \mathrm{d}t = 0$$

or

$$\delta \int_{t_1}^{t_2} (T^* - V)\, \mathrm{d}t = -\int_{t_1}^{t_2} (\sum_i F_i \delta x_i)\, \mathrm{d}t = \delta \int_{t_1}^{t_2} (-W)\, \mathrm{d}t \tag{3.5}$$

where δW is the virtual work done by non-conservative forces. This is *Hamilton's principle.* If momentum is a linear function of velocity then $T^* = T$. It is seen in section 3.4 that the quantity $(T^* - V)$ is in fact the Lagrangian.

If all the forces are derivable from potential functions then Hamilton's principle reduces to

$$\delta \int_{t_1}^{t_2} \mathfrak{L}\, \mathrm{d}t = 0 \tag{3.6}$$

All the comments made in the previous chapter regarding generalized co-ordinates apply equally well here so that $\mathfrak{L}$ is independent of the co-ordinate system.

3.3 Application of Hamilton's principle

In order to establish a general method for seeking a stationary value of the action integral we shall consider the simple mass/spring system with a single degree of freedom shown in Fig. 3.3. Figure 3.4 shows a plot of x versus t between two arbitrary times. The solid line is the actual plot, or path, and the dashed line is a varied path. The difference between the two paths is δx. This is made equal to $\varepsilon\eta(t)$, where η is an arbitrary function of time except that it is zero at the extremes. The factor ε is such that when it equals zero the two paths coincide. We can establish the conditions for a stationary value of the integral I by setting $\mathrm{d}I/\mathrm{d}\varepsilon = 0$ and then putting $\varepsilon = 0$.

From Fig. 3.4 we see that

$$\delta\,(x + \mathrm{d}x) = \delta x + \mathrm{d}(\delta x)$$

Therefore $\delta\,(\mathrm{d}x) = \mathrm{d}(\delta x)$ and dividing by $\mathrm{d}t$ gives

$$\delta\frac{\mathrm{d}x}{\mathrm{d}t} = \frac{\mathrm{d}}{\mathrm{d}t}(\delta x) \tag{3.7}$$

For the problem at hand the Lagrangian is

$$\mathfrak{L} = \frac{m\dot{x}^2}{2} - \frac{kx^2}{2}$$

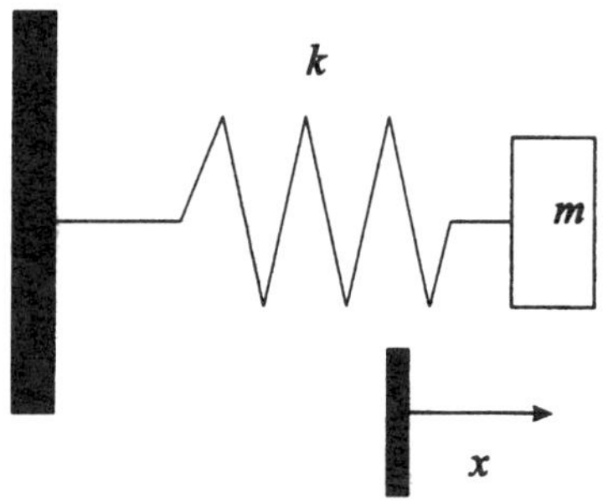

Fig. 3.3

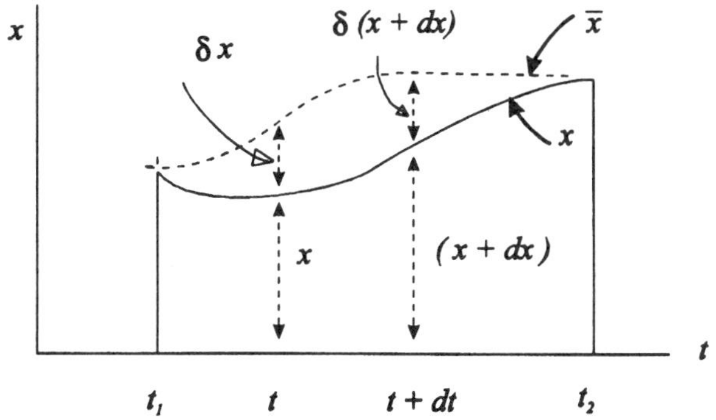

Fig. 3.4

Thus the integral to be minimized is

$$I = \int_{t_1}^{t_2} \left(\frac{m\dot{x}^2}{2} - \frac{kx^2}{2} \right) \mathrm{d}t$$

The varied integral with x replaced by $\bar{x} = x + \varepsilon\eta$ is

$$\bar{I} = \int_{t_1}^{t_2} \left(\frac{m\dot{\bar{x}}^2}{2} - \frac{k\bar{x}^2}{2} \right) \mathrm{d}t$$

$$= \int_{t_1}^{t_2} \left(\frac{m}{2} (\dot{x} + \varepsilon\dot{\eta})^2 - \frac{k}{2} (x + \varepsilon\eta)^2 \right) \mathrm{d}t$$

Therefore

$$\frac{\partial \bar{I}}{\partial \varepsilon_{\varepsilon \to 0}} = \int_{t_1}^{t_2} (m\dot{x}\dot{\eta} - kx\eta) \, \mathrm{d}t = 0$$

Integrating the first term in the integral by parts gives

$$m\dot{x}\eta \Big|_{t_1}^{t_2} - \int_{t_1}^{t_2} m\ddot{x}\eta \, \mathrm{d}t - \int_{t_1}^{t_2} kx\eta \, \mathrm{d}t = 0$$

By the definition of η the first term vanishes on account of η being zero at t_1 and at t_2, so

$$\int_{t_1}^{t_2} (m\ddot{x} + kx) \, \eta \, \mathrm{d}t = 0 \qquad (3.8)$$

Now η is an arbitrary function of time and can be chosen to be zero except for time $= t$ when it is non-zero. This means that the term in parentheses must be zero for any value of t, that is

$$m\ddot{x} + kx = 0 \qquad (3.9)$$

A quicker method, now that the exact meaning of variation is known, is as follows

$$\delta \int_{t_1}^{t_2} \left(\frac{m}{2} \dot{x}^2 - \frac{k}{2} x^2 \right) \mathrm{d}t = 0 \qquad (3.10)$$

Making use of equation (3.7), equation (3.10) becomes

$$\int_{t_1}^{t_2} (m\dot{x} \, \delta\dot{x} - kx \, \delta x) \, \mathrm{d}t = 0$$

Again, integrating by parts,

$$m\dot{x} \, \delta x \Big|_{t_1}^{t_2} - \int_{t_1}^{t_2} m\ddot{x} \, \delta x \, \mathrm{d}t - \int_{t_1}^{t_2} kx \, \delta x \, \mathrm{d}t = 0$$

or

$$-\int_{t_1}^{t_2} (m\ddot{x} + kx)\ \delta x\ \mathrm{d}t = 0$$

and because δx is arbitrary it follows that

$$m\ddot{x} + kx = 0 \tag{3.11}$$

3.4 Lagrange's equations derived from Hamilton's principle

For a system having n degrees of freedom the Lagrangian can be expressed in terms of the generalized co-ordinates, the generalized velocities and time, that is $\mathfrak{L} = \mathfrak{L}\ (q_i, \dot{q}_i, t)$. Thus with

$$I = \int_{t_1}^{t_2} \mathfrak{L}\ \mathrm{d}t \tag{3.12}$$

we have

$$\delta I = \int_{t_1}^{t_2} \left(\sum_i \frac{\partial \mathfrak{L}}{\partial q_i} \delta q_i + \sum_i \frac{\partial \mathfrak{L}}{\partial \dot{q}_i} \delta \dot{q}_i \right) \mathrm{d}t = 0$$

Note that there is no partial differentiation with respect to time since the variation applies only to the co-ordinates and their derivatives. Because the variations are arbitrary we can consider the case for all q_i to be zero except for q_j. Thus

$$\delta I = \int_{t_1}^{t_2} \left(\frac{\partial \mathfrak{L}}{\partial q_j} \delta q_j + \frac{\partial \mathfrak{L}}{\partial \dot{q}_j} \delta \dot{q}_j \right) \mathrm{d}t = 0$$

Integrating the second term by parts gives

$$\delta I = \int_{t_1}^{t_2} \frac{\partial \mathfrak{L}}{\partial q_j} \delta q_j\ \mathrm{d}t + \frac{\partial \mathfrak{L}}{\partial \dot{q}_j} \delta q_j \Bigg|_{t_1}^{t_2} - \int_{t_1}^{t_2} \frac{\mathrm{d}}{\mathrm{d}t} \left(\frac{\partial \mathfrak{L}}{\partial \dot{q}_j} \right) \delta q_j\ \mathrm{d}t = 0$$

Because $\delta q_j = 0$ at t_1 and at t_2

$$\int_{t_1}^{t_2} \left[\frac{\partial \mathfrak{L}}{\partial q_j} - \frac{\mathrm{d}}{\mathrm{d}t} \left(\frac{\partial \mathfrak{L}}{\partial \dot{q}_j} \right) \right] \delta q_j\ \mathrm{d}t = 0$$

Owing to the arbitrary nature of δq_j we have

$$\frac{\mathrm{d}}{\mathrm{d}t} \left(\frac{\partial \mathfrak{L}}{\partial \dot{q}_j} \right) - \frac{\partial \mathfrak{L}}{\partial q_j} = 0 \tag{3.13}$$

These are Lagrange's equations for conservative systems. It should be noted that $\mathfrak{L} = T^* - V$ because, with reference to Fig. 3.2, it is the variation of co-kinetic energy which is related to the momentum. But, as already stated, when the momentum is a linear function of velocity the co-kinetic energy $T^* = T$, the kinetic energy. The use of co-kinetic energy becomes important when particle speeds approach that of light and the non-linearity becomes apparent.

3.5 Illustrative example

One of the areas in which Hamilton's principle is useful is that of continuous media where the number of degrees of freedom is infinite. In particular it is helpful in complex problems for which approximate solutions are sought, because approximations in energy terms are often easier to see than they are in compatibility requirements.

As an example we shall look at wave motion in long strings under tension. The free-body diagram approach requires assumptions to be made in order that a simple equation of motion is generated; whilst the same is true for this treatment the implications of the assumptions are clearer.

Figure 3.5 shows a string of finite length. We assume that the stretching of the string is negligible and that no energy is stored owing to bending. We further assume that the tension τ in the string remains constant. This can be arranged by having a pre-tensioned constant-force spring at one end and assuming that $\partial u/\partial x$ is small. In practice the elasticity of the string and its supports is such that for small deviations the tension remains sensibly constant.

We need an expression for the potential energy of the string in a deformed state. If the string is deflected from the straight line then point B will move to the left. Thus the negative of the work done by the tensile force at B will be the change in potential energy of the system.

The length of the deformed string is

$$L = \int_{x=0}^{x=L-s} \sqrt{(\mathrm{d}x^2 + \mathrm{d}u^2)} = \int_{x=0}^{x=L-s} \sqrt{\left[1 + \left(\frac{\mathrm{d}u}{\mathrm{d}x}\right)^2\right]} \mathrm{d}x$$

If we assume that the slope $\mathrm{d}u/\mathrm{d}x$ is small then

$$L = \int_{x=0}^{x=L-s} \left[1 + \frac{1}{2}\left(\frac{\mathrm{d}u_2}{\mathrm{d}x}\right)\right] \mathrm{d}x = (L - s) + \int_{x=0}^{x=L-s} \frac{1}{2}\left(\frac{\mathrm{d}u}{\mathrm{d}x}\right)^2 \mathrm{d}x$$

For small deflections $s \ll L$ so the upper limit can be taken as L. Thus

$$s = \int_{x=0}^{x=L} \frac{1}{2}\left(\frac{\mathrm{d}u}{\mathrm{d}x}\right)^2 \mathrm{d}x$$

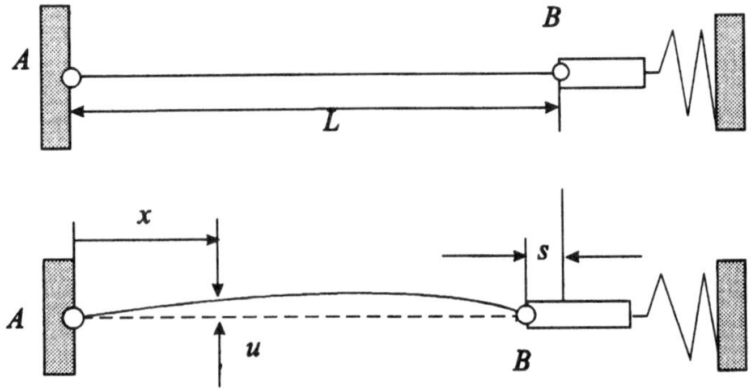

Fig. 3.5

The potential energy is $-\tau(-s) = \tau s$ giving

$$V = \tau s = \int_{x=0}^{x=L} \frac{\tau}{2}\left(\frac{du}{dx}\right)^2 dx \tag{3.14}$$

If u is also a function of time then du/dx will be replaced by $\partial u/\partial x$.

If ρ is the density and a is the cross-sectional area of the string then the kinetic energy is

$$T = \int_{x=0}^{x=L} \frac{\rho a}{2}\left(\frac{\partial u}{\partial x}\right)^2 dx \tag{3.15}$$

The Lagrangian is

$$\mathfrak{L} = \int_{x=0}^{x=L}\left[\frac{\rho a}{2}\left(\frac{\partial u}{\partial t}\right)^2 - \frac{\tau}{2}\left(\frac{\partial u}{\partial x}\right)^2\right] dx \tag{3.16}$$

According to Hamilton's principle we need to find the conditions so that

$$\delta\int_{t_1}^{t_2}\int_{x=0}^{x=L}\left[\frac{\rho a}{2}\left(\frac{\partial u}{\partial t}\right)^2 - \frac{\tau}{2}\left(\frac{\partial u}{\partial x}\right)^2\right] dx\, dt = 0 \tag{3.17}$$

Carrying out the variation

$$\int_{t_1}^{t_2}\int_{x=0}^{x=L}\left[\rho a\left(\frac{\partial u}{\partial t}\right)\delta\left(\frac{\partial u}{\partial t}\right) - \tau\left(\frac{\partial u}{\partial x}\right)\delta\left(\frac{\partial u}{\partial x}\right)\right] dx\, dt = 0 \tag{3.18}$$

To keep the process as clear as possible we will consider the two terms separately. For the first term the order of integration is reversed and then the time integral will be integrated by parts

$$\int_0^L\int_{t_1}^{t_2}\left[\rho a\left(\frac{\partial u}{\partial t}\right)\delta\left(\frac{\partial u}{\partial t}\right)\right] dt\, dx$$

$$= \int_0^L\left[\rho a\left(\frac{\partial u}{\partial t}\right)\delta u\,\Bigg|_{t_1}^{t_2} - \int_{t_1}^{t_2}\rho a\left(\frac{\partial^2 u}{\partial t^2}\right) dt\right] dx$$

$$= -\int_0^L\int_{t_1}^{t_2}\rho a\left(\frac{\partial^2 u}{\partial t^2}\right) dt\, dx \tag{3.19}$$

because $\delta u = 0$ at t_1 and t_2. The second term in equation (3.18) is

$$-\int_{t_1}^{t_2}\int_0^L\left[\tau\left(\frac{\partial u}{\partial x}\right)\delta\left(\frac{\partial u}{\partial x}\right)\right] dx\, dt$$

Integrating by parts gives

$$-\int_{t_1}^{t_2}\left[\tau\left(\frac{\partial u}{\partial x}\right)\delta u\,\Bigg|_0^L - \int_0^L \tau\left(\frac{\partial^2 u}{\partial x^2}\right)\delta u\, dx\right] dt$$

$$= \int_{t_1}^{t_2} \int_0^L \tau \left(\frac{\partial^2 u}{\partial x^2} \right) \delta u \, \mathrm{d}x \, \mathrm{d}t \tag{3.20}$$

The first term is zero provided that the ends are passive, that is no energy is being fed into the string after motion has been initiated. This means that either $\delta u = 0$ or $\partial u / \partial x = 0$ at each end. The specification of the problem indicated that $\delta u = 0$ but any condition that makes energy transfer zero at the extremes excludes the first term.

Combining equations (3.19) and (3.20) and substituting into equation (3.18) yields

$$\int_{t_1}^{t_2} \int_0^L \left[- \rho a \left(\frac{\partial^2 u}{\partial t^2} \right) + \tau \left(\frac{\partial^2 u}{\partial x^2} \right) \right] \delta u \, \mathrm{d}x \, \mathrm{d}t = 0$$

and because δu is arbitrary the integrand must sum to zero so that finally

$$\rho a \frac{\partial^2 u}{\partial t^2} = \tau \frac{\partial^2 u}{\partial x^2} \tag{3.21}$$

This is the well-known wave equation for strings. It is readily obtained from free-body diagram methods but this approach is much easier to modify if other effects, such as that of bending stiffness of the wire, are to be considered. Extra energy terms can be added to the above treatment without the need to rework the whole problem. This fact will be exploited in Chapter 6 which discusses wave motion in more detail.

4

Rigid Body Motion in Three Dimensions

4.1 Introduction

A rigid body is an idealization of a solid object for which no change in volume or shape is permissible. This means that the separation between any two particles of the body remains constant.

If we know the positions of three non-colinear points, i, j and k, then the position of the body in space is defined. However, there are three equations of constraint of the form $|r_i - r_j| =$ constant so the number of degrees of freedom is $3 \times 3 - 3 = 6$.

4.2 Rotation

If the line joining any two points changes its orientation in space then the body has suffered a rotation. If no rotation is taking place then all particles will be moving along parallel paths. If the paths are straight then the motion is described as rectilinear translation and if not the motion is curvilinear translation. From the definitions it is clear that a body can move along a circular path but there need be no rotation of the body.

It follows that for any pure translational motion there is no relative motion between individual particles. Conversely any relative motion must be due to some rotation.

The rotation of a rigid body can be described in terms of the motion of points on a sphere of radius a centred on some arbitrary reference point, say i. The body, shown in Fig. 4.1, is now reorientated so that the points j and k are moved, by any means, to positions j' and k'. The arc of the great circle joining j and k will be the same length as the arc joining j' and k', by definition of a rigid body. Next we construct the great circle through points j and j' and another through the points k and k'. We now draw great circles which are the perpendicular bisectors of arcs jj' and kk'. These two circles intersect at point N. The figure is now completed by drawing the four great circles through N and the points j, k, j' and k' respectively.

By the definition of the perpendicular bisector arc Nj = arc Nj' and arc Nk = arc Nk'. Also arc jk = arc $j'k'$ and thus it follows that the spherical triangle kNj is congruent with k'Nj'. Now the angle kNj = k'Nj' and the angle kNj' is common; therefore angle kNk' = jNj'.

With i as reference the line iN is an axis of rotation. Therefore we have proved that any displacement relative to i can be represented by a rotation of angle jNj' about the line iN.

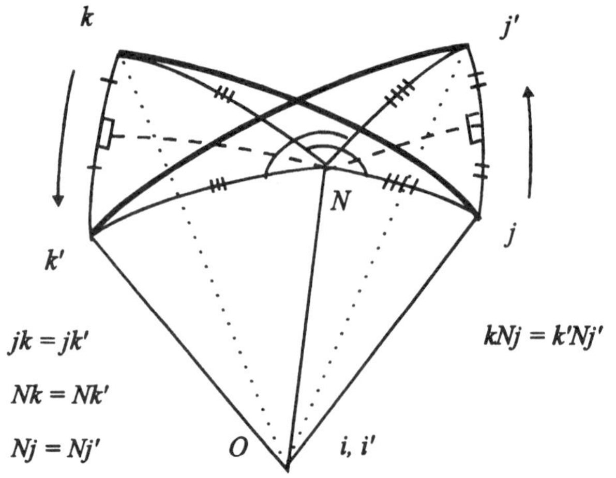

Fig. 4.1

> *In general we can state that any change in orientation can be achieved by a rotation about a single axis through any chosen reference point.*

This is often referred to as *Euler's theorem.*

It also follows directly that

> *any displacement of a rigid body can be obtained as the sum of the rectilinear displacement of some arbitrary point plus a rotation about an axis through that point.*

This is known as *Chasles's theorem.*

Note that the reference point is arbitrary so that the direction of the displacement is variable but the direction of the axis of rotation is constant. Indeed the reference point can be chosen such that the direction of the displacement is the same as the axis of rotation; this is known as screw motion.

The validity of the last statement can be justified by reference to Fig. 4.2(a). The body is moved by a rotation of θ about the OA axis and then translated along OO′. Alternatively the translation can be made first followed by a rotation about the O′A′ axis, which is parallel to OA. OA and OO′ define a plane and the view along arrow A is shown in Fig. 4.2(b). The point N is located such that ON = ON′ and angle ONO′ is also θ. This rotation will move the point O to O″ and a translation along the O′A′ axis will bring the body into the desired position.

It is worth noting that if the displacement of all particles is planar such that the rotation axis is normal to that plane then any change in position can be achieved by a rotation about a fixed axis. The case of pure translation may be thought of as a rotation about an axis at infinity.

The definition of rotation does not require the location of the axis to be specified – only its direction is needed. If the reference point is a fixed point then the axis of rotation can be regarded as a fixed axis through that point and points lying on the axis will not be displaced.

A corollary of Euler's theorem is that a rotation about axis 1 followed by a rotation about axis 2 can be replaced by a single rotation about axis 3. It should be noted that if the order

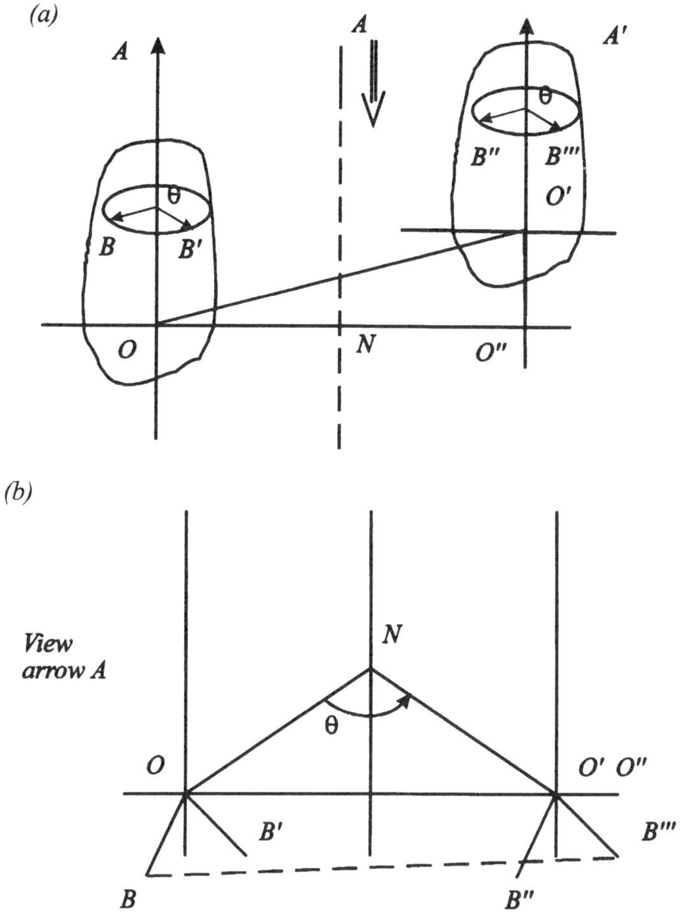

Fig. 4.2 (a) and **(b)**

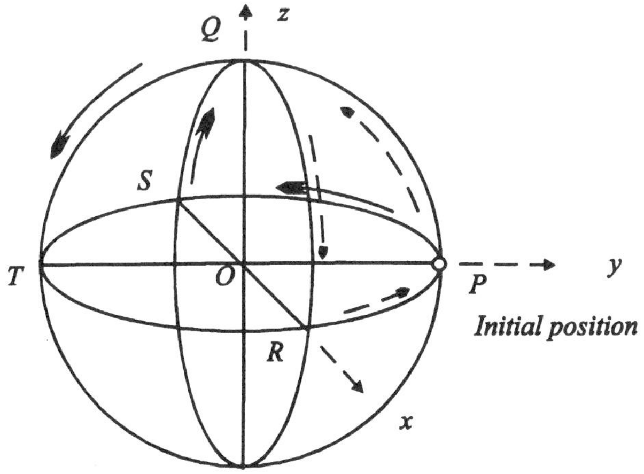

Fig. 4.3

of the first two rotations is interchanged then the equivalent third rotation will be different. Finite rotations do not obey the law of vector addition; this is discussed in detail in Chapter 8 which discusses robot dynamics. The fact is easily demonstrated by reference to Fig. 4.3, depicting three consecutive 90° rotations. The line OP is rotated 90° about the x axis to OQ, then the y axis to OR and then the z axis back to OP. Alternatively the line OP is rotated about the z axis to OS, then the y axis to OQ and finally about the x axis to OT. Clearly the results are different.

4.3 Angular velocity

Consider a small rotation $\mathrm{d}\theta$ about some axis Oz as shown in Fig. 4.4. A point on a sphere of radius a will move a distance

$$\mathrm{d}s = b\,\mathrm{d}\theta \tag{4.1}$$

Dividing by $\mathrm{d}t$, the time interval, gives

$$\dot{s} = b\dot{\theta} = a \sin \phi \, \dot{\theta} \tag{4.2}$$

The direction of the velocity is AA′ which is normal to the plane containing the radius vector and the axis of rotation. The angle ø is the angle between the radius vector and the axis of rotation so by definition of the vector product of two vectors

$$\boldsymbol{v} = \dot{s}\boldsymbol{e} = \dot{\theta}\boldsymbol{k} \times \boldsymbol{a} \tag{4.3}$$

where $\boldsymbol{k}$ is the unit vector in the z direction.

In general

$$\boldsymbol{v} = \boldsymbol{\omega} \times \boldsymbol{r} \tag{4.4}$$

where $\boldsymbol{\omega}$ is the angular velocity vector of magnitude $\dot{\theta}$ in a direction parallel to the axis of rotation and $\boldsymbol{r}$ is the position vector.

We still require to show that the angular velocity vector obeys the parallelogram law of vector addition. For small displacements on the surface of the sphere the surface tends to a flat surface. Thus the geometry is Euclidean and the order of the rotations may be reversed

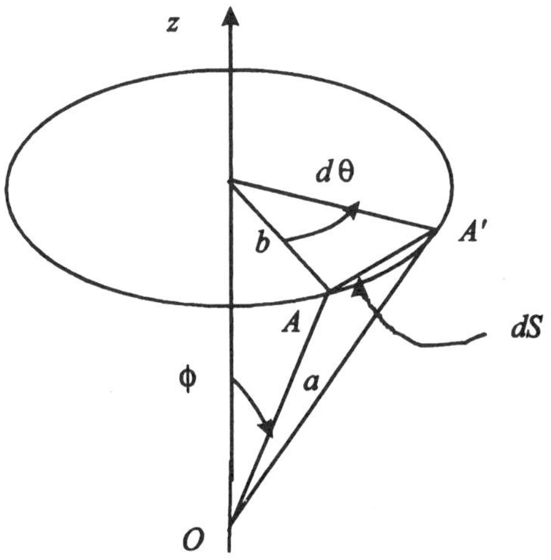

Fig. 4.4

$$\mathrm{d}\boldsymbol{s} = \mathrm{d}\boldsymbol{\theta}_1 \times \boldsymbol{a} + \mathrm{d}\boldsymbol{\theta}_2 \times (\boldsymbol{a} + \mathrm{d}\boldsymbol{s}) = (\mathrm{d}\boldsymbol{\theta}_1 + \mathrm{d}\boldsymbol{\theta}_2) \times \boldsymbol{a} \tag{4.5}$$

neglecting second-order terms. Dividing equation (4.5) by dt gives

$$\boldsymbol{v} = (\boldsymbol{\omega}_1 + \boldsymbol{\omega}_2) \times \boldsymbol{a} \tag{4.6}$$

Thus, although finite rotations do not obey the law of vector addition angular velocities do.

The velocity of point P on a rigid body may be written

$$\boldsymbol{v}_{\mathrm{P}} = \boldsymbol{v}_{\mathrm{A}} + \boldsymbol{\omega} \times \boldsymbol{r}_{\mathrm{P/A}} \tag{4.7}$$

where $\boldsymbol{v}_{\mathrm{A}}$ is the velocity of some reference point, $\boldsymbol{\omega}$ is the angular velocity and $\boldsymbol{r}_{\mathrm{P/A}}$ is the position vector of P relative to A.

4.4 Kinetics of a rigid body

From equation (1.48) we have that the linear momentum is the total mass times the velocity of the centre of mass. This is true whether the body is rigid or not, so equation (1.50) is valid

$$\sum \boldsymbol{F}_i = \frac{\mathrm{d}}{\mathrm{d}t}(m\dot{\boldsymbol{r}}_{\mathrm{G}}) = m\ddot{\boldsymbol{r}}_{\mathrm{G}} \tag{4.8}$$

Let us now consider the general space motion of a rigid body. From equation (1.53) the moment of momentum about some origin O is

$$\boldsymbol{L}_{\mathrm{O}} = \Sigma\, \boldsymbol{r}_i \times m_i\dot{\boldsymbol{r}}_i \tag{4.9}$$

From Fig. 4.5

$$\boldsymbol{r}_i = \boldsymbol{r}_{\mathrm{A}} + \boldsymbol{\rho}_i \tag{4.10}$$

where $\boldsymbol{\rho}_i$ is the position vector of particle i relative to A. For a rigid body equation (4.7) gives

$$\dot{\boldsymbol{r}}_i = \boldsymbol{v}_{\mathrm{A}} + \boldsymbol{\omega}_i \times \boldsymbol{\rho}_i \tag{4.11}$$

Substituting equations (4.10) and (4.11) into equation (4.9) gives

$$\begin{aligned}\boldsymbol{L}_{\mathrm{O}} &= \Sigma\, (\boldsymbol{r}_{\mathrm{A}} + \boldsymbol{\rho}_i) \times m_i\, (\boldsymbol{v}_{\mathrm{A}} + \boldsymbol{\omega} \times \boldsymbol{\rho}_i) \\ &= \boldsymbol{r}_{\mathrm{A}} \times m\boldsymbol{v}_{\mathrm{A}} + \boldsymbol{r}_{\mathrm{A}} \times (\boldsymbol{\omega} \times \Sigma\, m_i\boldsymbol{\rho}_i) + (\Sigma\, m_i\boldsymbol{\rho}_i) \times \boldsymbol{v}_{\mathrm{A}} \\ &\quad + \Sigma\, \boldsymbol{\rho}_i \times (\boldsymbol{\omega} \times m_i\boldsymbol{\rho}_i)\end{aligned} \tag{4.12}$$

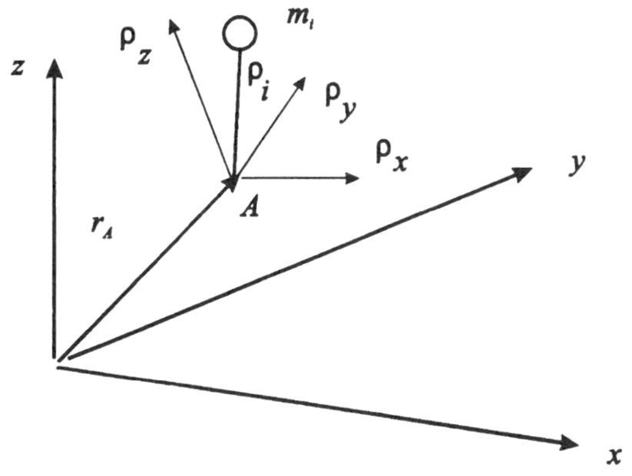

Fig. 4.5

From the definition of the centre of mass

$$\Sigma\, m_i\,(\boldsymbol{\rho}_i + \boldsymbol{r}_{\mathrm{A}}) = m\boldsymbol{r}_{\mathrm{G}}$$

or

$$\Sigma\, m_i\boldsymbol{\rho}_i = m\,(\boldsymbol{r}_{\mathrm{G}} - \boldsymbol{r}_{\mathrm{A}}) \tag{4.13}$$

Using equation (4.13), equation (4.12) becomes

$$\boldsymbol{L}_{\mathrm{O}} = \boldsymbol{r}_{\mathrm{A}} \times \boldsymbol{\omega} \times (m\boldsymbol{r}_{\mathrm{G}} - m\boldsymbol{r}_{\mathrm{A}}) + m\boldsymbol{r}_{\mathrm{G}} \times \boldsymbol{v}_{\mathrm{A}} + \Sigma\, \boldsymbol{\rho}_i \times (\boldsymbol{\omega} \times m_i\boldsymbol{\rho}_i) \tag{4.14}$$

This equation is cumbersome but it takes a simpler form for two special cases.

The first case is when the point A is fixed and is used as the origin, that is $\boldsymbol{r}_{\mathrm{A}} = 0$ and $\boldsymbol{v}_{\mathrm{A}} = 0$. Equation (4.14) is now

$$\boldsymbol{L}_{\mathrm{O}} = \Sigma\, \boldsymbol{\rho}_i \times (\boldsymbol{\omega} \times m_i\boldsymbol{\rho}_i) \tag{4.15}$$

The second case is when G is the reference point, that is A coincides with G. Equation (4.14) is now

$$\boldsymbol{L}_{\mathrm{O}} = \boldsymbol{r}_{\mathrm{G}} \times m\boldsymbol{v}_{\mathrm{G}} + \Sigma\, \boldsymbol{\rho}_i \times (\boldsymbol{\omega} \times m_i\boldsymbol{\rho}_i) \tag{4.16}$$

This may be further simplified if we take the origin to be coincident with G, in which case

$$\boldsymbol{L}_{\mathrm{O}} = \boldsymbol{L}_{\mathrm{G}} = \Sigma\, \boldsymbol{\rho}_i \times (\boldsymbol{\omega} \times m_i\boldsymbol{\rho}_i) \tag{4.17}$$

Note that $\boldsymbol{\rho}$ is measured from the reference point, that is the fixed point A in the first case and the centre of mass for the second.

From equation (1.53) we have that the moment of the external forces about the chosen origin O is equal to the time rate of change of the moment of momentum, or

$$\boldsymbol{M}_{\mathrm{O}} = \frac{\mathrm{d}}{\mathrm{d}t}\,\boldsymbol{L}_{\mathrm{O}} \tag{4.18}$$

Let us first consider the case of rotation about a fixed axis of symmetry, say the z axis, so $\boldsymbol{\omega} = \omega_z\boldsymbol{k}$ and $\boldsymbol{\rho} = x\boldsymbol{i} + y\boldsymbol{j} + z\boldsymbol{k}$. Equation (4.15) is now

$$\begin{aligned}\boldsymbol{L}_{\mathrm{O}} &= \Sigma\,(x_i\boldsymbol{i} + y_i\boldsymbol{j} + z_i\boldsymbol{k}) \times (\omega_z m_i x_i\boldsymbol{j} - \omega_z m_i y_i\boldsymbol{i})\\ &= \Sigma\,\omega_z m_i\,(x_i^2 + y_i^2)\,\boldsymbol{k} = \Sigma\, m_i b_i^2\,\omega_z\boldsymbol{k}\end{aligned} \tag{4.19}$$

where $b_i = \sqrt{(x_i^2 + y_i^2)}$ is the distance of the particle from the z axis and remains constant as the body rotates. The term $\Sigma\, m_i b_i^2$ is a constant of the body known as the moment of inertia about the z axis and is given the symbol I_z. Equation 4.19 can now be written

$$\boldsymbol{L}_{\mathrm{O}} = I_z\omega_z\boldsymbol{k} \tag{4.20}$$

Equation (4.18) gives

$$\boldsymbol{M}_{\mathrm{O}} = I_z\dot{\omega}_z\boldsymbol{k} \tag{4.21}$$

The differentiation is easy because the moment of inertia is a constant and shows that the moment of forces about O depends on the angular acceleration $\dot{\omega}_z$.

We now return to the case of rotation about the fixed point. If we express the moment of inertia in terms of the fixed co-ordinate system it will no longer be constant because the orientation of the body will be changing with time. To avoid the difficulty of coping with a variable moment of inertia it is convenient to choose a set of moving axes such that the moment of inertia is constant. For the general case these axes will be fixed to the body but for the common situation where the body has an axis of symmetry we can use any set of axes for which one axis coincides with the axis of symmetry.

This means that equation (4.18) will become (*see* equation (1.13))

$$\boldsymbol{M}_\mathrm{O} = \frac{\partial \boldsymbol{L}_\mathrm{O}}{\partial t} + \boldsymbol{\omega}_\mathrm{R} \times \boldsymbol{L}_\mathrm{O} \tag{4.22}$$

where $\boldsymbol{\omega}_\mathrm{R}$ is the angular velocity of the moving axes.

4.5 Moment of inertia

In the previous section we found an expression for the moment of inertia about a fixed axis. Clearly different axes will produce different values for this quantity. We need to look at the formula for moment of momentum in some detail. For rotation about a fixed point we have, by equation (4.15),

$$\boldsymbol{L}_\mathrm{O} = \Sigma\, \boldsymbol{\rho}_i \times (\boldsymbol{\omega} \times m_i \boldsymbol{\rho}_i) \tag{4.15}$$

Using the expansion formula for a triple vector product

$$\begin{aligned} \boldsymbol{L}_\mathrm{O} &= \boldsymbol{\omega}\, (\Sigma\, m_i \boldsymbol{\rho}_i \cdot \boldsymbol{\rho}_i) - \Sigma\, m_i \boldsymbol{\rho}_i\, (\boldsymbol{\omega} \cdot \boldsymbol{\rho}_i) \\ &= \boldsymbol{\omega}\, (\Sigma\, m_i \rho_i^2) - \boldsymbol{\omega} \cdot (\Sigma\, m_i \boldsymbol{\rho}_i \boldsymbol{\rho}_i) \end{aligned} \tag{4.23}$$

From the appendix on tensors and dyadics we recognize that the second term is the product of the vector $\boldsymbol{\omega}$ and a dyadic. The first term can be put in the same form by introducing the unit dyadic $\mathsf{1}$ so that equation (4.23) becomes

$$\boldsymbol{L}_\mathrm{O} = \boldsymbol{\omega} \cdot [\, \mathsf{1}\, (\Sigma\, m_i \rho_i^2) - \Sigma\, m_i \boldsymbol{\rho}_i \boldsymbol{\rho}_i)] \tag{4.24}$$

The terms in the square brackets are the moment of inertia. This quantity is not a scalar or a vector but a dyadic, or second-order tensor, and is given the symbol I so that equation (4.24) reads

$$\boldsymbol{L}_\mathrm{O} = \boldsymbol{\omega} \cdot \mathsf{I} \tag{4.25}$$

It is quite possible to expand the terms in the square brackets in equation (4.24) but we believe that it is clearer to obtain the expression for the components of $\boldsymbol{L}_\mathrm{O}$ directly by forming the dot product of equation (4.24) or (4.25) with the unit vectors

$$\begin{aligned} L_x &= \boldsymbol{L}_\mathrm{O} \cdot \boldsymbol{i} = \boldsymbol{\omega} \cdot \mathsf{I} \cdot \boldsymbol{i} \\ &= \boldsymbol{\omega} \cdot [\, (\Sigma\, m_i \rho_i^2)\, \boldsymbol{i} - \Sigma\, m_i \boldsymbol{\rho}_i x_i \,] \end{aligned} \tag{4.26}$$

Because $\boldsymbol{\rho}_i = x_i \boldsymbol{i} + y_i \boldsymbol{j} + z_i \boldsymbol{k}$ it follows that $\boldsymbol{\rho}_i \cdot \boldsymbol{i} = x_i$.

Expanding equation (4.26) we have

$$\begin{aligned} L_x &= \boldsymbol{\omega} \cdot \Sigma\, [\, m_i\, (x_i^2 + y_i^2 + z_i^2)\, \boldsymbol{i} - m_i x_i^2 \boldsymbol{i} - m_i y_i x_i \boldsymbol{j} - m_i z_i x_i \boldsymbol{k} \,] \\ &= \boldsymbol{\omega} \cdot \Sigma\, [\, m_i\, (y_i^2 + z_i^2)\, \boldsymbol{i} - m_i y_i x_i \boldsymbol{j} - m_i z_i x_i \boldsymbol{k} \,] \\ &= \omega_x \Sigma\, m_i\, (y_i^2 + z_i^2) - \omega_x \Sigma\, m_i y_i x_i - \omega_z \Sigma\, m_i z_i x_i \\ &= \omega_x I_{xx} + \omega_y I_{xy} + \omega_z I_{xz} \end{aligned} \tag{4.27}$$

where

$I_{xx} = \Sigma\, m_i\, (y_i^2 + z_i^2)$, moment of inertia about x axis
$I_{xy} = -\, \Sigma\, m_i y_i x_i$, product moment of inertia, $= I_{yx}$
$I_{xz} = -\, \Sigma\, m_i z_i x_i$, product moment of inertia, $= I_{zx}$

Some texts define the product moment of inertia as the negative of the above.

Similar expressions for L_y and L_z can be found and the results written as a matrix equation as follows

$$\begin{bmatrix} L_x \\ L_y \\ L_z \end{bmatrix} \quad \begin{bmatrix} I_{xx} & I_{xy} & I_{xz} \\ I_{yx} & I_{yy} & I_{yz} \\ I_{zx} & I_{zy} & I_{zz} \end{bmatrix} \begin{bmatrix} \omega_x \\ \omega_y \\ \omega_z \end{bmatrix}$$

or, in short form,

$$(L_{\mathrm{O}}) = [I_{\mathrm{O}}]\,(\omega) \tag{4.28}$$

where the symmetric square matrix $[I_{\mathrm{O}}]$ is the moment of inertia matrix with respect to point O.

An alternative method of obtaining the moment of inertia matrix is to use the vector–matrix algebra shown in Appendix 1.

Equation (4.15),

$$\boldsymbol{L}_{\mathrm{O}} = -\sum \boldsymbol{\rho}_i \times (m_i \boldsymbol{\rho}_i \times \boldsymbol{\omega})$$

may be written

$$L_{\mathrm{O}} = (e)^{\mathrm{T}}\,(L_{\mathrm{O}}) = (e)^{\mathrm{T}}\,[\,\Sigma\,[\rho]_i^{\mathrm{x}}\,[m\rho]_i^{\mathrm{x}}\,]\,(\omega)$$

$$= (e)^{\mathrm{T}}\,\Sigma \begin{bmatrix} 0 & -z_i & y_i \\ z_i & 0 & -x_i \\ -y_i & x_i & 0 \end{bmatrix} \begin{bmatrix} 0 & -m_i z_i & m_i y_i \\ m_i z_i & 0 & -m_i x_i \\ -m_i y_i & m_i x_i & 0 \end{bmatrix} \begin{bmatrix} \omega_x \\ \omega_y \\ \omega_z \end{bmatrix} \tag{4.29}$$

Carrying out the matrix multiplication yields the same result as equation (4.28).

The co-ordinate axes have been chosen arbitrarily so we now ask the question whether there are any preferred axes. In general the moment of momentum vector will not be parallel to the angular velocity vector. It can be seen that if a body is spinning about an axis of symmetry then $\boldsymbol{L}$ will be parallel to $\boldsymbol{\omega}$. Can this also be true for the general case?

We seek a vector $\boldsymbol{\omega}$ such that

$$\boldsymbol{L}_{\mathrm{O}} = \lambda \boldsymbol{\omega}$$

where λ is a scalar constant. Thus

$$(e)^{\mathrm{T}}\,[I_{\mathrm{O}}](\omega) = \lambda\,(e)^{\mathrm{T}}\,(\omega)$$

or

$$\{\,[I_{\mathrm{O}}] - \lambda\,[1]\,\}\,(\omega) = 0 \tag{4.30}$$

This is the classical eigenvalue problem in which λ is an *eigenvalue* and the corresponding (ω) is an *eigenvector*; note that for the eigenvector it is only the direction which is important – the magnitude is arbitrary. Writing equation (4.30) in full gives

$$\begin{bmatrix} (I_{xx}-\lambda) & I_{xy} & I_{xz} \\ I_{yx} & (I_{yy}-\lambda) & I_{yz} \\ I_{zx} & I_{zy} & (I_{zz}-\lambda) \end{bmatrix} \begin{bmatrix} \omega_x \\ \omega_y \\ \omega_z \end{bmatrix} = \begin{bmatrix} 0 \\ 0 \\ 0 \end{bmatrix} \tag{4.31}$$

From the theory of homogeneous linear equations a non-trivial result is obtained when the determinant of the square matrix is zero. This leads to a cubic in λ and therefore there are three roots (λ_1, λ_2 and λ_3) and three corresponding vectors (ω_1, ω_2 and ω_3). Each pair of eigenvalues and eigenvectors satisfy equation (4.30). There are, therefore, three equations

$$[I_O](\omega_1) - \lambda_1(\omega_1) = (0) \quad (4.32)$$
$$[I_O](\omega_2) - \lambda_2(\omega_2) = (0) \quad (4.33)$$
$$[I_O](\omega_3) - \lambda_3(\omega_3) = (0) \quad (4.34)$$

If we premultiply equation (4.32) by $(\omega_2)^T$ and subtract equation (4.33) premultiplied by $(\omega_1)^T$ the resulting scalar equation is

$$(\omega_2)^T [I_O](\omega_1) - (\omega_1)^T [I_O](\omega_2) - \lambda_1(\omega_2)^T(\omega_1) + \lambda_2(\omega_1)^T(\omega_2) = 0$$

Because $[I_O]$ is symmetrical the second term is the transpose of the first and as they are scalar they cancel. Since the product of the two vectors is independent of the order of multiplication we are left with

$$(\lambda_2 - \lambda_1)(\omega_2)^T(\omega_1) = 0 \quad (4.35)$$

and if λ_1 does not equal λ_2 then ω_2 is orthogonal to ω_1. The same argument is true for the other two pairings of vectors, which means that the eigenvectors form an orthogonal set of axes.

From equation (4.32) it follows that if $(\omega_2)^T(\omega_1) = 0$ then

$$(\omega_2)^T [I_O](\omega_1) = (\omega_1)^T [I_O](\omega_2) = 0 \quad (4.36)$$

and similarly for the other two equations.

We shall now construct a square matrix such that the columns are the three eigenvectors, that is

$$[A] = [(\omega_1)\ (\omega_2)\ (\omega_3)] \quad (4.37)$$

We now use this matrix to transform the moment of inertia matrix to give

$$[I_{PO}] = [A]^T [I_O][A] \quad (4.38)$$

A typical element of the transformed matrix is $(\omega_i)^T [I_O](\omega_j)$ which, by virtue of the orthogonality condition in equation (4.36), is zero if i does not equal j. The matrix is therefore diagonal with the diagonal elements equal to $(\omega_i)^T [I_O](\omega_i)$.

We have shown that for any body and for any arbitrary reference point there exists a set of axes for which the moment of inertia matrix is diagonal. These axes are called the *principal axes* and the elements of the matrix are the *principal moments* of inertia. These axes are unique except for the degenerate case when two of the eigenvalues are identical. From equation (4.35) if $\lambda_1 = \lambda_2$ then the eigenvectors are not unique but they must both be orthogonal to λ_3. With this proviso they may be chosen at will.

An example is that for a right circular cylinder the axis of symmetry is a principal axis; clearly any pair of axes normal to the axis of symmetry will be a principal axis. Although it is not obvious a prism of square cross-section will satisfy the same criteria as the previous case. In fact any prism whose cross-section is a regular polygon has degenerate principal axes. Another useful property of symmetry is that for a body which has a plane of symmetry one principal axis will be normal to that plane.

The above argument is true for any reference point in the body. We now seek a relationship between the moment of inertia about some arbitrary point O and that about the centre of mass G. If $\boldsymbol{R}$ is the position vector of G relative to O then the position of mass m_i can be written $\boldsymbol{\rho}_i = \boldsymbol{R} + \bar{\boldsymbol{\rho}}_i$, where $\bar{\boldsymbol{\rho}}$ is the position relative to G. Substitution into equation (4.24) yields

$$\mathsf{I}_O = \Sigma m_i [(R^2 + \bar{\rho}_i^2 + 2\boldsymbol{R}\cdot\bar{\boldsymbol{\rho}}_i)\,\mathsf{1} - (\boldsymbol{R}\boldsymbol{R} + \bar{\boldsymbol{\rho}}_i\bar{\boldsymbol{\rho}}_i + \boldsymbol{R}\bar{\boldsymbol{\rho}}_i + \bar{\boldsymbol{\rho}}_i\boldsymbol{R})]$$

Remembering that $\Sigma\, m_i \bar{\boldsymbol{\rho}}_i = 0$ we obtain

$$\mathsf{I}_O = m\,(R^2\mathsf{1} - \boldsymbol{RR}) + \Sigma\, m_i\,(\bar{\boldsymbol{\rho}}_i^2\mathsf{1} - \bar{\boldsymbol{\rho}}_i\bar{\boldsymbol{\rho}}_i) = m\,(R^2\mathsf{1} - \boldsymbol{RR}) + \mathsf{I}_G \tag{4.39}$$

It should be pointed out that the principal axes of I_O are only parallel to those of I_G if the shift of the reference point is along one of the principal axes. However, if we wish to find just the component about a given axis, say the x axis, by forming $\boldsymbol{i}\cdot\mathsf{I}_O\cdot\boldsymbol{i} = I_{Ox} = m\,(R^2 - x^2) + I_{Gx} = m\,(y^2 + z^2) + I_{Gx}$, we see that it becomes the familiar *parallel axes theorem*.

The other well-known theorem relating to moment of inertia is the *perpendicular axes theorem*. For a thin lamina one principal axis is normal to the plane of the lamina, the other two being in the plane.

Taking the axis normal to the plane as the z axis, the moment of inertia with reference to point O is

$$I_{Oz} = \Sigma\, m_i\,(x_i^2 + y_i^2) = \Sigma\, m_i x_i^2 + \Sigma\, m_i y_i^2$$

which, for a lamina, gives

$$I_{Oz} = I_{Oy} + I_{Ox} \tag{4.39a}$$

If the x and y axes are chosen to be the principal axes

$$I_3 = I_2 + I_1 \tag{4.39b}$$

4.6 Euler's equation for rigid body motion

For the rotation of a rigid body about a fixed point equation (4.22) tells us that

$$\boldsymbol{M}_O = \frac{\partial \boldsymbol{L}_O}{\partial t} + \boldsymbol{\omega}_R \times \boldsymbol{L}_O \tag{4.22}$$

and because I_O is symmetrical equation (4.25) reads

$$\boldsymbol{L}_O = \boldsymbol{\omega}\cdot\mathsf{I}_O = \mathsf{I}_O\cdot\boldsymbol{\omega} \tag{4.25}$$

In the general case for which the axes are fixed to the body $\boldsymbol{\omega}_R = \boldsymbol{\omega}$. Hence combining equations (4.22) and (4.25) gives

$$\boldsymbol{M}_O = \frac{\partial\,(\mathsf{I}_O\cdot\boldsymbol{\omega})}{\partial t} + \boldsymbol{\omega}\times(\mathsf{I}_O\cdot\boldsymbol{\omega}) = \mathsf{I}_O\dot{\boldsymbol{\omega}} + \boldsymbol{\omega}\times(\mathsf{I}_O\cdot\boldsymbol{\omega}) \tag{4.40}$$

Here we have replaced $\dfrac{\partial\boldsymbol{\omega}}{\partial t}$ by $\dot{\boldsymbol{\omega}}$ without ambiguity because

$$\frac{\mathrm{d}\boldsymbol{\omega}}{\mathrm{d}t} = \frac{\partial\boldsymbol{\omega}}{\partial t} + \boldsymbol{\omega}\times\boldsymbol{\omega} = \frac{\partial\boldsymbol{\omega}}{\partial t} = \dot{\boldsymbol{\omega}}$$

Choosing principal axes the moment of inertia dyadic (*see* Appendix 1) may be written

$$\mathsf{I}_O = I_1\boldsymbol{ii} + I_2\boldsymbol{jj} + I_3\boldsymbol{kk}$$

or

$$[I_O] = \begin{bmatrix} I_1 & 0 & 0 \\ 0 & I_2 & 0 \\ 0 & 0 & I_3 \end{bmatrix}$$

In matrix form equation (4.40) is

$$(e)^{\mathrm{T}}(M_{\mathrm{O}}) = (e)^{\mathrm{T}}[I_{\mathrm{O}}](\dot{\omega}) + (e)^{\mathrm{T}}[\omega]^{\mathrm{x}}[I_{\mathrm{O}}](\omega)$$

$$\begin{bmatrix} M_x \\ M_y \\ M_z \end{bmatrix} = \begin{bmatrix} I_1 & 0 & 0 \\ 0 & I_2 & 0 \\ 0 & 0 & I_3 \end{bmatrix}\begin{bmatrix} \dot{\omega}_x \\ \dot{\omega}_y \\ \dot{\omega}_z \end{bmatrix} + \begin{bmatrix} 0 & -\omega_z & \omega_y \\ \omega_z & 0 & -\omega_x \\ -\omega_y & \omega_x & 0 \end{bmatrix}\begin{bmatrix} I_1 & 0 & 0 \\ 0 & I_2 & 0 \\ 0 & 0 & I_3 \end{bmatrix}\begin{bmatrix} \omega_x \\ \omega_y \\ \omega_z \end{bmatrix}$$

which, on carrying out the multiplication, yields

$$\begin{aligned} M_x &= I_1\dot{\omega}_x - (I_2 - I_3)\,\omega_y\omega_z \\ M_y &= I_2\dot{\omega}_y - (I_3 - I_1)\,\omega_z\omega_x \\ M_z &= I_3\dot{\omega}_z - (I_1 - I_2)\,\omega_x\omega_y \end{aligned} \tag{4.41}$$

These last three equation are *Euler's equations.*

For the case where the z axis is an axis of symmetry $I_1 = I_2$ the x and y axes can take any position, provided that the set is still orthogonal. Equation (4.40) can be now modified to take account of the angular velocity of the axes being

$$\boldsymbol{\omega}_{\mathrm{R}} = \Omega_x\boldsymbol{i} + \Omega_y\boldsymbol{j} + \Omega_z\boldsymbol{k}$$

Hence we have

$$\begin{aligned} M_x &= I_1\frac{\partial\omega_x}{\partial t} - \Omega_z I_2\omega_y + \Omega_y I_3\omega_z \\ M_y &= I_2\frac{\partial\omega_y}{\partial t} - \Omega_x I_3\omega_z + \Omega_z I_1\omega_x \\ M_z &= I_3\frac{\partial\omega_z}{\partial t} - \Omega_y I_1\omega_x + \Omega_x I_2\omega_y \end{aligned} \tag{4.42}$$

It is still imperative that from the moving axes the moment of inertia is constant. Either the axes are fixed to the body or at least one axis is an axis of symmetry. The latter implies that if one axis is an axis of symmetry then the other two are equal. If the body has point symmetry then all principal moments of inertia are equal. Notice also that we have reverted to the partial differential for the first term on the right hand side of equation (4.42).

The above analysis has been developed on the assumption that the reference point is fixed. However, the same formulation is applicable if the reference point is the centre of mass and is independent of any motion of the centre of mass. We need, of course, to evaluate the moment of inertia with respect to the centre of mass.

4.7 Kinetic energy of a rigid body

The total kinetic energy is $\frac{1}{2}\Sigma\, m_i v_i^2$ and for a rigid body the velocity relative to the reference point is $\boldsymbol{v}_i = \boldsymbol{\omega} \times \boldsymbol{r}_i$. Therefore the kinetic energy is

$$T = \frac{1}{2}\,\Sigma\, m_i(\boldsymbol{\omega} \times \boldsymbol{r}_i)\cdot(\boldsymbol{\omega} \times \boldsymbol{r}_i) \tag{4.43}$$

$$= \frac{1}{2}\,\Sigma\, m_i(\boldsymbol{r}_i \times \boldsymbol{\omega})\cdot(\boldsymbol{r}_i \times \boldsymbol{\omega})$$

In vector–matrix notation

$$T = \frac{1}{2}\,\Sigma\, m_i\,\{[r]^{\times}(\omega)\}^{\mathrm{T}}\,[r]^{\times}(\omega) \tag{4.44}$$

Now the transpose of a cross-matrix (*see* Appendix 1) is its negative so

$$T = -\frac{1}{2}\,\Sigma\, m_i\,(\omega)^{\mathrm{T}}\,[r]^{\times}\,[r]^{\times}\,(\omega)$$

$$= -\frac{1}{2}\,(\omega)^{\mathrm{T}}\,(\,\Sigma\, m_i\,[r]^{\times}\,[r]^{\times}\,)\,(\omega) \tag{4.45}$$

Also from Appendix 1

$$[r]^{\times}\,[r]^{\times} = (r)\,(r)^{\mathrm{T}} - \mathrm{r}^2\,[1]$$

and therefore equation (4.45) is

$$T = \frac{1}{2}\,(\omega)^{\mathrm{T}}\left(\Sigma\; m_i\,(r^2\,[1]\; - \;(r)\,(r)^{\mathrm{T}}\,)\right)(\omega) \tag{4.46}$$

The term in the large parentheses is recognized as the moment of inertia matrix $[I]$. Thus

$$T = \frac{1}{2}\,(\omega)^{\mathrm{T}}\,[I]\,(\omega) \tag{4.47}$$

If the rotation is about the z axis equation (4.43) reduces to

$$T = \frac{1}{2}\,\Sigma\, m_i\,(\omega_z x\boldsymbol{j}\; - \;\omega_z y\boldsymbol{i})^2$$

$$= \frac{1}{2}\,\omega^2\,\Sigma\, m_i\,(x^2 + y^2)$$

$$= \frac{1}{2}\,\omega^2\, I_{zz} \tag{4.48}$$

Since the choice of axis was arbitrary it follows that the kinetic energy is half the angular speed squared times the moment of inertia about the axis of rotation. If we write $\boldsymbol{\omega} = \omega\,\boldsymbol{e}$, where

$$\boldsymbol{e} = l\boldsymbol{i} + m\boldsymbol{j} + n\boldsymbol{k} \tag{4.49}$$

is the unit vector in the direction of rotation, equation (4.47) gives

$$T = \frac{1}{2}\,\omega^2\,(e)^{\mathrm{T}}\,[I](e) \tag{4.50}$$

Hence

$$(e)^{\mathrm{T}}\,[I](e) = I \tag{4.51}$$

where I is the moment of inertia about the axis of rotation.
With $(e) = (l\;m\;n)^{\mathrm{T}}$ and noting that $[I]$ is symmetrical, equation (4.50) expands to

$$l^2 I_{xx} + m^2 I_{yy} + n^2 I_{zz} + 2lmI_{xy} + 2mnI_{yz} + 2nlI_{zx} \tag{4.52}$$

If principal axes are chosen then

$$l^2 I_1 + m^2 I_2 + n^2 I_3 = I \tag{4.53}$$

The equation $\frac{x^2}{a^2} + \frac{y^2}{b^2} + \frac{z^2}{c^2} = 1$ is the equation for an ellipsoid where a, b and c are the semi-major and semi-minor axes. Equation (4.52) can be put in this form by taking the magnitude of a radius vector in the direction of $\boldsymbol{e}$ as $1/\sqrt{I}$ as shown in Fig. 4.6. It is seen that $l = x\sqrt{I}$, $m = y\sqrt{I}$ and $n = z\sqrt{I}$.

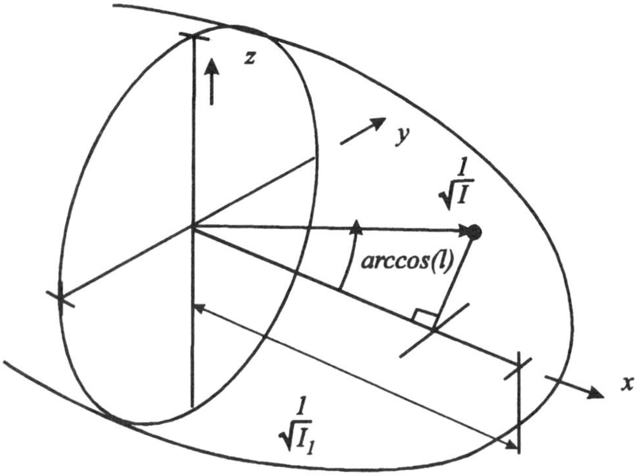

Fig. 4.6

Substituting into equation (4.53) and dividing through by I gives

$$I_1 x^2 + I_2 y^2 + I_3 z^2 = 1 \tag{4.54}$$

This is the equation of an ellipsoid with semi-axes $1/\sqrt{I_1}$, $1/\sqrt{I_2}$ and $1/\sqrt{I_3}$. This is known as *the moment of inertia ellipsoid.*

4.8 Torque-free motion of a rigid body

From equation (4.18) we have that if the torque $\boldsymbol{M}_O$ is zero then the moment of momentum is constant. We shall take, as before, the reference point to be either a fixed point or the centre of mass. The only difference is that the appropriate moment of inertia has to be used. Therefore

$$\boldsymbol{L}_O = \mathbf{I}_O \cdot \boldsymbol{\omega} = \text{constant} \tag{4.55}$$

Since for a rigid body there can be no internal energy losses and because there is no external work being done the kinetic energy will be constant

$$T = \frac{1}{2}\boldsymbol{\omega} \cdot \mathbf{I}_O \cdot \boldsymbol{\omega} = \frac{1}{2}\boldsymbol{\omega} \cdot \boldsymbol{L}_O = \text{constant} \tag{4.56}$$

We can write $2T = |\boldsymbol{\omega}|\,|\boldsymbol{L}_O| \cos\alpha$, where α is the acute angle between the angular velocity vector and the moment of momentum vector. Therefore

$$|\boldsymbol{\omega}| \cos\alpha = 2T/|\boldsymbol{L}_O| = \text{constant} \tag{4.57}$$

$$\cos\alpha = \frac{2T}{|\boldsymbol{L}_{\rm O}|\,|\boldsymbol{\omega}|} \tag{4.58}$$

This says that the component of the angular velocity parallel to the fixed moment of momentum is constant. Another way of expressing this result is to note that the tip of the angular velocity vector lies on a fixed plane normal to the moment of momentum vector known as the invariable plane.

We choose the body axes to correspond to the principal axes so that

$$\boldsymbol{L}_{\rm O} = I_1\omega_x\boldsymbol{i} + I_2\omega_y\boldsymbol{j} + I_3\omega_z\boldsymbol{k} \tag{4.59}$$

and

$$2T = I_1\omega_x^2 + I_2\omega_y^2 + I_3\omega_z^2 \tag{4.60}$$

Also

$$\boldsymbol{\omega} = \omega_x\boldsymbol{i} + \omega_y\boldsymbol{j} + \omega_z\boldsymbol{k} \tag{4.61}$$

The angle between the moment of momentum vector and the z axis, β, can be found from

$$\cos\beta = \frac{I_3\omega_z}{|\boldsymbol{L}_{\rm O}|} \tag{4.62}$$

and the angle between the angular velocity vector and the z axis, γ, is found from

$$\cos\gamma = \frac{\omega_z}{|\boldsymbol{\omega}|} \tag{4.63}$$

Without loss of generality we can choose the sense of the z axis such that ω_z is positive, in which case α, β and γ are all acute angles.

Expanding equation (4.62)

$$\cos\beta = \frac{I_3\omega_z}{\sqrt{(I_1^2\omega_x^2 + I_2^2\omega_y^2 + I_3^2\omega_z^2)}} \tag{4.64}$$

Expanding equation (4.63) and multiplying the numerator and denominator by I_3

$$\cos\gamma = \frac{I_3\omega_z}{\sqrt{(I_3^2\omega_x^2 + I_3^2\omega_y^2 + I_3^2\omega_z^2)}} \tag{4.65}$$

Now if I_3 is the largest principal moment of inertia $\cos\beta > \cos\gamma$ and therefore $\gamma > \beta$. If I_3 is the smallest principal moment of inertia $\beta > \gamma$.

Expanding equation (4.58)

$$\cos\alpha = \frac{I_1\omega_x^2 + I_2\omega_y^2 + I_3\omega_z^2}{|\boldsymbol{L}_{\rm O}|\,|\boldsymbol{\omega}|} \tag{4.66}$$

From the expressions for $\cos\alpha$, $\cos\beta$ and $\cos\gamma$ it can be shown that α can never be the largest of the three angles.

The justification of the last statement is dependent on the fact that not all combinations of the principal moment of inertia are possible. The perpendicular axes theorem as given in equation (4.39b) shows that for a lamina in the xy plane $I_z = I_x + I_y$ and as one principal axis is in the z direction we have for the principal axes that $I_3 = I_1 + I_2$. Thus

$$1 = \frac{I_1}{I_3} + \frac{I_2}{I_3} \tag{4.67}$$

Figure 4.7 shows a plot of I_1/I_3 against I_2/I_3 on which equation (4.67) plots as line *ab*. For a long slender rod with its axis along the 3 axis $I_3 = 0$ and $I_1 = I_2$ and is shown as point *e*. It is easy to show that rods in the *y* and *x* axes respectively are represented by points *a* and *b*. Lines *ac* and *bd* are for laminae in the *yz* and *xz* planes. The region enclosed by *cabd* is the allowable region for moments of inertia.

Figure 4.8 shows the relative positions of the **L**, **ω** and **k** vectors for the case where I_3 is smallest so that $\beta > \gamma$. Because α can never be the largest of the three angles the arrangement must be as shown. The three vectors need not be coplanar. Figure 4.9 is similar to Fig. 4.8 except that I_3 is the largest so that $\gamma > \beta$. If I_3 has the intermediate value then either pattern is possible.

An interesting geometrical interpretation was put forward by Poinsot in 1834. He discovered that the body could be represented by its inertia ellipsoid touching the invariable plane and with its centroid at a fixed distance from the plane as shown in Fig. 4.10.

From the discussion leading to equation (4.54) the radius vector $\rho = 1/\sqrt{I}$ but as $2T = I\omega^2$ we have that $\rho = \omega/\sqrt{(2T)}$.

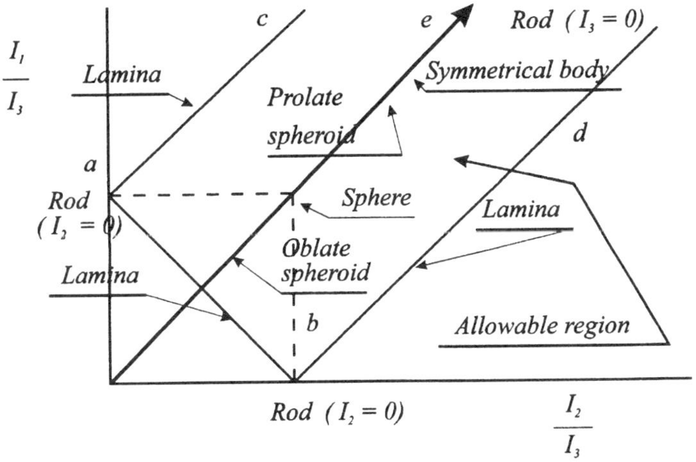

Fig. 4.7 Moment of inertia bounds

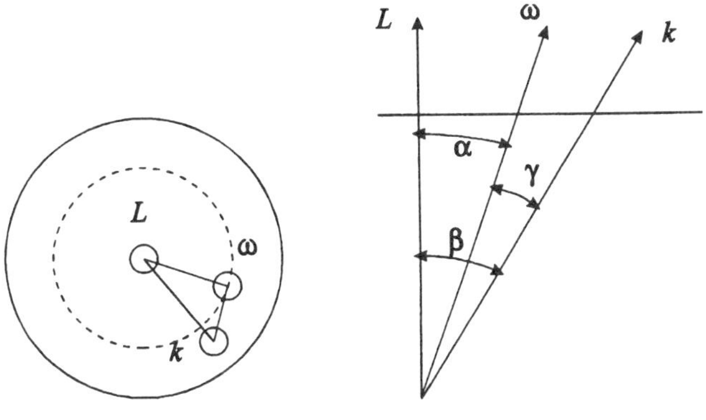

Fig. 4.8

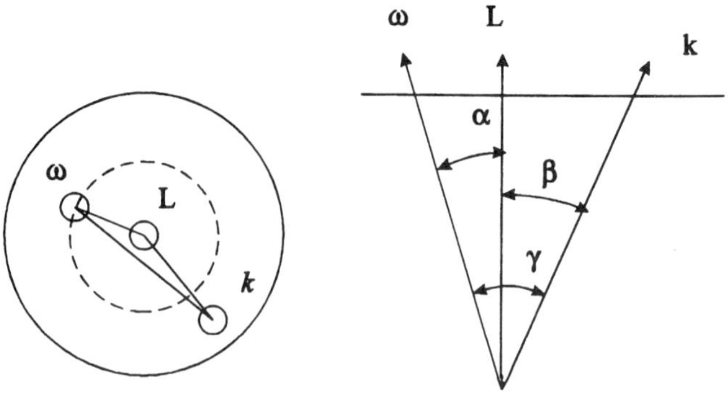

Fig. 4.9

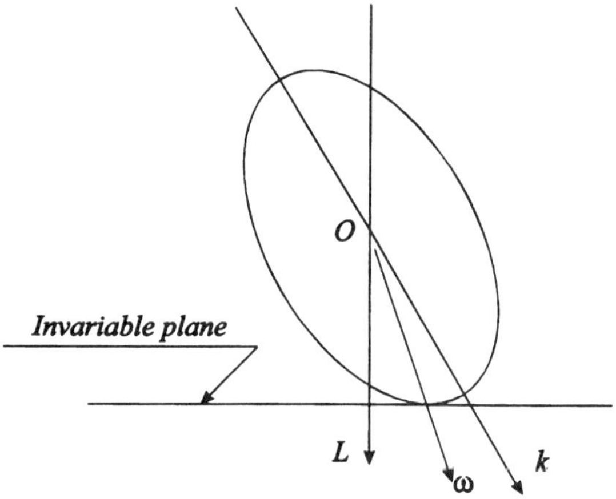

Fig. 4.10 Poinsot's ellipsoid

The equation for the ellipsoid can be written as

$$I_1x^2 + I_2y^2 + I_3z^2 = f$$

where f is a scalar function; $f = 1$ is the case for the inertia ellipsoid. The gradient of f gives a vector which points in the direction of the normal to the surface at given value of x, y, z, and hence $\boldsymbol{\omega}$. Thus

$$\text{grad}\, f = 2I_1x\boldsymbol{i} + 2I_2y\boldsymbol{j} + 2I_3z\boldsymbol{k}$$

Now

$$x = l\rho = \frac{l\omega}{\sqrt{(2T)}} = \frac{\omega_x}{\sqrt{(2T)}}$$

etc., so

$$\text{grad}\, f = \frac{2}{\sqrt{(2T)}}\boldsymbol{L}$$

This means that the normal to the ellipsoid surface at the contact point is parallel to $\boldsymbol{L}$ and is therefore normal to the invariable plane.

All the above is consistent for a body with the shape of the inertia ellipsoid rolling without slip on an invariable plane. The curve that the $\boldsymbol{\omega}$ vector traces out on the body is known as the *polhode* and the curve which it traces out on the invariable plane is known as the *herpolhode.*

For a torque-free symmetrical body with $I_1 = I_2$ we have, from Euler's equation (4.41), that $I_3\omega_z =$ constant. Equation (4.62) shows that β is constant and thus the inclination of the moment of inertia ellipsoid is constant. Since the distance of the origin of the ellipsoid to the invariable plane is constant it follows that the radius vector from the origin to the contact point is also constant and therefore the magnitude of the angular velocity is constant, its direction of course not being constant. From equation (4.58) we see that α is constant and from equation (4.63) γ is constant. By forming the triple scalar product of $\boldsymbol{L}_{\mathrm{O}}$, $\boldsymbol{\omega}$ and $\boldsymbol{k}$ it is seen that the value is zero when $I_1 = I_2$ showing that for this case the three vectors are coplanar.

Referring to Figs 4.8 and 4.9 it is clear that because all three angles are constant and the three vectors are coplanar both the polhode and the herpolhode are circles. These circles can be thought of as the bases of cones centred on the origin, the one with semi-angle α being the space cone and the one with semi-angle γ the body cone. When I_3 is the smallest the outside of the body cone rolls on the outside of the space cone, see Fig. 4.11, and when I_3 is the greatest the *inside* of the body cone rolls on the outside of the space cone, as shown in Fig. 4.12.

For any given starting values of ω the constants L_{O} and T are determined and from these the constant angles α, β and γ can be found. The precession of the body axis around the moment of momentum vector can be evaluated from the kinematics of the space and body cones as shown in Figs 4.11 and 4.12.

Letting the precession rate about the L_{O} axis be Ω we can write expressions for the velocity of the point $\mathrm{C_b}$ as

$$\boldsymbol{\omega} \times \overrightarrow{(\mathrm{OC_b})} = \boldsymbol{\Omega} \times \overrightarrow{(\mathrm{OC_b})}$$

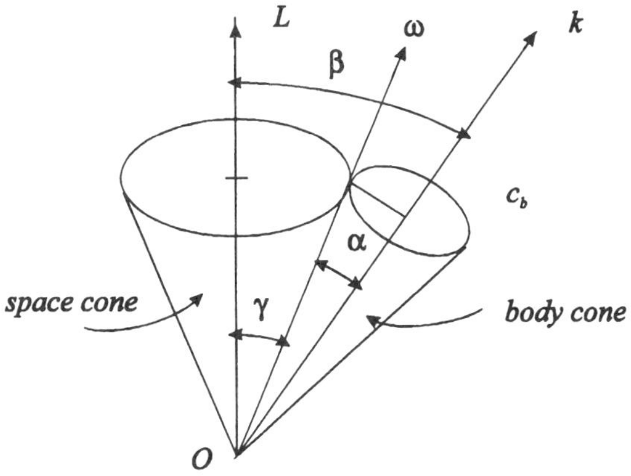

Fig. 4.11 $I_3 < (I_2 = I_1)$

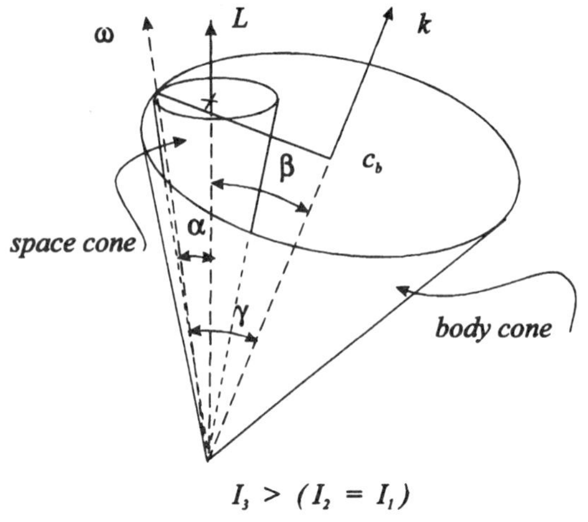

Fig. 4.12 $I_3 > (I_2 = I_1)$

Equating the magnitudes, since directions are the same, gives

$$|\omega|\,|\overrightarrow{OC_b}|\sin\gamma = |\Omega|\,|\overrightarrow{OC_b}|\sin\beta$$

giving

$$|\Omega| = |\omega|\frac{\sin\gamma}{\sin\beta} = |\omega|\frac{\sqrt{(\omega_x^2+\omega_y^2)}}{|\omega|}\;\frac{|L_O|}{\sqrt{[I_1^2(\omega_x^2+\omega_y^2)]}}$$

$$= \frac{|L_O|}{I_1} = \omega_z\sqrt{(\omega_x^2/\omega_z^2 + \omega_y^2/\omega_z^2 + I_3^2/I_1^2)} \quad (4.68)$$

Now the angular velocity of the body cone relative to the precessing plane containing $\boldsymbol{L}_O$, $\boldsymbol{\omega}$ and $\boldsymbol{k}$ vectors is $\boldsymbol{\omega} - \boldsymbol{\Omega}$. The component in the z directon is

$$(\boldsymbol{\omega} - \boldsymbol{\Omega})\cdot\boldsymbol{k} = \omega_z - |\Omega|\cos\beta$$

$$= \omega_z - \frac{|L_O|}{I_1}\cos\beta$$

$$= \omega_z - I_3\omega_z/I_1$$

$$= \omega_z(1 - I_3/I_1)$$

This is the rotation of the body relative to the frame containing the $\boldsymbol{L}_O$, $\boldsymbol{\omega}$ and $\boldsymbol{k}$ vectors. Therefore its negative will be the precession of the ω axis relative to the body, and thus

$$\Omega_{\omega/b} = \omega_z(I_3/I_1 - 1) \quad (4.68a)$$

4.9 Stability of torque-free motion

The investigation of the stability of torque-free motion can be carried out using standard mathematical techniques, but the semi-graphical method which follows gives all the essential information.

For torque-free motion the moment of momentum vector is constant, as is the kinetic energy. The square of the magnitude of the moment of momentum expressed in terms of its components along the principal axes is

$$L^2 = L_1^2 + L_2^2 + L_3^2 \tag{4.69}$$

and as $L_1 = I_1\omega_x$ etc. the kinetic energy T can be written

$$2T = \frac{L_1^2}{I_1} + \frac{L_2^2}{I_2} + \frac{L_3^2}{I_3} \tag{4.70}$$

Taking L_1, L_2 and L_3 as the co-ordinate axes equation (4.69) plots as a sphere of radius L and equation (4.70) plots as an ellipsoid with semi-axes $\sqrt{(2TI_1)}$, $\sqrt{(2TI_2)}$ and $\sqrt{(2TI_3)}$ and is known as *Binet's ellipsoid.*

For convenience we choose the principal axes such that $I_3 > I_2 > I_1$. By multiplying equation (4.69) by I_1 we have

$$2TI_1 = L_1^2 + \frac{I_1}{I_2} L_2^2 + \frac{I_1}{I_3} L_3^2 \tag{4.71}$$

which is less than L^2, since I_1 is the smallest moment of inertia. Therefore $\sqrt{(2TI_1)} < L$. Similarly by multiplying equation (4.69) by I_3 we show that $\sqrt{(2TI_3)} > L$. This means that the sphere always intersects the ellipsoid.

Figures 4.13 to 4.16 show the form of the surfaces for various conditions. The intersection curve is the locus of $\boldsymbol{L}$ relative to the body axes as only points on this curve will satisfy both equation (4.69) and equation (4.70). The curve is obviously closed so it follows that the curve traced out by the angular velocity vector will also be closed, that is the polhode is a closed curve. We cannot say the same for the herpolhode but it can be seen that this falls within a fixed band.

In Fig. 4.13 the moment of momentum vector is close to the 1 axis (smallest moment of inertia) and the curve tends to a circle, so for a rigid body this is a stable condition. In Fig. 4.16 the moment of momentum vector is close to the 3 axis (the maximum moment of inertia) and again the intersection curve tends to a circle and stable motion. When the moment of momentum vector is close to the 2 axis (the intermediate moment of inertia) the intersection curve is long and wanders over a large portion of the surface, as shown in Figs 4.14 and 4.15. This indicates an unstable motion.

So far we have considered the body to be rigid and torque free. In the case of a non-rigid body any small amount of flexing will dissipate energy whilst the moment of momentum

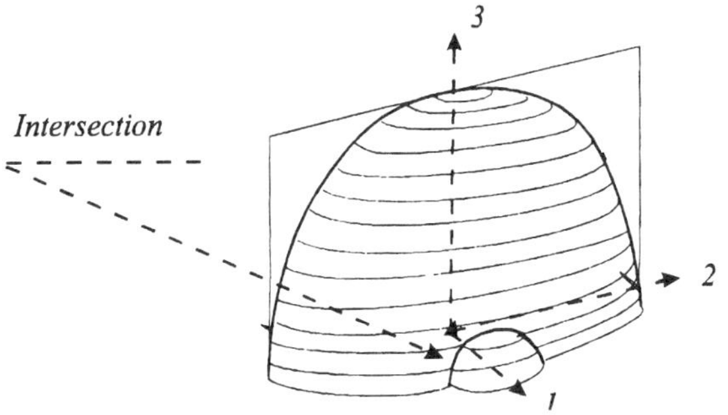

Fig. 4.13 Binet diagram: ellipsoid principal axes 1,2,3, sphere radius 1.1

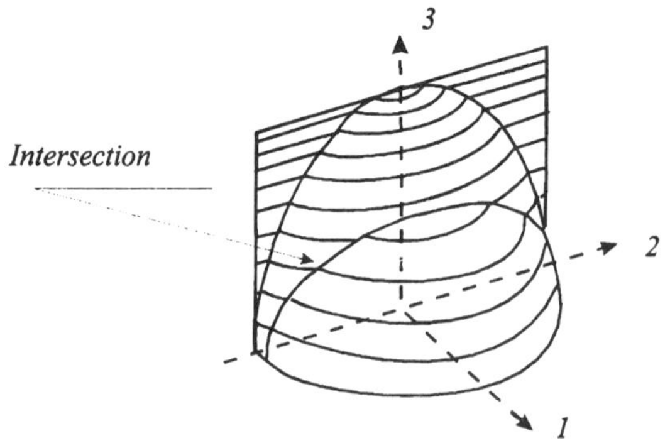

Fig. 4.14 Binet diagram: ellipsoid principal axes 1,2,3, sphere radius 1.95

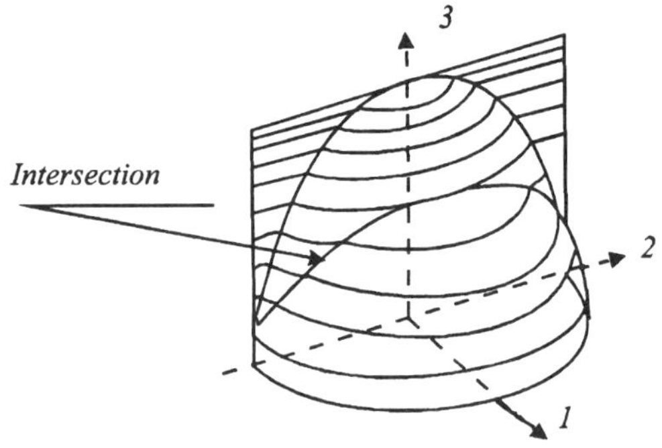

Fig. 4.15 Binet diagram: ellipsoid principal axes 1,2,3, sphere radius 2.05

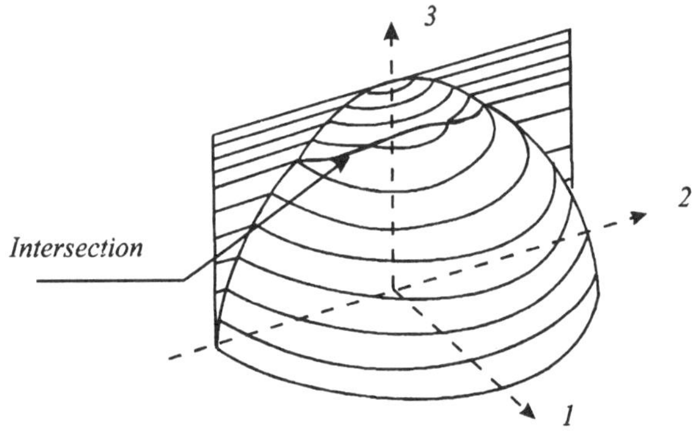

Fig. 4.16 Binet diagram: ellipsoid principal axes 3,2,1, sphere radius 2.6

remains constant. A similar situation occurs when tidal effects are present. In both cases it is assumed that variations from the nominal shape are small.

The general effect is that Binet's ellipsoid will slowly shrink. For the case of rotation about the 3 axis the intersection curve reduces and the motion remains stable, but in the case of rotation close to the 1 axis the intersection curve will slowly increase in size leading to an unstable condition. Rotation close to the 2 axis is unstable under all conditions.

4.10 Euler's angles

The previous sections have been concerned, for the most part, with setting up the equations of motion and looking at the properties of a rigid body. Some insight to the solution of these equations was gained by means of Poinsot's construction for the case of torque-free motion.

The equations obtained involved the components of angular velocity and acceleration but they cannot be integrated to yield angles because the co-ordinate axes are changing in direction so that finite rotation about any of the body axes has no meaning.

We are now going to express the angular velocity in terms of angles which can uniquely define the orientation of the body. Such a set are Euler's angles which we now define.

Figure 4.17 shows a body rotating about a fixed point O (or its centre of mass). The *XYZ* axes are an inertial set with origin O. The *xyz* axes are, in the general case, attached to the body. If the body has an axis of symmetry then this is chosen to be the *z* axis. Starting with *XYZ* and the *xyz* coincident we impose a rotation of ø about the *Z* axis. There then follows a rotation of θ about the new *x* axis (the *x'* axis) and finally we give a rotation of θ about the final *z* axis.

The angular velocity vector is

$$\boldsymbol{\omega} = \dot{ø}\boldsymbol{K} + \dot{\theta}\,\boldsymbol{i}' + \dot{\psi}\boldsymbol{k} \tag{4.72}$$

where $\boldsymbol{K}$ is the unit vector in the Z direction, $\boldsymbol{i}'$ is the unit vector in the x' direction and $\boldsymbol{k}$ is the unit vector in the z direction.

From the figure we see that

$$\boldsymbol{K} = \cos(\theta)\,\boldsymbol{k} + \sin(\theta)\,\boldsymbol{j}'' \tag{4.73}$$

$$\boldsymbol{j}'' = \cos(\psi)\,\boldsymbol{j} + \sin(\psi)\,\boldsymbol{i} \tag{4.74}$$

$$\boldsymbol{i}' = \cos(\psi)\,\boldsymbol{i} - \sin(\psi)\,\boldsymbol{j} \tag{4.75}$$

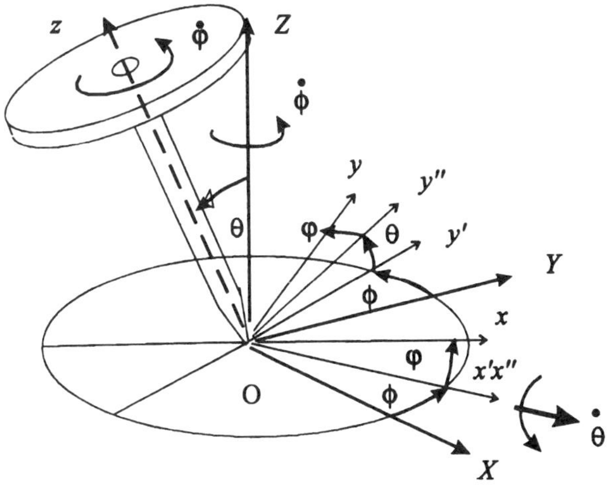

Fig. 4.17

Thus

$$\boldsymbol{K} = \cos(\theta)\,\boldsymbol{k} + \sin(\theta)\cos(\psi)\,\boldsymbol{j} + \sin(\theta)\sin(\psi)\,\boldsymbol{i} \tag{4.76}$$

Writing

$$\boldsymbol{\omega} = \omega_x \boldsymbol{i} + \omega_y \boldsymbol{j} + \omega_z \boldsymbol{k} \tag{4.77}$$

and substituting for $\boldsymbol{K}$ and $\boldsymbol{i}'$ in equation (4.72) gives

$$\begin{aligned} \omega_x &= \dot{\phi}\sin\theta\sin\psi + \dot{\theta}\cos\psi + 0 \\ \omega_y &= \dot{\phi}\sin\theta\cos\psi - \dot{\theta}\sin\psi + 0 \\ \omega_z &= \dot{\phi}\cos\theta + 0 + \dot{\psi} \end{aligned} \tag{4.78}$$

4.11 The symmetrical body

The equation for angular velocity given in the previous section, even when used in conjunction with principal co-ordinates, leads to lengthy expressions when substituted into Euler's equations or Lagrange's equations. For the body whose axis of symmetry is the z axis (so that $I_2 = I_1$) the xyz axes need not be rotated about the z axis. This means that for the axes ψ is zero. Nevertheless the body still has an angular velocity component $\dot{\psi}$ about the z axis.

For the axes

$$\boldsymbol{\omega}_R = \Omega_x \boldsymbol{i} + \Omega_y \boldsymbol{j} + \Omega_z \boldsymbol{k}$$

where

$$\begin{aligned} \Omega x &= \dot{\theta} \\ \Omega y &= \dot{\phi}\sin\theta \\ \Omega z &= \dot{\phi}\cos\theta \end{aligned} \tag{4.79}$$

For the body

$$\begin{aligned} \omega_x &= \dot{\theta} \\ \omega_y &= \dot{\phi}\sin\theta \\ \omega_z &= \dot{\phi}\cos\theta + \dot{\psi} \end{aligned} \tag{4.80}$$

These terms may now be inserted into the modified Euler's equations (4.42) to give

$$\begin{aligned} M_x &= I_1\ddot{\theta} - (I_1 - I_3)\,\dot{\phi}^2\sin\theta\cos\theta + I_3\dot{\phi}\dot{\psi}\sin\theta \\ M_y &= I_1(\ddot{\phi}\sin\theta + \dot{\phi}\dot{\theta}\cos\theta) - (I_3 - I_1)\,\dot{\phi}\dot{\theta}\cos\theta - I\,\dot{\theta}\dot{\psi} \\ M_z &= I_3\frac{\partial}{\partial t}(\dot{\phi}\cos\theta + \dot{\psi}) \end{aligned} \tag{4.81}$$

Alternatively we can write an expression for the kinetic energy

$$T = \frac{1}{2}I_1\omega_x^2 + \frac{1}{2}I_1\omega_y^2 + \frac{1}{2}I_3\omega_z^2$$

and substituting the angular velocities from equation (4.80) gives

$$T = \frac{1}{2}I_1\dot{\theta}^2 + \frac{1}{2}I_1\dot{\phi}^2\sin^2\theta + \frac{1}{2}I_3(\dot{\phi}\cos\theta + \dot{\psi})^2 \tag{4.82}$$

It is interesting to note that use of equation (4.78) gives the same result because $(\omega_x^2 + \omega_y^2)$ does not contain ψ.

We now consider the classic case of the symmetric top in a gravitational field $-g\boldsymbol{K}$. Figure 4.18 shows the relevant data.

The torque is

$$\boldsymbol{M}_O = mgh \sin(\theta)\, \boldsymbol{i} \tag{4.83}$$

and the potential energy

$$V = mgh \cos\theta \tag{4.84}$$

We choose to use Lagrange's equations because they yield some first integrals in a convenient form. The Lagrangian ($\mathfrak{L} = T - V$) is

$$\mathfrak{L} = \frac{1}{2} I_1 \dot{\theta}^2 + \frac{1}{2} I_1 \dot{\phi}^2 \sin^2\theta + \frac{1}{2} I_3 (\dot{\phi} \cos\theta + \dot{\psi})^2 - mgh \cos\theta \tag{4.85}$$

from which we see that neither ψ nor ϕ appear explicitly in the Lagrangian (they are cyclic or ignorable). Therefore we have two first integrals of the motion in the form of constant generalized momenta

$$p_\psi = \frac{\partial \mathfrak{L}}{\partial \dot{\psi}} = I_3 (\dot{\psi} + \dot{\phi} \cos\theta) = \text{constant} \tag{4.86}$$

and

$$p_\phi = \frac{\partial \mathfrak{L}}{\partial \dot{\phi}} = I_1 \dot{\phi} \sin^2\theta + I_3 (\dot{\psi} + \dot{\phi} \cos\theta) \cos\theta$$

$$= I_1 \dot{\phi} \sin^2\theta + p_\psi \cos\theta = \text{constant} \tag{4.87}$$

Because time does not appear explicitly in the Lagrangian the energy, E, is constant

$$E = \frac{1}{2} I_1 \dot{\theta}^2 + \frac{1}{2} I_1 \dot{\phi}^2 \sin^2\theta + \frac{1}{2} I_3 (\dot{\phi} \cos\theta + \dot{\psi})^2 + mgh \cos\theta \tag{4.88}$$

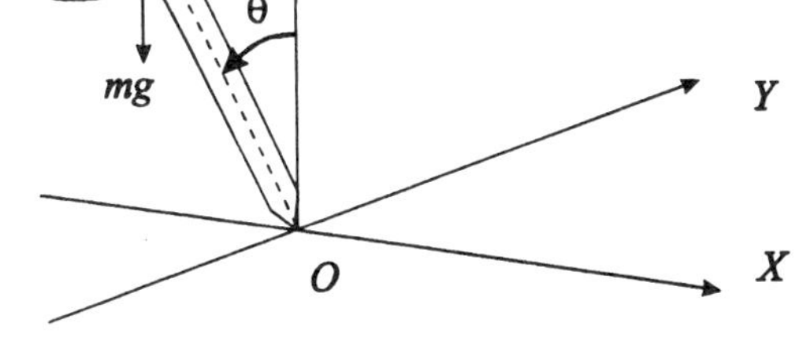

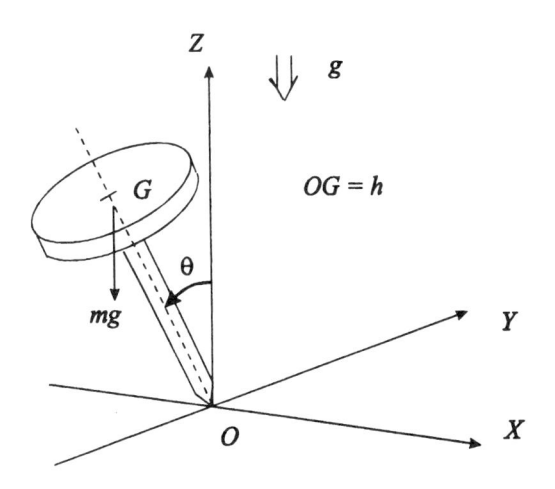

Fig. 4.18

Substituting equations (4.86) and (4.87) into (4.88) gives

$$E = \frac{1}{2} I_1\dot{\theta}^2 + \frac{(p_ø - p_\psi \cos\theta)^2}{2I_1 \sin^2\theta} + \frac{p_\psi^2}{2I_3} + mgh \cos\theta \tag{4.89}$$

Equation (4.89) can be rearranged in the following form

$$E' = \frac{1}{2} I_1\dot{\theta}^2 + V'(\theta)$$

where the constant

$$E' = E - \frac{p_\psi^2}{2I_3} \text{ is the effective energy and} \tag{4.90}$$

$$V' = \frac{(p_ø - p_\psi \cos\theta)^2}{2I_1 \sin^2\theta} + mgh \cos\theta \tag{4.91}$$

which is a function of θ only and may be considered to be a 'pseudo' potential energy. A typical plot of V' and E' against θ is shown in Fig. 4.19. In this case the shaded area is the region where $\dot{\theta}^2$ is positive and is therefore the only possible values of θ for the given initial conditions. It is seen that θ oscillates between levels θ_1 and θ_2 whilst θ_3 is the value of θ where $\ddot{\theta} = 0$.

We shall next generate Lagrange's equation for θ as the generalized co-ordinate

$$\frac{\mathrm{d}}{\mathrm{d}t}\left(\frac{\partial \mathfrak{L}}{\partial \dot{\theta}}\right) - \frac{\partial \mathfrak{L}}{\partial \theta} = 0$$

The right hand side is zero because it is assumed that there is no friction or other forces applied; the effect of gravity is covered by the potential energy term.

Thus

$$I_1\ddot{\theta} - [\, I_1\dot{ø}^2 \sin\theta \cos\theta - I_3(\dot{ø}\cos\theta + \dot{\psi})(\dot{ø}\sin\theta) + mgh\sin\theta \,] \tag{4.92}$$

Now from equation (4.86)

$$I_3(\dot{ø}\cos\theta + \dot{\psi}) = I_3\omega_z$$

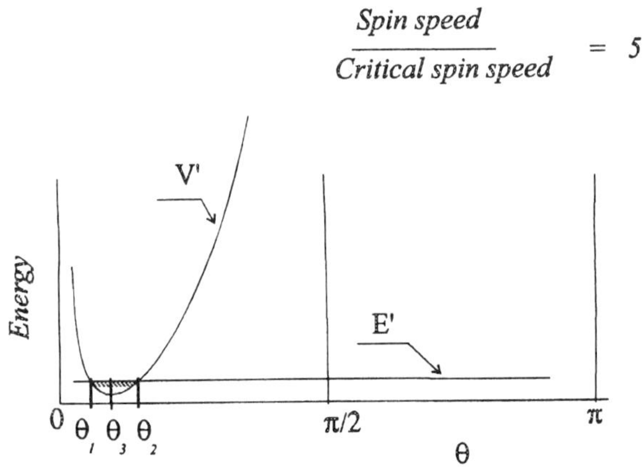

Fig. 4.19

Therefore

$$I_1\ddot{\theta} = (-I_3\omega_z\dot{ø} + I_1\dot{ø}^2 \cos\theta + mgh) \sin\theta = 0 \tag{4.93}$$

For $\ddot{\theta} = 0$ and θ not equal to zero

$$-I_3\omega_z\dot{ø} + I_1\dot{ø}^2 \cos\theta + mgh = 0 \tag{4.94}$$

This expression is valid for $\ddot{\theta} = 0$ and is independent of $\dot{\theta}$ so it is true for the case of steady precession where $\dot{\theta} = 0$.

It is convenient to rewrite equation (4.94) as

$$\cos\theta = \frac{I_3\omega_z}{I_1}\left(\frac{1}{\dot{ø}}\right) - \frac{mgh}{I_1}\left(\frac{1}{\dot{ø}}\right)^2 \tag{4.95}$$

A plot of cos θ against $1/\dot{ø}$ is shown in Fig. 4.20. The maximum value for cos θ is found by equating the slope of the curve to zero,

$$\frac{d\cos\theta}{d(1/\dot{ø})} = \frac{I_3\omega_z}{I_1} - \frac{2mgh}{I_1}\left(\frac{1}{\dot{ø}}\right) = 0 \tag{4.96}$$

so the maximum occurs when

$$I/\dot{ø} = I_3\omega_z / 2mgh \tag{4.97}$$

and the maximum value for cos θ is

$$\cos\theta_{\max} = \frac{I_3^2\omega_z^2}{4mghI_1} \tag{4.98}$$

For the special case of $\theta = 0$ the minimum value of ω_z that will maintain stable motion with the axis vertical is

$$\omega_{z,\text{crit}} = \sqrt{(4mgI_1 / I_3^2)} \tag{4.99}$$

This condition is that of the 'sleeping top'.

The motion of the *z* axis can be found by numerically integrating equation (4.93) in conjunction with equation (4.87) to generate θ and ø as functions of time. Some typical results are shown in Figs 4.21 to 4.24. If the initial precessional speed is that corresponding to those for steady precession then a circular motion is achieved. The time for one revolution about

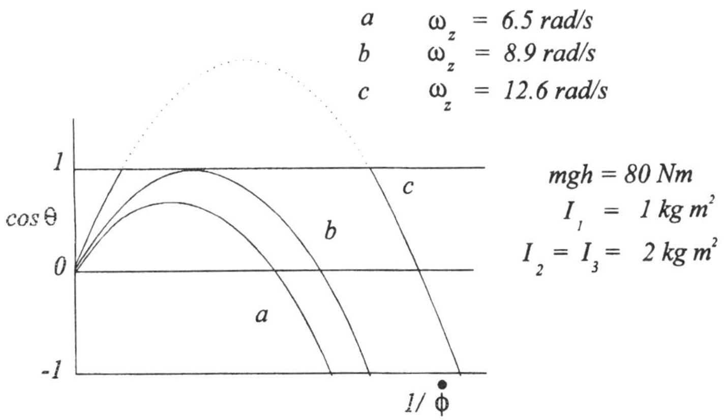

Fig. 4.20

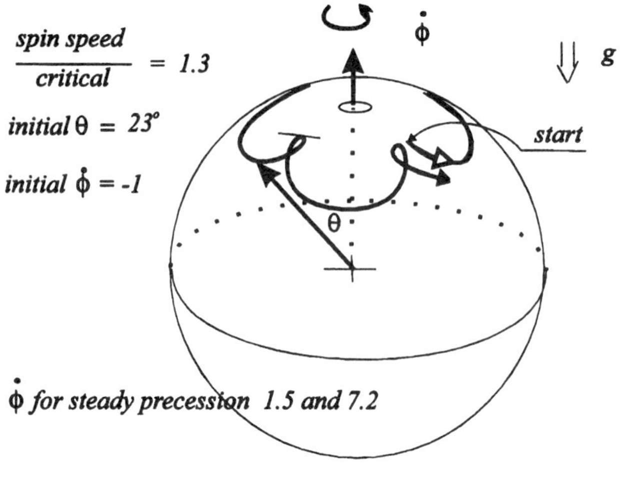

Fig. 4.21

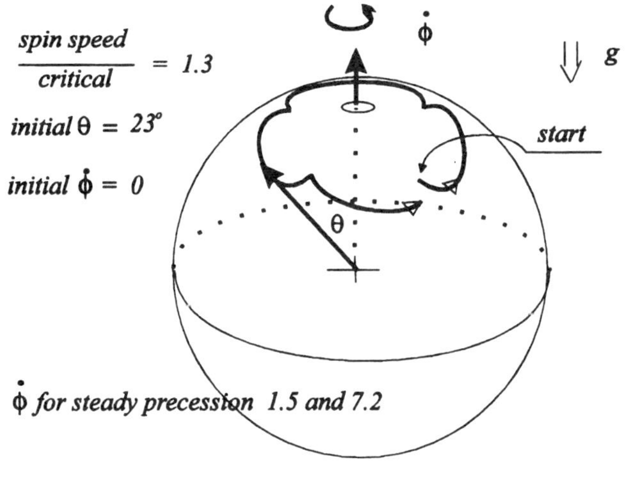

Fig. 4.22

the Z axis for initial speeds not equal to a steady precessional speed varies slightly for small oscillations of θ and when $\dot{\phi}$ is in the range 0 to a little above the slow precessional speed. A plot of the ratio of precessional time to time for steady precession against initial precessional speed is shown in Fig. 4.25.

4.12 Forced precession

So far we have considered the body to be free to respond to applied torques, or the absence of torque. A much simpler problem is to determine the torques required to give a prescribed motion to a body. We shall tackle a specific problem and find the solution by the direct application of first principles.

Figure 4.26 depicts a rigid symmetrical wheel W which runs, with negligible friction, on axle A. The axle is freely pivoted to a block which is free to rotate about the vertical Z axis. The wheel is rotating relative to the axle at a speed $\dot{\psi}$ and the whole assembly is rotated

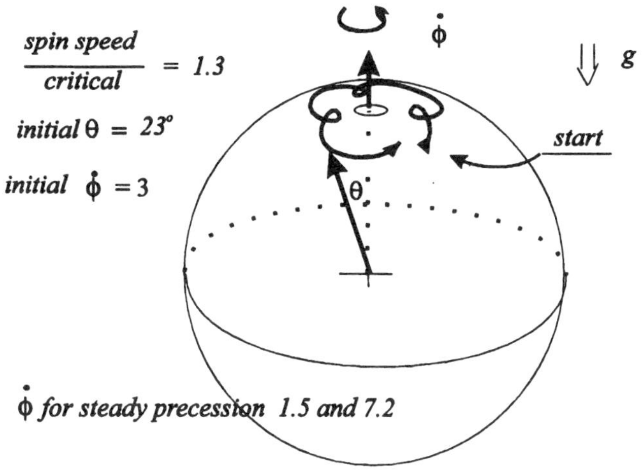
spin speed / critical = 1.3
initial θ = 23°
initial $\dot{\phi}$ = 3
$\dot{\phi}$
g
start
θ
$\dot{\phi}$ for steady precession 1.5 and 7.2

Fig. 4.23

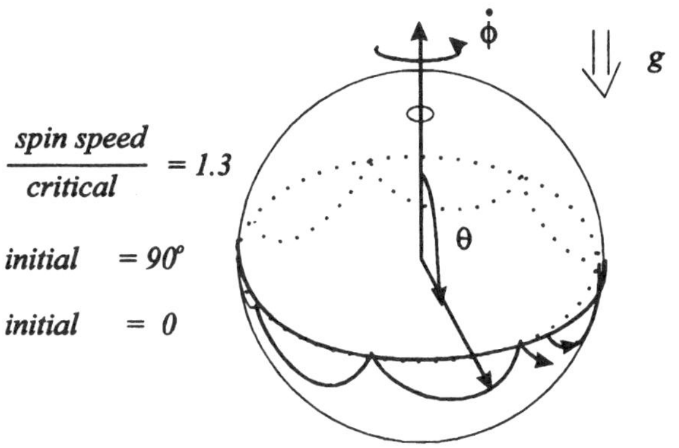
$\dot{\phi}$
g
spin speed / critical = 1.3
initial = 90°
initial = 0
θ

Fig. 4.24

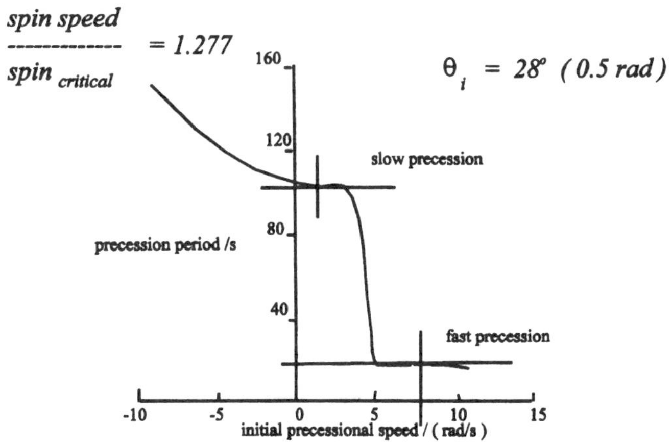
spin speed / spin critical = 1.277
θ i = 28° (0.5 rad)
160
120
80
40
slow precession
fast precession
precession period /s
-10
-5
0
5
10
15
initial precessional speed / (rad/s)

Fig. 4.25

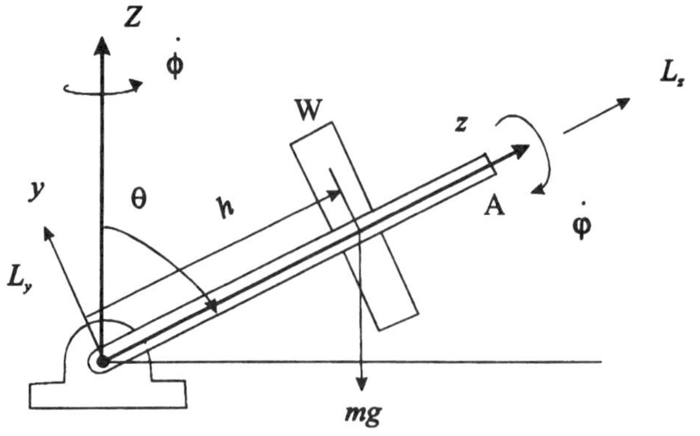

Fig. 4.26

about the Z axis at a constant angular speed of $\dot{\phi}$. The moment of inertia of the wheel about its axis, the z axis, is I_3 and the moment of inertia of the wheel and axle about the y and x axes is I_1.

EXAMPLE

Determine the torque which must be applied to the axle about the x axis so that the angle θ is maintained constant.
The angular velocity about the z axis is

$$\dot{\psi} + \dot{\phi} \cos \theta = \omega_z \tag{i}$$

and the angular velocity about the y axis is $\dot{\phi} \sin \theta$.

The moment of momentum vector is

$$\boldsymbol{L}_{\mathrm{O}} = I_1 \dot{\phi} \sin(\theta) \boldsymbol{j} + I_3 \omega_z \boldsymbol{k} \tag{ii}$$

From Figs 4.26 and 4.27 we see that the change in the moment of momentum vector is

$$I_3 \omega_z \sin \theta \, \mathrm{d}\phi \, \boldsymbol{i} - I_1 \dot{\phi} \sin \theta \cos \theta \, \mathrm{d}\phi \, \boldsymbol{i} = \mathrm{d}\boldsymbol{L}_{\mathrm{O}}$$

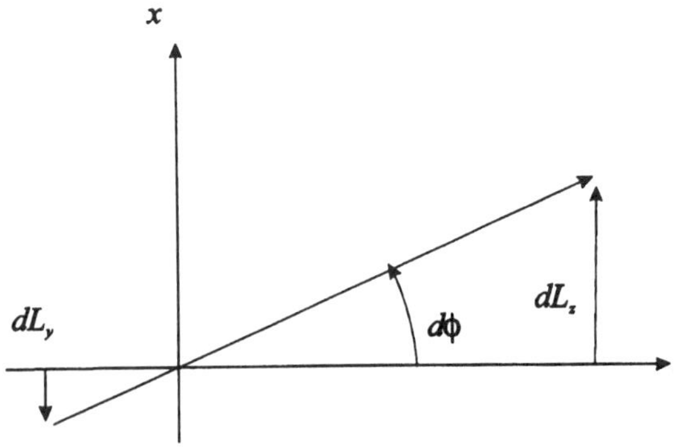

Fig. 4.27

Thus

$$M_O = \frac{dL_O}{dt} = (I_3\omega_z\dot{\varnothing} - I_1\dot{\varnothing}^2\cos\theta)\,\sin(\theta)\;\boldsymbol{i} \qquad \text{(iii)}$$

but the torque about O is

$$M_O = (mgh\sin\theta + Q_x)\,\boldsymbol{i} \qquad \text{(iv)}$$

Equating torques from equations (iii) and (iv) gives

$$Q_x = (I_3\omega_z\dot{\varnothing} - I_1\dot{\varnothing}^2\cos\theta - mgh)\sin\theta \qquad \text{(v)}$$

which is the required holding torque. Since there is no torque about the z axis $I_3\omega_z = I_3(\dot{\psi} + \dot{\varnothing}\cos\theta)$ = constant.

If Q_x is zero we replicate equation (4.94).

4.13 Epilogue

The reader may well feel at this point that some new basic principle has been uncovered owing to the somewhat unexpected behaviour of rotating rigid bodies. We appear to come a long way from Newton's laws of motion with notions such as the moment of inertia tensor and the need for three-dimensional rotating axes. The fact that torques do not just produce angular accelerations in a straightforward analogy with particle dynamics seems to require reconciliation.

A simple example will serve to illustrate the origins of gyroscopic behaviour. Figure 4.28 shows two identical satellites in circular orbit about a massive central body, the satellites being diametrically opposed. At the same instant the satellites receive impulses, ΔJ, normal to the plane of the orbit but in opposite senses. The effect of these impulses is to give each satellite a velocity of $\Delta J/m$ in the same direction as the impulse. If the initial tangential velocity is V then the change in direction of the path $d\varnothing = \Delta J/(mV)$. This is the simple particle dynamic solution. However, we can regard the system as originally rotating about the

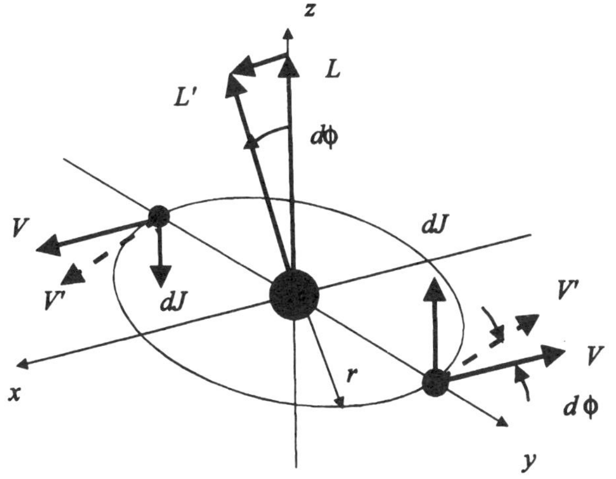

Fig. 4.28

z axis; the pair of impulses constitute a couple about the x axis yet the plane of rotation has rotated about the y axis.

We could solve the problem by equating the impulsive couple to the change in moment of momentum, that is

$$\boldsymbol{M}_{\mathrm{O}}\mathrm{d}t = \mathrm{d}\boldsymbol{L}_{\mathrm{O}}$$

$$\Delta J(2r)\boldsymbol{i} = I_z\omega_z\mathrm{d}ø\ \boldsymbol{i}$$

Thus

$$\Delta J 2r = (2mr^2)(V/r)\ \mathrm{d}ø$$

giving

$$\mathrm{d}ø = \Delta J/(mV)$$

as expected.

5
Dynamics of Vehicles

5.1 Introduction

A vehicle in this chapter is taken to be one which travels on land, in the air or in space. The purpose of the chapter is to bring out some of the characteristic dynamics in the particular domain.

Satellite motion is typified by the motion of a small body about a large body under the action of a central force. However, for the two-body problem the restriction of one body being significantly smaller than the other is not restrictive because when considering the *relative* motion of two bodies the equations for relative motion are the same as those pertaining to one small body in orbit about a large one.

5.2 Gravitational potential

Before considering the motion of bodies under gravity it is necessary to look at the distinction between centre of mass and centre of gravity. If the gravitational field is uniform then the two centres will coincide. One important result which, so far, we have taken for granted is that for a uniform, spherical body the centre of gravity is at the geometric centre; as is the centre of mass.

The quickest method of proving the last statement is to utilize the concept of gravitational potential. The change in potential is defined to be the negative of the work done by the force of gravity acting on a unit mass at some point P. Figure 5.1 shows a mass m and a unit mass at point P. Therefore the change in potential is

$$\mathrm{d}\bar{V} = -F\mathrm{d}r = -\left(-\frac{Gm}{r^2}\right)\mathrm{d}r \tag{5.1}$$

Thus

$$\bar{V} = \int \frac{Gm}{r^2}\mathrm{d}r = -\frac{Gm}{r} + \text{constant} \tag{5.2}$$

Because ultimately only the differences in potential are required the additive constant can take any convenient value. In this case we shall make the constant equal to zero.

Note that here $\bar{V}$ is the potential energy for a unit mass at the point P (or, the work done divided by the mass at P) and is therefore described as the *gravitational potential* at the point P. In many texts $-\bar{V}$ is called the gravitational potential.

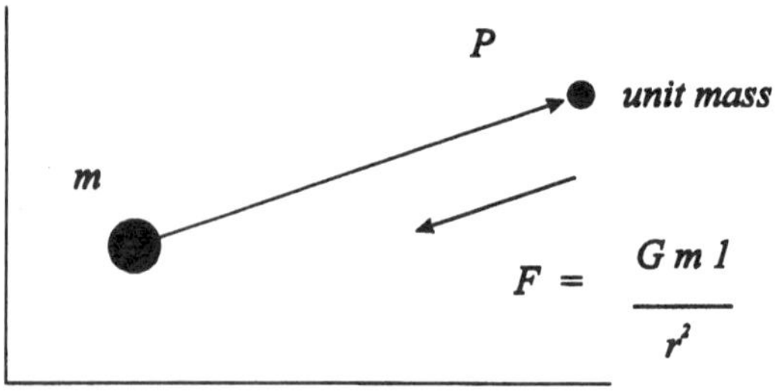

Fig. 5.1

If there are many masses then, since the potential is a scalar function of position, the total potential is

$$\bar{V} = -\sum \frac{Gm_i}{r_i} \tag{5.3}$$

From the definition of the potential the force in the r direction is

$$F_r = -\frac{\mathrm{d}\bar{V}}{\mathrm{d}r}$$

and using the chain rule for differentiation

$$F_r = -\left(\frac{\partial \bar{V}}{\partial x}\frac{\mathrm{d}x}{\mathrm{d}r} + \frac{\partial \bar{V}}{\partial y}\frac{\mathrm{d}y}{\mathrm{d}r} + \frac{\partial \bar{V}}{\partial z}\frac{\mathrm{d}z}{\mathrm{d}r}\right)$$

The unit vector in the r direction, $\boldsymbol{e}_r$, can be expressed in terms of the Cartesian unit vectors as

$$\boldsymbol{e}_r = \frac{\mathrm{d}x}{\mathrm{d}r}\boldsymbol{i} + \frac{\mathrm{d}y}{\mathrm{d}r}\boldsymbol{j} + \frac{\mathrm{d}z}{\mathrm{d}r}\boldsymbol{k}$$

so

$$F_r = -\left(\frac{\partial \bar{V}}{\partial x}\boldsymbol{i} + \frac{\partial \bar{V}}{\partial y}\boldsymbol{j} + \frac{\partial \bar{V}}{\partial z}\boldsymbol{k}\right)\cdot \boldsymbol{e}_r$$

or

$$F_r = -\boldsymbol{F}\cdot\boldsymbol{e}_r$$

We define the gravitational field strength, $\boldsymbol{g}$, as the force acting on a unit mass. Therefore $\boldsymbol{g} = \boldsymbol{F}$ and thus

$$\boldsymbol{g} = -\left(\frac{\partial \bar{V}}{\partial x}\boldsymbol{i} + \frac{\partial \bar{V}}{\partial y}\boldsymbol{j} + \frac{\partial \bar{V}}{\partial z}\boldsymbol{k}\right)$$

or, using the definition of the gradient of a scalar,

$$\boldsymbol{g} = -\nabla\bar{V} = -\mathrm{grad}\bar{V} \tag{5.4}$$

We shall now use the potential to find the field due to a thin hollow shell as shown in Fig. 5.2. The density of the material is ρ, the shell thickness is t and its radius is R. The point P is situated a distance x from the centre of the shell and all points on the elemental annular ring are a distance r from P.

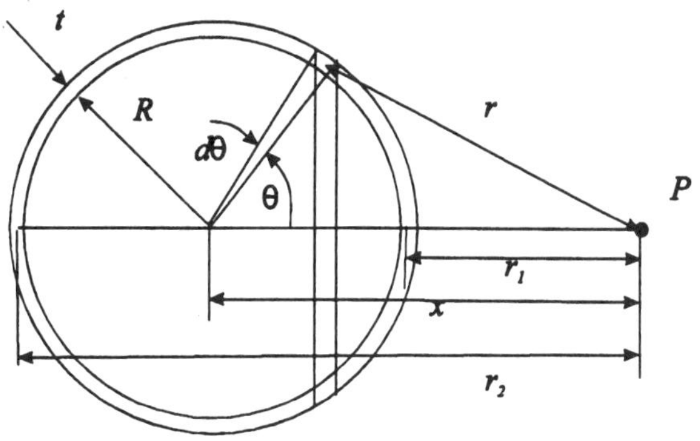

Fig. 5.2

The mass of the annular ring is

$$dm = \rho 2\pi R \sin\theta\, R\, d\theta\, t$$

so the potential at P due to the ring is

$$d\overline{V} = -\frac{G\rho 2\pi R^2 t \sin\theta\, d\theta}{r}$$

Now the mass of the whole shell is

$$m = \rho 4\pi R^2 t$$

Therefore

$$d\overline{V} = -\frac{Gm \sin\theta\, d\theta}{2r} \tag{5.5}$$

Using the cosine rule

$$r^2 = x^2 + R^2 - 2Rx\cos\theta \tag{5.6}$$

and differentiating gives

$$2r\, dr = 2Rx \sin\theta\, d\theta \tag{5.7}$$

Substituting equation (5.7) into equation (5.5) gives

$$d\overline{V} = -\frac{Gm\, dr}{2xR}$$

Integration produces

$$\overline{V} = -\int_{r_1}^{r_2} \frac{Gm}{2xR}\, dr = -\frac{Gm}{2xR}\left[r_2 - r_1\right] \tag{5.7a}$$

For the case as shown $r_2 = x + R$ and $r_1 = x - R$. Thus

$$\overline{V} = -\frac{Gm}{x} \tag{5.8}$$

which is identical to the result for a point mass of m at the centre. Applying equation (5.4) to equation (5.8)

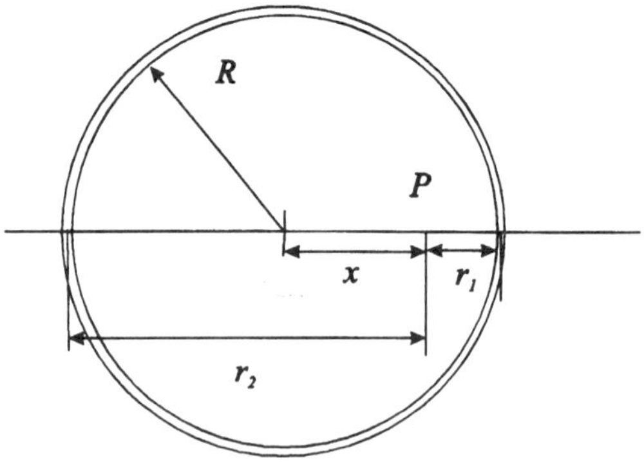

Fig. 5.3

$$g = -\frac{Gm}{x^2}\boldsymbol{i} \tag{5.9}$$

If the point P is inside the shell we must re-examine the limits. By the definition of potential r must be positive so, from Fig. 5.3, we have that

$$r_2 = R + x \text{ and } r_1 = R - x$$

which, when substituted into equation (5.7a), gives

$$\overline{V} = -\frac{Gm}{R} \tag{5.10}$$

which is constant for any point inside the shell. From equation (5.4) it follows that the field is zero.

For a body comprising concentric spherical shells the field outside the body will be as if all the mass is concentrated at the centre. The field inside the body will be due only to the mass which is at a radius smaller than the distance of P from the centre, the outer shells making no contribution. If a body, of radius R, has a uniform density then the field at radius a (a < R) will be

$$g = G\rho\frac{4}{3}\pi a^3\frac{1}{a^2} = G\rho\frac{4}{3}\pi a \tag{5.11}$$

From the above arguments it can be seen that two spherically symmetric bodies will attract each other as if each was a point mass concentrated at their respective centres.

5.3 The two-body problem

Figure 5.4 shows two spherical bodies under the action of equal but opposite central forces F_{12} and F_{21}. The centre of mass lies along the line joining the centres at a position such that

$$m_1r_1 = m_2r_2$$

With the separation of the bodies being s then we can define a mass μ so that

$$m_1r_1 = m_2r_2 = \mu s \tag{5.12}$$

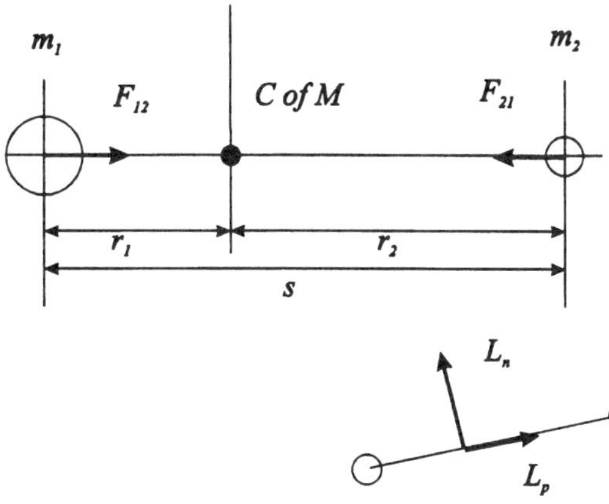

Fig. 5.4

Now

$$s = r_1 + r_2 = \frac{\mu s}{m_1} + \frac{\mu s}{m_2} = \mu s\left(\frac{1}{m_1} + \frac{1}{m_2}\right)$$

Therefore

$$\frac{1}{\mu} = \frac{1}{m_1} + \frac{1}{m_2} \tag{5.13}$$

By conservation of momentum the centre of mass will be unaccelerated and this will initially be chosen as the origin. By conservation of moment of momentum we can fix a direction for one of the inertial axes. The moment of momentum vector can be written as two components, one along the line joining the centres and one normal to it. The component along the line will be due to the spin of the bodies and this will have no effect on the moment of momentum normal to the line. It follows that the motion of the two bodies will be confined to the plane which contains the line of centres and has L_n as its normal.

We can now write the equations of motion for the two masses in polar co-ordinates. Resolving radially we have

$$|\boldsymbol{F}_{12}| = m_1\omega^2 r_1 - m_1\ddot{r}_1 = m_2\omega^2 r_2 - m_2\ddot{r}_2 \tag{5.14}$$

and taking moments about the centre of mass (C of M) gives

$$0 = m_1(r_1\dot{\omega} + 2\omega\dot{r}_1) = m_2(r_2\dot{\omega} + 2\omega\dot{r}_2)$$

or

$$0 = \frac{1}{r_1}\frac{\mathrm{d}}{\mathrm{d}t}\left(m_1 r_1^2 \omega\right) = \frac{1}{r_2}\frac{\mathrm{d}}{\mathrm{d}t}\left(m_2 r_2^2 \omega\right) \tag{5.15}$$

Substituting equation (5.12) into equation (5.14) gives

$$|\boldsymbol{F}_{12}| = \mu\omega^2 s + \mu\ddot{s} \tag{5.16}$$

Now

$$L_n = m_1 r_1^2 \omega + m_2 r_2^2 \omega$$

and again using equation 5.12

$$L_n = \mu s r_1 \omega + \mu s r_2 \omega = \mu s \omega (r_1 + r_2)$$
$$= \mu s^2 \omega \tag{5.17}$$

So we see that the motion is identical to that of a body of mass μ at a distance *s* from a fixed body. The quantity μ is known as the *reduced mass*. We shall now consider the central force problem because any two-mass system can be replaced by a single mass under the action of a central force.

5.4 The central force problem

Figure 5.5 shows a body of mass *m* at a distance *r* from the origin of a central force *F*(*r*) acting towards the origin. As already discussed the motion is in a plane so we can write the equations of motion directly in polar co-ordinates.

In the radial direction

$$-F(r) = m\ddot{r} - mr\dot{\theta}^2 \tag{5.18}$$

and normal to the radius

$$0 = mr\ddot{\theta} + m2\dot{r}\dot{\theta}$$
$$= \frac{1}{r}\frac{\mathrm{d}}{\mathrm{d}t}\left(mr^2\dot{\theta}\right) \tag{5.19}$$

Now $mr^2\dot{\theta} = L$ is the moment of momentum and therefore constant. Eliminating $\dot{\theta}$ in equation (5.18) leads to

$$-F(r) = m\ddot{r} - \frac{L^2}{mr^3}$$

Dividing through by *m* we have

$$-f(r) = \ddot{r} - \frac{L^{*2}}{r^3} \tag{5.20}$$

where L^* is the moment of momentum per unit mass. *f*(*r*) is the central force per unit mass; let this force be Kr^n. Noting that $\ddot{r} = \dot{r}\,\mathrm{d}\dot{r}/\mathrm{d}r$, integrating equation (5.20) with respect to *r* provides

$$-\frac{Kr^{n+1}}{(n+1)} = \frac{\dot{r}^2}{2} + \frac{L^{*2}}{2r^2} - E^* \tag{5.21}$$

or, if $n = -1$,

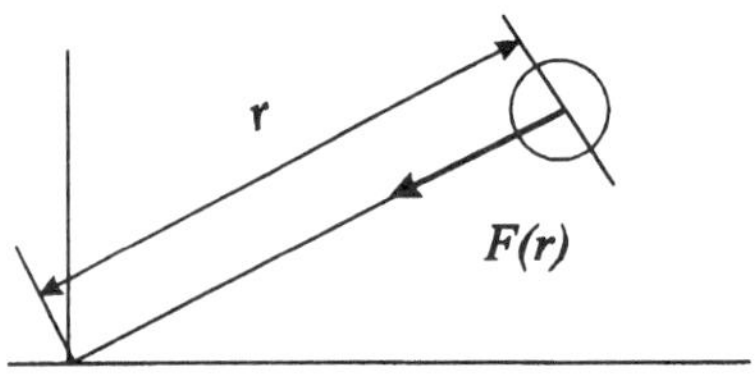

Fig. 5.5

$$-K\log(r) = \frac{\dot{r}^2}{2} + \frac{L^{*2}}{2r^2} - E^* \tag{5.21a}$$

where the constant E^* has the dimensions of energy per unit mass.

Equations (5.21) and (5.21a) may be written

$$\frac{\dot{r}^2}{2} = -\left(\frac{Kr^{n+1}}{(n+1)} + \frac{L^{*2}}{2r^2}\right) + E^*$$

or, with $n = -1$,

$$\frac{\dot{r}^2}{2} = -\left(K\log(r) + \frac{L^{*2}}{2r^2}\right) + E^*$$

$$= -\text{'}V\text{'} + E^* \tag{5.22}$$

The terms in the large parentheses can be regarded as a pseudo potential 'V'. A plot of 'V' versus radius for various integer values of n is given in Fig. 5.6. It is seen that for certain values of n the curve exhibits a minimum and for others a maximum. Simple differentiation of 'V' with respect to r reveals that for $n > -3$ the curve has a minimum and for $n < -3$ the curve has a maximum. The curves are drawn for a fixed value of moment of momentum and all (bar $n = -3$) show that for some range of values of E^* there are two value of r for $\dot{r} = 0$. In the cases where 'V' exhibits a minimum the value of $\dot{r}^2 > 0$ giving a real value for $\dot{r}$ and thus stable motion occurs between the inner and outer bounds. The motion is said to be bounded.

We now wish to know which values of n lead to closed orbits, that is orbits which repeat themselves after an integer number of orbits. A useful change of variable is $r = 1/u$.

Differentiation with respect to t gives

$$\frac{\mathrm{d}r}{\mathrm{d}t} = -\frac{1}{u^2}\frac{\mathrm{d}u}{\mathrm{d}\theta}\dot{\theta}$$

Now

$$L^* = r^2\dot{\theta} = \dot{\theta}/u^2$$

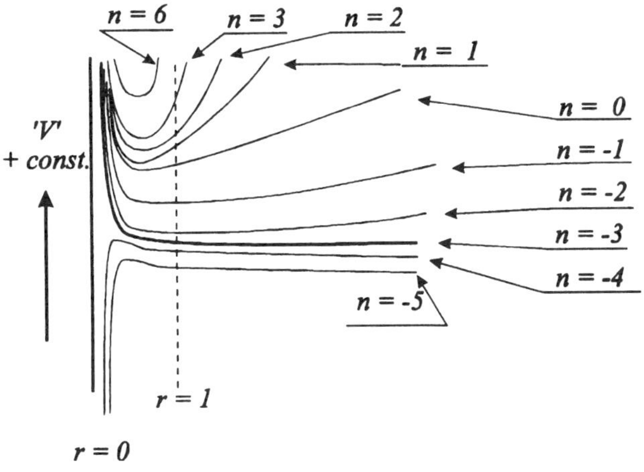

Fig. 5.6 Constant chosen to give non-intersecting curves.

and therefore

$$\frac{dr}{dt} = -\frac{du}{d\theta}L^*$$

and

$$\frac{d^2r}{dt^2} = -\frac{d^2u}{d\theta^2}L^*\dot{\theta}$$

$$= -\frac{d^2u}{d\theta^2}L^{*2}u^2 \tag{5.23}$$

Substitution of equation (5.23) into equation (5.20) gives

$$-Ku^{-n} = -\frac{d^2u}{d\theta^2}L^{*2}u^2 - L^{*2}u^3$$

or

$$\frac{d^2u}{d\theta^2} + u = \frac{K}{L^{*2}}u^{-(n+2)} \tag{5.24}$$

If u is constant, that is a circular orbit of inverse radius u_0, then

$$u = u_0 = \frac{K}{L^{*2}}u_0^{-(n+2)}$$

Let us now assume that $u = u_0 + \varepsilon(\theta)$ where ε is small. Substitution into equation (5.24) leads to

$$\frac{d^2\varepsilon}{d\theta^2} + \varepsilon = \frac{K}{L^{*2}}(u_0 + \varepsilon)^{-(n+2)} - u_0$$

Expanding the right hand side by the binomial theorem and neglecting terms above the first order gives

$$\frac{d^2\varepsilon}{d\theta^2} + \varepsilon = \frac{K}{L^{*2}}u_0^{-(n+2)}\left[1 - (n+2)\varepsilon/u_0\right] - u_0$$

$$= -(n+2)\varepsilon$$

so

$$\frac{d^2\varepsilon}{d\theta^2} + (3 + n)\varepsilon = 0 \tag{5.25}$$

This is the equation of simple harmonic motion (SHM) with the solution

$$\varepsilon = \varepsilon_0 \cos(p\theta + \text{ø}) \tag{5.26}$$

where $p = \sqrt{(3 + n)}$, ø is a constant of integration and ε_0 is a small amplitude. From this it is clear that n must be greater than -3 for sinusoidal error motion, which agrees with the result obtained from the pseudo potential. We may further argue that if the orbit is closed after one cycle then p must be an integer so that

$$p^2 = 3 + n$$

or

$$n = p^2 - 3$$

Some of the values are

$p = 1 \quad n = -2$
$p = 2 \quad n = 1$
$p = 3 \quad n = 6$
$p = 4 \quad n = 13$

Notice that the only negative value of n is -2, which is the inverse square law. The second value with $n = 1$ is Hooke's law.

If we let p be the ratio of two integers then a further series of non-integer values of n are generated which indicate paths which close after a finite number of orbits. However, more detailed analysis carried out by Bertrand in 1873 leads to the conclusion that for large deviations from the circular orbit only the inverse square law and Hooke's law generate closed orbits. Numerical integration of equation (5.24) supports this theory. See Figs 5.7(a) and 5.7(b).

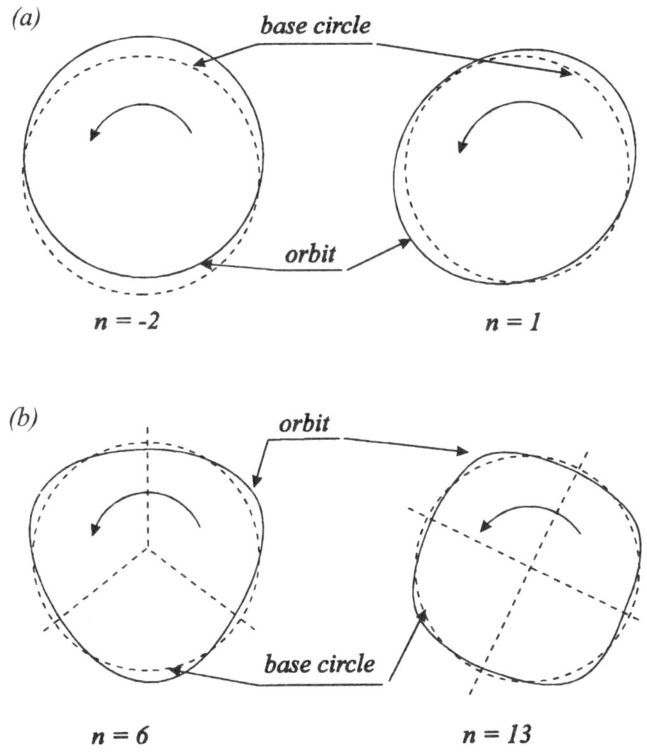

Fig. 5.7 (a) Apsides stationary. **(b)** Apsides precess slowly.

Astronomical observations have to date revealed only closed orbits other than deviations due to extra bodies or the effects of Einstein's theory of relativity. This lends weight to the belief that the inverse square law of gravitational attraction is universal.

5.5 Satellite motion

We shall now consider the case for $n = -2$, that is the inverse square law of attraction. Equation (5.24) becomes

$$\frac{d^2u}{d\theta^2} + u = \frac{K}{L^{*2}} \tag{5.27}$$

the solution to which is

$$u = A\cos(\theta + \phi) + \frac{K}{L^{*2}}$$

where A and ϕ are constants. Choosing the constant ϕ to be zero we have

$$\frac{1}{r} = A\cos\theta + \frac{K}{L^{*2}} \tag{5.28}$$

The locus definition of a conic is that the distance from some point known as the focus is a fixed multiple of the distance of that point from a line called the directrix. From Fig. 5.8 we have

$$r = ed \tag{5.29}$$

where the positive constant e is known as the eccentricity. Also

$$r\cos(\theta) + d = D \tag{5.30}$$

at $\theta = \pi/2$

$$r = l = eD \tag{5.31}$$

where the length l is the distance to a point called the latus rectum. Substituting equation (5.31) into equation (5.30) and rearranging gives

$$\frac{1}{r} = \frac{1}{l} + \frac{e}{l}\cos\theta$$

or

$$\frac{l}{r} = 1 + e\cos\theta \tag{5.32}$$

At $\theta = 0$

$$r = r_1 = \frac{l}{1 + e} \tag{5.33}$$

and at $\theta = \pi$

$$r = r_2 = \frac{l}{1 - e} \tag{5.34}$$

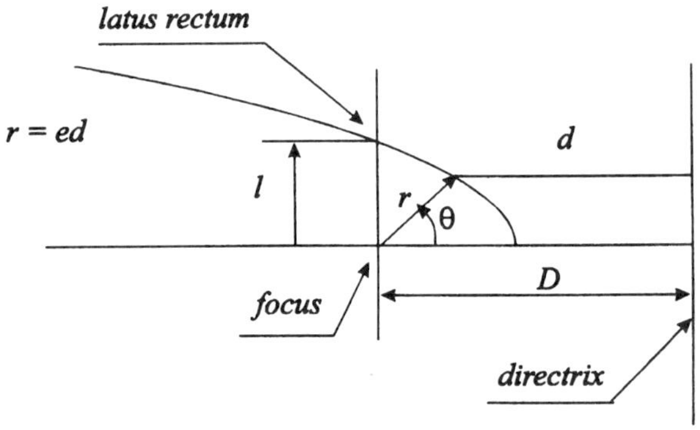

Fig. 5.8

Since r is positive this expression is only valid for $e < 1$.
So for $e < 1$

$$r_1 + r_2 = \frac{2l}{1 - e^2} \tag{5.35}$$

The type of conic is determined by the value of the eccentricity. If $e = 0$ then $r_1 = r_2 = l$ is the radius of a circle. If $e = 1$ then r_2 goes to infinity and the curve is a parabola. For $0 < e < 1$ the curve is an ellipse and for $e > 1$ an hyperbola is generated.

For an ellipse, as shown in Fig. 5.9, $r_1 + r_2 = 2a$ where a is the semi-major axis. From equation (5.35)

$$a = \frac{l}{1 - e^2} \tag{5.36}$$

The length CF is

$$a - r_1 = \frac{l}{1 - e^2} - \frac{l}{1 + e} = ae$$

We notice that if $\cos\theta = -e$ equation (5.32) gives

$$\frac{l}{r} = 1 - e^2$$

which by inspection of equation (5.36) shows that $r = a$.
From Fig. 5.9 it follows that triangle FCB is a right-angled triangle with b the semi-minor axis. Therefore

$$b = \sqrt{(a^2 - e^2a^2)} = a\sqrt{(1 - e^2)} \tag{5.37}$$

Comparing equation (5.28) with equation (5.32) we see that

$$l = \frac{L^{*2}}{K} \tag{5.38}$$

The energy equation for a unit mass in an inverse square law force field (*see* equation (5.21)) is

$$E^* = \frac{\dot{r}}{2} + \frac{L^{*2}}{2r^2} - \frac{K}{r} \tag{5.39}$$

and when the radial component of the velocity is zero ($\dot{r} = 0$) equation (5.39) becomes the quadratic

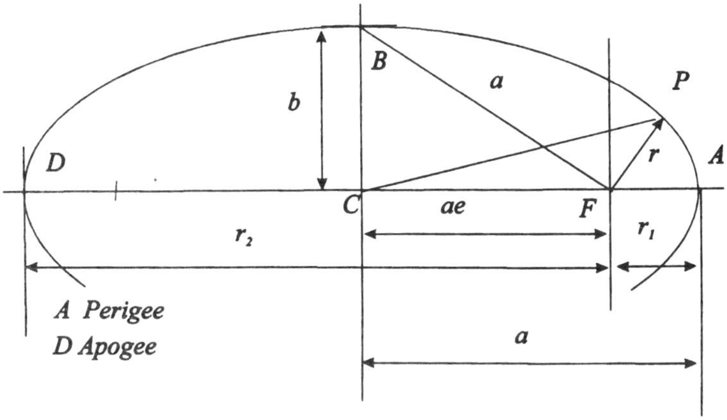

Fig. 5.9

$$2E^*r^2 + 2Kr - L^{*2} = 0$$

The values of r satisfying this equation are

$$r = \frac{-2K \pm \sqrt{(4K^2 + 8E^*L^{*2})}}{4E^*}$$

For real roots $4K^2 + 8E^*L^{*2} > 0$ or

$$E^* > -\frac{K^2}{2L^{*2}} \tag{5.40}$$

The sum of the two roots, r_1 and r_2, is

$$r_1 + r_2 = -\frac{K}{E^*} \tag{5.41}$$

If both roots are positive, as they are for elliptic motion, then E^* must be negative since K is a positive constant. For circular motion the roots are equal and thus

$$E^* = -\frac{K^2}{2L^{*2}} \tag{5.42}$$

Using equations (5.36) and (5.38)

$$r_1 + r_2 = 2a = \frac{2l}{1 - e^2} = \frac{2L^{*2}}{K(1 - e^2)}$$

Therefore equating expressions for the sum of the roots from equation (5.41)

$$-\frac{K}{E^*} = \frac{2L^{*2}}{K(1 - e^2)}$$

giving

$$e_2 = 1 + \frac{2L^{*2}E^*}{K^2} \tag{5.43}$$

Figure 5.10 summarizes the relationship between eccentricity and energy.

Our next task is to find expressions involving time. Starting with equation (5.32) and the fact that the moment of momentum is constant we write

$$\frac{l}{r} = 1 + e\cos\theta$$

and

$$L^* = r^2\dot{\theta} = \text{constant}$$

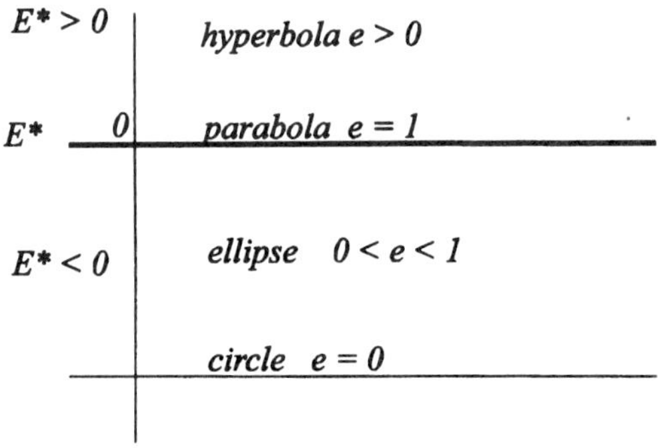

Fig. 5.10

Therefore

$$\frac{d\theta}{dt} = \frac{L^*(1 + e\cos\theta)^2}{l^2}$$

$$t = \frac{l^2}{L^*}\int_0^{\theta} \frac{d\theta}{(1 + e\cos\theta)^2} \tag{5.44}$$

Evaluation of this integral will give time as a function of angle after which equation (5.32) will furnish the radius.

For elliptic orbits a graphical construction leads to a simple solution of the problem. In Fig. 5.11 a circle of radius a, the semi-major axis, is drawn centred at the centre of the ellipse. The line PQ is normal to a. The area FQA = area CQA − area CFQ

$$\text{area FQA} = \frac{a^2}{2}\phi - \frac{1}{2}ae\,a\sin\phi$$

Now

$$\text{area FPA} = A = \text{area FQA} \times b/a$$

Thus the area swept out by the radius r is

$$A = \frac{ba}{2}(\phi - e\sin\phi) \tag{5.45}$$

Now

$$\frac{L^*}{2} = \frac{1}{2}r^2\dot{\theta} = \frac{dA}{dt} \tag{5.46}$$

This is Kepler's second law of planetary motion, which states that the rate at which area is being swept by the radius vector is constant. Combining equations (5.45) and (5.46) and integrating gives

$$\frac{L^*}{2}t = A = \frac{ba}{2}(\phi - e\sin\phi)$$

Using equations (5.37) and (5.38)

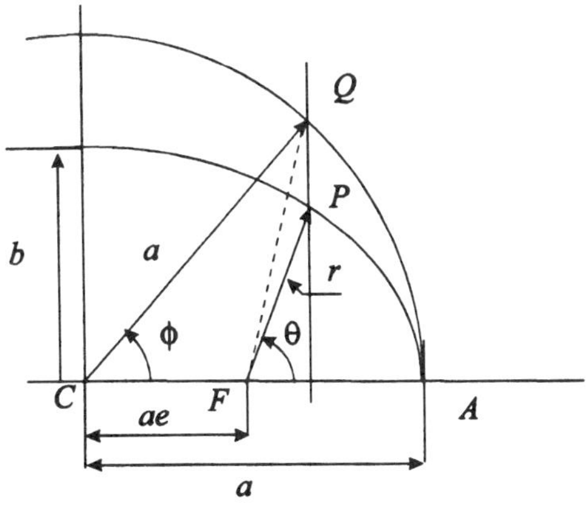

Fig. 5.11

$$t = \frac{a^{3/2}}{\sqrt{K}}(\phi - e \sin\phi) \tag{5.47}$$

From Fig. 5.11 we have

$$(a \sin\phi) b/a = r \sin\theta$$

Substituting for r from equation (5.32)

$$\sin\phi = \frac{r \sin\theta}{b} = \frac{l \sin\theta}{b(1 + e \cos\theta)}$$

and finally, combining equations (5.36) and (5.37) gives

$$l/b = \sqrt{(1 - e^2)}$$

so that

$$\sin\phi = \frac{\sqrt{(1 - e^2)} l \sin\theta}{(1 + e \cos\theta)} \tag{5.48}$$

Equations (5.47) and (5.48) are sufficient to calculate t as a function of θ but it is more accurate to use half-angle format.

Let $\tau = \tan(\theta/2)$ so that

$$\sin\theta = \frac{2\tau}{1 + \tau^2}$$

and

$$\cos\theta = \frac{1 - \tau^2}{1 + \tau^2}$$

Substituting into equation 5.48 gives

$$\sin\phi = \frac{\sqrt{(1 - e^2)} 2\tau}{(1 + \tau^2) + e(1 - \tau^2)}$$

$$= \frac{\sqrt{(1 - e^2)} 2\tau}{(1 + e) + \tau^2(1 - e)}$$

$$= \frac{2\tau\sqrt{[(1 - e)/(1 + e)]}}{1 + \{\tau\sqrt{[(1 - e)/(1 + e)]}^2} \tag{5.49}$$

Now

$$\sin\phi = \frac{2 \tan(\phi/2)}{1 + \tan^2(\phi/2)} \tag{5.50}$$

Comparison of equations (5.49) and (5.50) shows that

$$\tan(\phi/2) = \sqrt{\left(\frac{(1 - e)}{(1 + e)}\right)} \tan(\theta/2) \tag{5.51}$$

Equation (5.47) may now be written as

$$t = \frac{a^{3/2}}{\sqrt{K}}\left\{2 \arctan\left[\tan(\theta/2)\sqrt{\left(\frac{(1 - e)}{(1 + e)}\right)}\right] - \frac{e\sqrt{(1 - e^2)} \sin(\theta)}{(1 + e \cos\theta)}\right\} \tag{5.52}$$

which holds for $0 \le e < 1$. Figure 5.12 shows plots of θ versus a non-dimensional time for various values of eccentricity.

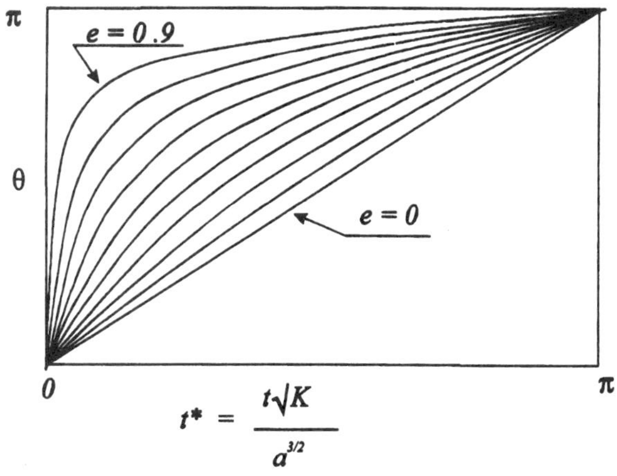

Fig. 5.12

From equation (5.47), since ø ranges from 0 to 2π, the time for one orbit is

$$T = \frac{2\pi a^{3/2}}{\sqrt{K}} \tag{5.53}$$

from which $T^2 \alpha\, a^3$; this is Kepler's third law. The first law was that the orbits of the planets about the Sun are ellipses. The second law is true for any central force problem whilst the first and third require that the law be an inverse square. The closure of the orbits also strongly supports the inverse square law as previously discussed.

For a parabolic path, $e = 1$, we return to equation (5.44) and note that $l = L^{*2}/K$ so that

$$t = \frac{l^{3/2}}{\sqrt{K}} \int_0^\theta \frac{d\theta}{(1 + e\cos\theta)^2}$$

Making a substitution of $\tau = \tan(\theta/2)$ leads to

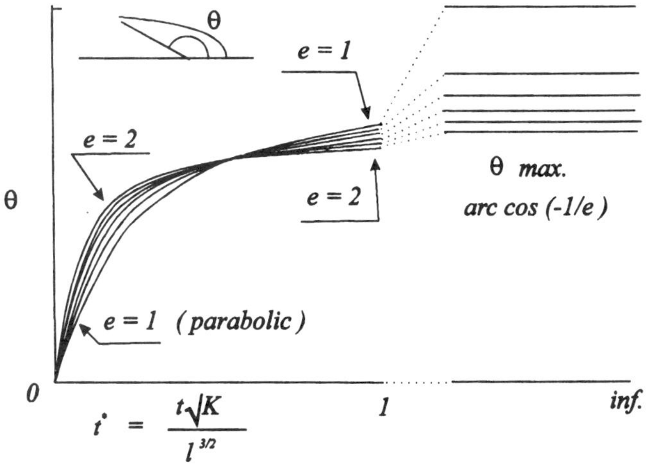

Fig. 5.13 Time for parabolic and hyperbolic trajectories

$$t = \frac{l^{3/2}}{\sqrt{K}} \int \frac{d\tau}{2/(1 + \tau^2)} = (\tau/2 + \tau^3/6) \frac{l^{3/2}}{\sqrt{K}}$$

$$= \frac{l^{3/2}}{\sqrt{K}} \left[\frac{1}{2} \tan(\theta/2) + \frac{1}{6} \tan^3(\theta/2) \right] \tag{5.54}$$

For hyperbolic orbits, $e > 1$, the integration follows the method as above but is somewhat longer. The result of the integration is

$$t = \frac{l^{3/2}}{\sqrt{K}(e^2 - 1)^{3/2}} \left[\frac{e\sqrt{(e^2 - 1)}\sin\theta}{1 + e\cos\theta} - \ln\left(\frac{\sqrt{(e+1)} + \sqrt{(e-1)}\tan(\theta/2)}{\sqrt{(e+1)} - \sqrt{(e-1)}\tan(\theta/2)} \right) \right] \tag{5.55}$$

Plots of equation (5.55), including equation (5.54), for different values of e are shown in Fig. 5.13.

5.6 Effects of oblateness

In the previous section we considered the interaction of two objects each possessing spherical symmetry. The Earth is approximately an oblate spheroid such that the moment of inertia about the spin axis is greater than that about a diameter. This means that the resultant attractive force is not always directed towards the geometric centre so that there may be a component of force normal to the ideal orbital plane.

For a satellite that is not spherical the centre of gravity will be slightly closer to the Earth than its centre of mass thereby causing the satellite's orientation to oscillate.

We first consider a general group of particles, as shown in Fig. 5.14, and use equation (5.3) to find the gravitational potential at point P. From the figure $\boldsymbol{R} = \boldsymbol{\rho}_i + \boldsymbol{r}_i$ where $\boldsymbol{\rho}_i$ is the position of mass m_i from the centre of mass. Thus

$$V = -\frac{Gm_i}{|\boldsymbol{R} - \boldsymbol{\rho}_i|}$$

$$= -\frac{Gm_i}{\sqrt{(R^2 + \rho_i^2 - 2\boldsymbol{\rho}_i \cdot \boldsymbol{R})}} \tag{5.56}$$

The binomial theorem gives

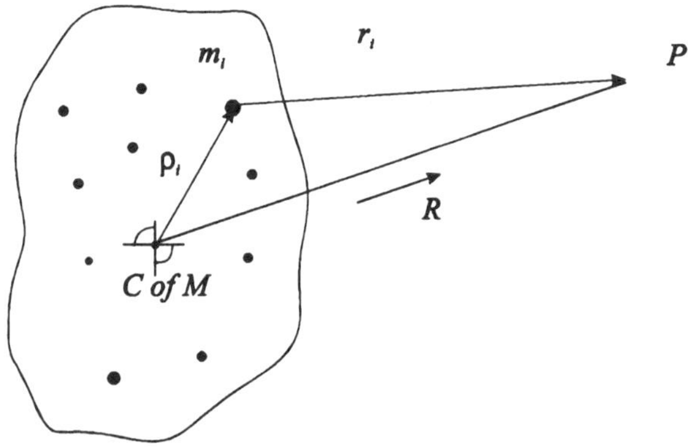

Fig. 5.14

$$(1 + x)^{-1/2} = 1 - \frac{1}{2}x + \frac{1}{2}\frac{3}{2}\frac{x^2}{2} + \cdots \tag{5.57}$$

so equation (5.56), assuming $R \gg \rho$, can be written

$$V = -\frac{Gm_i}{R}\left[1 - \frac{\rho_i^2}{2R^2} + \frac{2\boldsymbol{\rho}_i\cdot\boldsymbol{R}}{2R^2} + \frac{3}{8}\left(\frac{\rho_i^2}{R^2} - \frac{2\boldsymbol{\rho}_i\cdot\boldsymbol{R}}{R^2}\right)^2 + \cdots\right]$$

As a further approximation we shall ignore all terms which include ρ to a power greater than 2. We now sum for all particles in the group and note that, by definition of the centre of mass, $\Sigma m_i\boldsymbol{\rho}_i = 0$. Thus

$$V = -\frac{G}{R}\left[\Sigma m_i - \frac{\Sigma m_i\rho_i^2}{2R^2} + \frac{3}{8}\left(\frac{4(\boldsymbol{R}\cdot\boldsymbol{\rho})(\boldsymbol{R}\cdot\boldsymbol{\rho})}{R^4}\right)\right]$$

$$= -\frac{Gm}{R} + \frac{G}{2R^3}\left(3\Sigma m_i\ (\boldsymbol{e}\cdot\boldsymbol{\rho}_i)(\boldsymbol{\rho}_i\cdot\boldsymbol{e}) - \Sigma m_i\rho_i^2\right) \tag{5.58}$$

where the unit vector $\boldsymbol{e} = \boldsymbol{R}/R$.

The term in the large parentheses may be written

$$\boldsymbol{e}\cdot\left(3\Sigma m_i\boldsymbol{\rho}_i\boldsymbol{\rho}_i - 2\Sigma m_i\rho_i^2\mathsf{1}\right)\cdot\boldsymbol{e}$$

By equation (4.24) the moment of inertia dyadic is

$$I = \Sigma(m_i\rho_i^2\mathsf{1} - m_i\boldsymbol{\rho}_i\boldsymbol{\rho}_i)$$

and by definition if $\boldsymbol{\rho}_i = x_i\boldsymbol{i} + y_i\boldsymbol{j} + z_i\boldsymbol{k}$ then

$$I_x = \Sigma m_i(y_i^2 + z_i^2)$$

$$I_y = \Sigma m_i(z_i^2 + x_i^2)$$

$$I_z = \Sigma m_i(x_i^2 + y_i^2)$$

so that

$$I_x + I_y + I_z = 2\Sigma(x_i^2 + y_i^2 + z_i^2) = 2\Sigma\rho_i^2$$

This is a scalar and is therefore invariant under the transformation of axes. Thus it will also equal the sum of the principal moments of inertia.

Using this information equation (5.58) becomes

$$V = -\frac{Gm}{R} + \frac{G}{2R^3}\,\boldsymbol{e}\cdot[3\mathsf{I} - (I_1 + I_2 + I_3)\mathsf{1}]\cdot\boldsymbol{e}$$

$$= -\frac{Gm}{R} + \frac{G}{2R^3}\,[3I_r - (I_1 + I_2 + I_3)] \tag{5.59}$$

where I_r is the moment of inertia about the centre of mass and in the direction of R.

Let us now consider the special case of a body with an axis of symmetry, that is $I_1 = I_2$. Taking $\boldsymbol{e} = l\boldsymbol{i} + m\boldsymbol{j} + n\boldsymbol{k}$ where l, m and n are the direction cosines of R relative to the principal axes, in terms of principal axes the inertia dyadic is

$$\mathsf{I} = \boldsymbol{i}I_1\boldsymbol{i} + \boldsymbol{j}I_2\boldsymbol{j} + \boldsymbol{k}I_3\boldsymbol{k}$$

Thus

$$\mathbf{e}.\mathsf{I}.\mathbf{e} = l^2 I_1 + m^2 I_2 + n^2 I_3$$

$$= (1 - n^2)I_1 + n^2 I_3$$

Finally equation (5.59) is

$$V = -\frac{Gm}{R} + \frac{G}{2R^3}[3(1 - n^2)I_1 + 3n^2 I_3 - 2I_1 - I_3]$$

$$= -\frac{Gm}{R} + \frac{G}{2R^3}(3n^2 - 1)(I_3 - I_1) \tag{5.60}$$

Referring to Fig. 5.15 we see that $n = \cos\gamma$ where γ is the angle between the figure axis and R. Also from Fig. 5.15 we have that

$$\cos\gamma = \mathbf{e}.\mathbf{e}_3 = [\cos(\psi)\mathbf{i} + \sin(\psi)\mathbf{j}]\cdot[\sin(\theta)\mathbf{i} + \cos(\theta)\mathbf{k})$$
$$= \cos\psi\sin\theta \tag{5.61}$$

Substituting equation (5.61) into equation (5.60) gives

$$V = -\frac{Gm}{R} + \frac{G}{2R^3}(3\sin^2\theta\cos^2\psi - 1)(I_3 - I_1) \tag{5.62}$$

The first term of equation (5.62) is the potential due to a spherical body and the second term is the approximate correction for oblateness. It is assumed that this has only a small effect on the orbit so that we may take an average value for $\cos^2\psi$ over a complete orbit which is 1/2. Also, by replacing $\sin^2\theta$ with $1 - \cos^2\theta$ equation (5.62) may be written as

$$V = -\frac{Gm}{R} + \frac{G}{2R^3}\left(\frac{1}{2} - \frac{3}{2}\cos^2\theta\right)(I_3 - I_1) \tag{5.62a}$$

We shall consider a ring of satellites with a total mass of μ on the assumption that motion of the ring will be the same as that for any individual satellite. Also the motion of the ring is identical to the motion of the orbit. The potential energy will then be

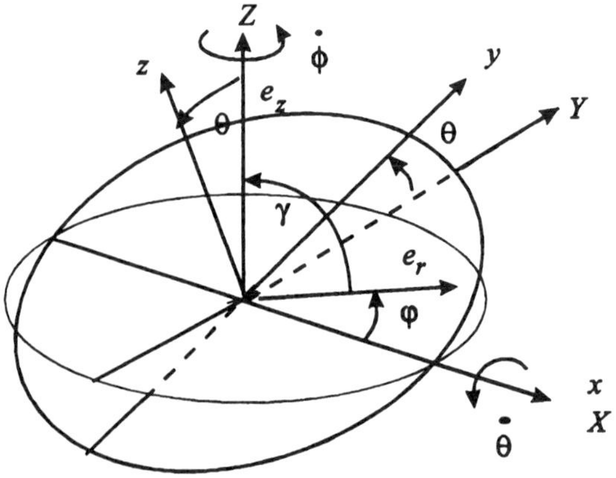

Fig. 5.15

$$V = \mu\mathcal{V} = -\frac{\mu Gm}{R} + \frac{\mu G}{2R^3}\left(\frac{1}{2} - \frac{3}{2}\cos^2\theta\right)(I_3 - I_1) \tag{5.63}$$

We can study the motion of the satellite ring in the same way as we treated the precession of a symmetrical rigid body in section 4.11. The moment of inertia of the ring about its central axis is μR^2 and that about the diameter is $\mu R^2/2$. Thus, referring to Fig. 5.15, the kinetic energy is

$$T = \frac{\mu R^2}{4}\dot{\theta}^2 + \frac{\mu R^2}{4}\left(\dot{ø}\sin\theta^2\right) + \frac{\mu R^2}{2}\left(\dot{\psi} + \dot{ø}\cos\theta\right)^2 \tag{5.64}$$

and the Lagrangian $\mathfrak{L} = T - V$.

Because ψ is an ignorable co-ordinate

$$\frac{\partial T}{\partial\dot{\psi}} = \mu R^2(\dot{\psi} + \dot{ø}\cos\theta) = \text{constant}$$

so that

$$(\dot{\psi} + \dot{ø}\cos\theta) = \omega_3 = \text{constant}$$

With θ as the generalized co-ordinate

$$\mu R^2\ddot{\theta} + \mu R^2(\dot{\psi} + \dot{ø}\cos\theta)\dot{ø}\sin\theta - \frac{\mu R^2\dot{ø}^2}{2}\sin\theta\cos\theta$$

$$+ \frac{\mu G}{2R^3}(3\sin\theta\cos\theta)(I_3 - I_1) = 0$$

For steady precession $\ddot{\theta} = 0$ and neglecting $\dot{ø}^2$, since we assume that $\dot{ø}$ is small, we obtain, after dividing through by $\mu\sin\theta$,

$$R^2\omega_3\dot{ø} + \frac{G}{2R^3}3\cos(\theta)(I_3 - I_1) = 0$$

or

$$\dot{ø} = -\frac{G}{2R^5\omega_3}3\cos(\theta)(I_3 - I_1) \tag{5.65}$$

This precession is the result of torque applied to the satellite ring, or more specifically a force acting normal to the radius R. There is, of course, the equal and opposite torque applied to the Earth which in the case of artificial satellites is negligible. However, the effect of the Moon is sufficient to produce small but significant precession of the Earth.

5.7 Rocket in free space

We shall now study the dynamics of a rocket in a gravitational field but without any aerodynamic forces being applied. The rocket will be assumed to be symmetrical and not rotating about its longitudinal axis. Under these circumstances the motion will be planar. Referring to Fig. 5.16 the XYZ axes are inertial with Y vertical. The xyz axes are fixed to the rocket body with the origin being the current centre of mass. Because of the large amount of fuel involved the centre of mass will not be a fixed point in the body. However, Newton's

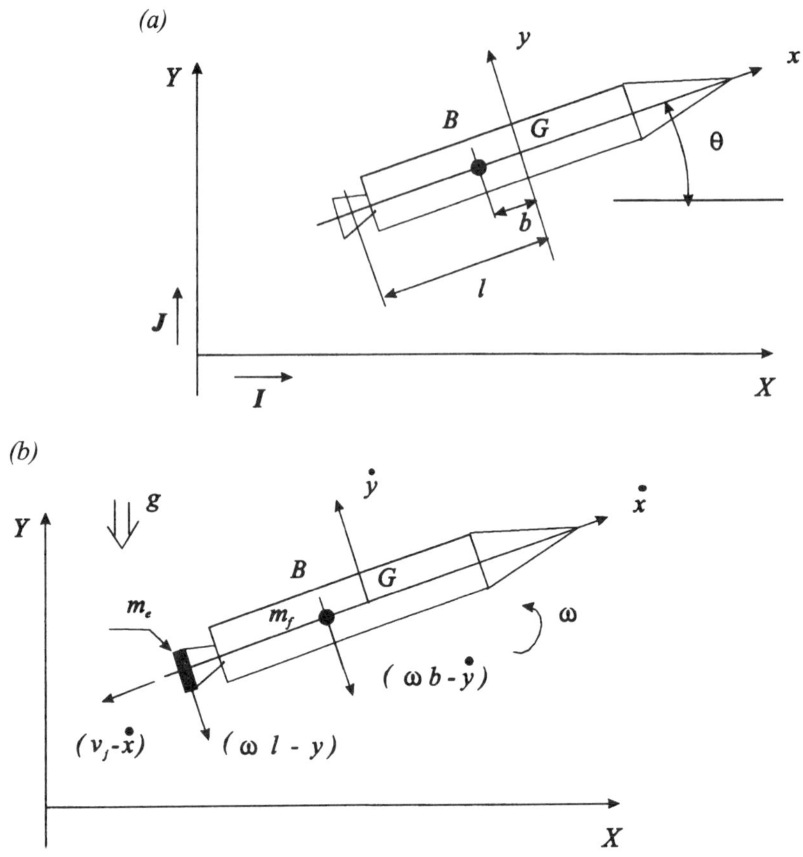

Fig. 5.16(a) and **(b)**

laws, in the form of equations (1.51) and (1.53b), apply to a constant amount of matter so great care is needed in setting up the model.

At a given time we shall consider that fuel is being consumed at a fixed rate, $\dot{m}$, from a location B a distance b from the centre of mass and ejected at a nozzle located l from the centre of mass. Let the mass of a small amount of fuel at location B be m_f and the total mass of the rocket at that time be m. The mass of the rocket structure is $m_0 = \mathrm{m} - m_{\mathrm{f}}$. The mass of burnt fuel in the exhaust is m_{e} and is taken to be vanishingly small, its rate of generation being, of course, $\dot{m}$. The speed of the exhaust *relative to the rocket* is v_j.

The angle that the rocket axis, the x axis, makes with the horizontal is θ and its time rate of change is ω. The angular velocity of the xyz axes will be $\omega \boldsymbol{k}$. If the linear momentum is $\boldsymbol{p}$ then

$$\frac{\mathrm{d}\boldsymbol{p}}{\mathrm{d}t} = \frac{\partial \boldsymbol{p}}{\partial t} + \omega \boldsymbol{k} \times \boldsymbol{p} = m\boldsymbol{g} \tag{5.66}$$

where $\boldsymbol{g} = -g\boldsymbol{J}$ is the gravitational field strength.

Now

$$\boldsymbol{p} = [m_0\dot{x} + m_{\mathrm{f}}\dot{x} + m_{\mathrm{e}}(\dot{x} - v_j)]\boldsymbol{i} + [m_0\dot{y} + m_{\mathrm{f}}\dot{y} + m_{\mathrm{e}}\dot{y}]\boldsymbol{j} \tag{5.67}$$

so

$$\frac{\mathrm{d}\boldsymbol{p}}{\mathrm{d}t}_{m_e \Rightarrow 0} = [m_0\ddot{x} + m_f\ddot{x} - \dot{m}\dot{x} + \dot{m}(\dot{x} - v_j)]\boldsymbol{i} + [m_0\ddot{y} + m_f\ddot{y} - \dot{m}\dot{y} + \dot{m}\dot{y}]\boldsymbol{j}$$

$$+ \omega[m_0\dot{x} + m_f\dot{x}]\boldsymbol{j} - \omega[m_0\dot{y} + m_f\dot{y}]\boldsymbol{i}$$

$$= [m\ddot{x} - \dot{m}v_j - \omega m\dot{y}]\boldsymbol{i} + [m\ddot{y} + \omega m\dot{x}]\boldsymbol{j}$$

$$= -gm\sin(\theta)\boldsymbol{i} - gm\cos(\theta)\boldsymbol{j}$$

or in scalar form, after dividing through by m,

$$\ddot{x} - \frac{\dot{m}}{m}v_j - \omega\dot{y} = -g\sin\theta \tag{5.68}$$

and

$$\ddot{y} + \omega\dot{x} = -g\cos\theta \tag{5.69}$$

By writing the moment of momentum equation using the centre of mass as the origin only motion relative to the centre of mass is involved. Because the fuel flow is assumed to be axial the only relative motion which has a moment about the centre of mass will be that due to rotation. We will use the symbol I_G' to signify the moment of momentum about the centre of mass of the rocket less that due to the small amount of fuel at B. Hence, the moment of momentum of the complete rocket is

$$\boldsymbol{L}_G = [I_G'\omega + m_f b^2\omega + m_e l^2\omega]\boldsymbol{k} \tag{5.70}$$

so

$$\frac{\mathrm{d}\boldsymbol{L}_G}{\mathrm{d}t}_{m_e \Rightarrow 0} = [(I_G' + m_f b^2)\dot{\omega} - \dot{m}b^2\omega + \dot{m}l^2\omega]\boldsymbol{k} + \omega\boldsymbol{k} \times \boldsymbol{L}_G$$

$$= [I_G\dot{\omega} + \omega\dot{m}(l^2 - b^2)]\boldsymbol{k}$$

$$= 0$$

in the absence of aerodynamic forces.

The scalar moment equation is

$$0 = I_G\dot{\omega} + \omega\dot{m}(l^2 - b^2) \tag{5.71}$$

The second term in the above equation provides a damping effect known as jet damping, provided that $l > b$.

Because the position of the centre of mass is not fixed in the body both l and b will vary with time. They are regarded as constants in the differentiation since $m_e \Rightarrow 0$ and m_f may also be regarded as small because it need not be any larger than m_e.

If the distribution of fuel is such that the radius of gyration of the complete rocket is constant then equation (5.70) is

$$\boldsymbol{L}_G = [mk_G^2\omega + m_e\omega l^2]\boldsymbol{k} \tag{5.70a}$$

and equation (5.71) becomes

$$0 = mk_G^2\dot{\omega} + \dot{m}\omega(l^2 - k_G^2) \tag{5.71a}$$

This last equation has a simple solution, we can write

$$\frac{\mathrm{d}\omega}{\mathrm{d}t} = -\frac{\mathrm{d}m}{\mathrm{d}t}\frac{\omega}{m}(l^2/k_G^2 - 1)$$

or

$$\frac{d\omega}{\omega} = -\frac{dm}{m}(l^2/k_G^2 - 1)$$

and the solution is

$$\ln(\omega/\omega_i) = -(l^2/k_G^2 - 1)\ln(m/m_i)$$

or

$$\omega/\omega_i = (m/m_i)^{-(l^2/k_G^2 - 1)}$$

If the initial mass is m_i then $m = m_i - \dot{m}t$ and thus

$$\omega/\omega_i = \left[1 - (\dot{m}/m_i)t\right]^{-(l^2/k_G^2 - 1)} \tag{5.72}$$

5.8 Non-spherical satellite

A non-spherical satellite will have its centre of gravity displaced relative to its centre of mass. The sense of the torque produced will depend on both the shape and its orientation.

Consider first a body with an axis of symmetry such that the moment of inertia about that axis is the greatest ($I_3 > I_1$). From equation (5.60) we obtain the potential energy of a non-spherical satellite, of mass m, and an assumed spherical Earth of mass M.

$$V = -\frac{GMm}{R} + \frac{GMm}{2R^3}(3\cos^2\gamma - 1)(I_3 - I_1)$$

With γ as the generalized co-ordinate the associated torque is

$$Q_\gamma = -\frac{\partial V}{\partial \gamma} = \frac{GMm}{2R^3}3(2\cos\gamma\sin\gamma)(I_3 - I_1)$$

$$= \frac{GMm}{R^3}3(\sin(2\gamma)/2)(I_3 - I_1) \tag{5.73}$$

If the figure axis is pointing towards the Earth (γ is small) then when $I_3 > I_1$ the torque is proportional to γ and is therefore unstable. When $I_3 < I_1$ the torque is proportional to $-\gamma$ and is stable. The satellite will then exhibit a pendulous motion with a period of

$$T = \frac{1}{2\pi}\left(\frac{GMm}{R^3}3(1 - I_3/I_1)\right) \tag{5.74}$$

For the case when γ is close to $\pi/2$ so that $\gamma = \pi/2 + \beta$ then the torque becomes

$$Q_\gamma = -\frac{GMm}{R^3}3(\sin(2\beta)/2)(I_3 - I_1) \tag{5.73a}$$

so that the configuration is stable when $I_3 > I_1$ and the period will be

$$T = \frac{1}{2\pi}\left(\frac{GMm}{R^3}3(I_3/I_1 - 1)\right) \tag{5.74a}$$

5.9 Spinning satellite

If the satellite is spinning about its figure axis then the torque described in the previous section will produce precession of the figure axis. The kinetic energy of a spinning symmetrical body is, by equation (4.82),

$$T = \frac{1}{2}I_1\dot{\theta}^2 + \frac{1}{2}I_1\dot{\phi}^2\sin^2\theta + \frac{1}{2}I_3(\dot{\phi}\cos\theta + \dot{\psi})^2$$

Lagrange's equation with θ as the generalized co-ordinate may be written

$$\frac{\mathrm{d}}{\mathrm{d}t}\left[\frac{\partial T}{\partial\dot{\theta}}\right] - \frac{\partial T}{\partial\theta} + \frac{\partial V}{\partial\theta} = 0$$

When the figure axis is along the radius to the Earth $\gamma \to \theta$ and thus equation (5.73), with γ replaced by θ, gives $\frac{\partial V}{\partial\theta}$.

Applying Lagrange's equation leads to

$$I_1\ddot{\theta} - \left(-I_3\omega_z\dot{\phi} + I_1\dot{\phi}^2\cos\theta\right)\sin\theta + \frac{GMm}{R^3}3\,\frac{(\sin(2\theta))}{2}\,(I_3 - I_1) = 0$$

For steady precession, $\ddot{\theta} = 0$,

$$I_3\omega_z\dot{\phi} - I_1\dot{\phi}^2\cos\theta + \frac{GMm}{R^3}\,3\,(\cos(\theta)(I_3 - I_1) = 0$$

Assuming that $\omega_z \gg \dot{\phi}$ and that θ is small

$$\dot{\phi} = -\frac{3GMm}{\omega_z R^3}(1 - I_1/I_3) \tag{5.75}$$

The effect of making $I_1 > I_3$ is simply to change the sign of the precessional velocity. If the figure axis is at 90° to the radius then the signs of $\dot{\phi}$ are reversed. Thus all configurations are stable but with differing precession rates.

5.10 De-spinning of satellites

An interesting method of stopping the spin of satellites is shown in Fig. 5.17. Two equal masses are attached to cables which are wrapped around the outside of the satellite shell. When it is required to stop the spin the masses are released so that they unwind. Relative to the satellite the masses follow an involute curve so that the velocity of the mass relative to the satellite is always normal to the cable and has the value $s\dot{\gamma}$, where s is the length of unwound cable and γ is unwrap angle. From the geometry $s = R\gamma$. The moment of inertia of the satellite about its spin axis is I and the angular velocity is ω. The kinetic energy of the system is

$$T = \frac{1}{2}I\omega^2 + m\left[(s\dot{\gamma} + s\omega)^2 + (R\omega)^2\right]$$

$$= \frac{1}{2}I\omega^2 + mR^2\left[\gamma^2(\dot{\gamma} + \omega)^2 + \omega^2\right]$$

which in the absence of external forces is constant. The constant may be equated to the initial conditions when $\omega = \omega_0$. Thus

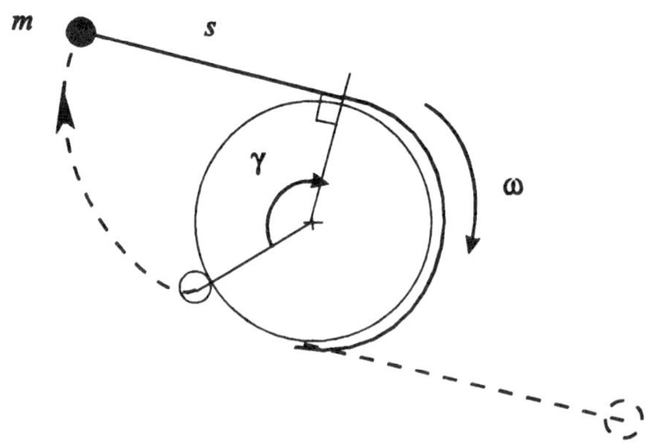

Fig. 5.17

$$\frac{1}{2}I\omega^2 + mR^2\left[\gamma^2(\dot{\gamma} + \omega)^2 + \omega^2\right] = \frac{1}{2}I\omega_0^2 + mR^2\omega_0^2 \tag{5.76}$$

Because the angle of rotation of the satellite is a cyclic co-ordinate

$$\frac{\partial T}{\partial \omega} = \text{constant}$$

Hence

$$I\omega + 2mR^2[\gamma^2(\dot{\gamma} + \omega) + \omega] = \text{constant} = I\omega_0 + 2mR^2\omega_0 \tag{5.77}$$

Equation (5.76) can be written as

$$(I + mR^2)(\omega_0^2 - \omega^2) = 2mR^2\gamma^2(\dot{\gamma} + \omega)^2 \tag{5.76a}$$

and equation (5.77) becomes

$$(I + mR^2)(\omega_0 - \omega) = 2mR^2\gamma^2(\dot{\gamma} + \omega) \tag{5.77a}$$

Dividing equation (5.76a) by equation (5.77a) gives

$$\omega_0 + \omega = \dot{\gamma} + \omega$$

and therefore

$$\dot{\gamma} = \omega_0 = \text{constant}$$

So we see that the rate of unwinding is constant.

If we require that the final spin rate is zero then putting $\omega = 0$ in equation (5.77) yields

$$-I\omega_0 + 2mR^2(\gamma^2\omega_0 - \omega_0) = 0$$

or

$$\gamma^2 - 1 = I/(2mR^2)$$

Thus the required length of cable is

$$s = \gamma R = R\sqrt{\left(\frac{I}{2mR^2} + 1\right)} \tag{5.78}$$

5.11 Stability of aircraft

In this section we shall examine the stability of an aircraft in steady horizontal flight. The general equations will be set up but only the stability requirements for longitudinal motion, that is motion in the vertical plane, will be studied. Since most aircraft are symmetrical with respect to the vertical plane motion in this plane will not be coupled to out-of-plane motion or to roll and yaw.

Referring to Fig. 5.18 we choose x to be positive forwards, y to be positive to the right and z positive downwards. The origin will be at the centre of mass. The motions in these directions are sometimes referred to as surge, sway and heave respectively. Rotations about the axes are referred to as roll, pitch and yaw.

The symbols used for the physical quantities are as follows

Axis	Displacement	Velocity / increment of		Force
x	x (longitudinal)	U	u	X
y	y (transverse)	V	v	Y
z	z (normal)	W	w	Z

(m = mass.)

Axis	Ang. disp.	Ang. vel.	M of I	Couple
x	ϕ (roll)	p	A	L
y	θ (pitch)	q	B	M
z	ψ (yaw)	r	C	N

Note that in many cases L is used for lift and V is used for forward velocity. In the present section we will use L for lift and use L_x to signify rolling moment.

By symmetry the y axis is a principal axis of inertia and we shall assume that the x axis is also a principal axis; therefore the z axis is a principal axis.

The momentum vector is

$$\boldsymbol{p} = mU\boldsymbol{i} + mV\boldsymbol{j} + mW\boldsymbol{k} \tag{5.79}$$

The angular velocity of the axes is

$$\boldsymbol{\omega} = p\boldsymbol{i} + q\boldsymbol{j} + r\boldsymbol{k} \tag{5.80}$$

Hence the time rate of change of momentum is

$$\begin{aligned}\dot{\boldsymbol{p}} &= \frac{\partial \boldsymbol{p}}{\partial t} + \boldsymbol{\omega} \times \boldsymbol{p} \\ &= m\dot{u}\boldsymbol{i} + m\dot{v}\boldsymbol{j} + m\dot{w}\boldsymbol{k} + m(qW - rV)\boldsymbol{i} + m(rU - pW)\boldsymbol{j} \\ &\quad + m(pV - qU)\boldsymbol{k}\end{aligned} \tag{5.81}$$

and the force is

$$\boldsymbol{F} = X\boldsymbol{i} + Y\boldsymbol{j} + Z\boldsymbol{k} \tag{5.82}$$

Therefore

$$X = m(\dot{u} + qW - rV) \tag{5.83}$$

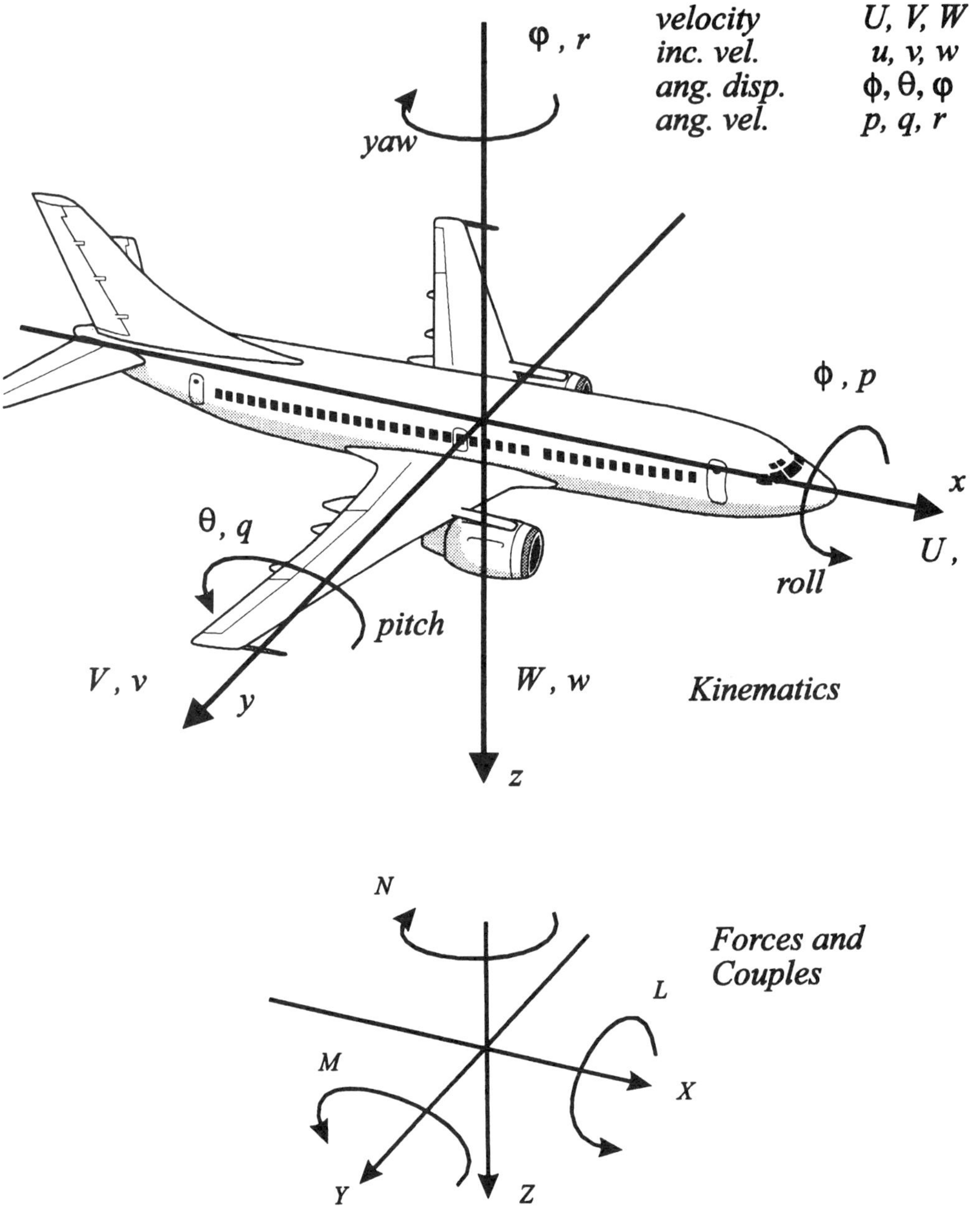

Fig. 5.18

$$Y = m(\dot{v} + rU - pW) \tag{5.84}$$

$$Z = m(\dot{w} + pV - qU) \tag{5.85}$$

The moment of momentum relative to the centre of mass is

$$\boldsymbol{L}_G = Ap\boldsymbol{i} + Bq\boldsymbol{j} + Cr\boldsymbol{k} \tag{5.86}$$

Hence the rate of change of moment of momentum is

$$\dot{\boldsymbol{L}}_G = \frac{\partial \boldsymbol{L}_G}{\partial t} + \boldsymbol{\omega} \times \boldsymbol{L}_G$$

$$= A\dot{p}\boldsymbol{i} + B\dot{q}\boldsymbol{j} + C\dot{r}\boldsymbol{k} + (Crq - Bqr)\boldsymbol{i} + (Apr - Crp)\boldsymbol{j}$$

$$+ (Bqp - Apq)\boldsymbol{k} \tag{5.87}$$

The moment of forces about G is

$$\boldsymbol{M}_G = L\boldsymbol{i} + M\boldsymbol{j} + N\boldsymbol{k} \tag{5.88}$$

Therefore

$$L_x = A\dot{p} + (C - B)qr \tag{5.89}$$

$$M = B\dot{q} + (A - C)rq \tag{5.90}$$

$$N = C\dot{r} + (B - A)pq \tag{5.91}$$

We now restrict the motion to the vertical xz plane so that $V = v = 0$, $p = 0$ and $r = 0$, and by symmetry $Y = 0$, $L_x = 0$ and $N = 0$. From this it follows that $\dot{v} = 0$, $\dot{p} = 0$ and $\dot{r} = 0$.

The equations of motion reduce to

$$X = m(\dot{u} + qW) = m\dot{u} \quad \text{as } W \to 0 \tag{5.92}$$

$$Z = m(\dot{w} - qU) = m(\dot{w} - \dot{\theta}U) \tag{5.93}$$

$$M = B\dot{q} \qquad = I_y\ddot{\theta} \tag{5.94}$$

where $q = \dot{\theta}$ and $B = I_y$.

Consider first the aircraft in straight and level flight. Figure 5.19 shows the major aerodynamic forces and gravity.The lift, L, is the aerodynamic force acting on the wing normal to the direction of airflow. It is related to the wing area S, the air density ρ, and the airspeed U by the following equation

$$L = C_L \frac{1}{2} \rho U^2 S \tag{5.95}$$

where C_L is known as the *lift coefficient*. The drag, D, is

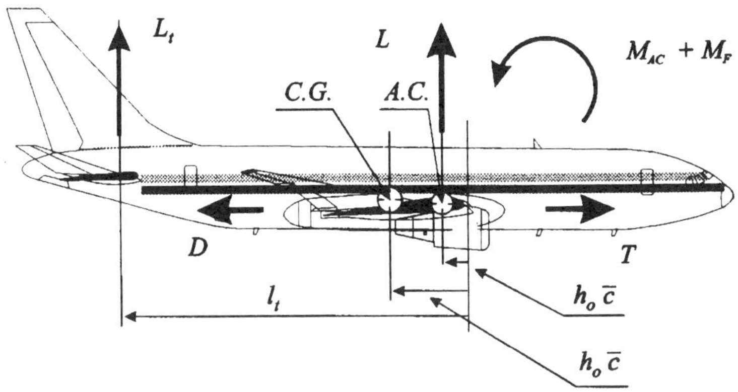

Fig. 5.19 Aerodynamic forces

$$D = C_D \frac{1}{2} \rho U^2 S \tag{5.96}$$

where C_D is the *drag coefficient.*

The drag coefficient is the sum of two parts, the first being the sum of the skin friction coefficient and form drag coefficient which will be assumed to be sensibly constant for this discussion. The second depends on the generation of lift and is known as vortex drag or induced drag. Texts on aerodynamics show that the theoretical value is $C_{Dv} = C_L^2/(\pi(AR))$, where (AR) is the aspect ratio, that is the ratio of wing span to the mean chord. Thus

$$C_D = C_{D0} + C_L^2/(\pi(AR)) \tag{5.96a}$$

T is the thrust generated by the engines and is assumed to be constant. The weight is, of course, mg.

The aerodynamic mean chord (amc), symbol $\bar{c}$, is the chord of the equivalent constant-chord wing. The line of action of the lift (centre of pressure) moves fore and aft as the angle between the chord line and the air (angle of incidence α) changes. This is taken into account by choosing a reference point and giving the moment of the lift force about this point. The pitching moment is given by

$$M = C_M \frac{1}{2} \rho U^2 S \bar{c} \tag{5.97}$$

where C_M is the pitching moment coefficient. It is found that a point exists along the chord such that the pitching moment coefficient remains sensibly constant with angle of incidence. This point is called the aerodynamic centre (ac) and typically is located at the quarter chord point. The tailplane lift is

$$L_t = C_{Lt} \frac{1}{2} \rho U^2 S_t \tag{5.98}$$

where the suffix t refers to the tailplane. The airspeed over the tailplane will be slightly less than the speed of the air relative to the wing because of the effect of drag. We shall ignore this effect but it can be included at a later stage by the introduction of a tailplane efficiency. Another effect of the wing is to produce a downwash at the tailplane. This is related to the lift of the wing which has the effect of reducing the gradient of the tailplane lift to incidence curve.

The lift and drag of the fuselage could be included by modifying the lift and drag coefficients but the pitching moment will be kept separate because in general it will vary with angle of incidence.

For steady horizontal flight with the x axis horizontal the only non-zero velocity is U, the forward speed. In this case the thrust is equal to the drag and the sum of the wing lift and the tailplane lift is equal to the weight. Only a small error will be introduced if the tailplane lift is neglected when compared with the wing lift; hence

$$T - D = 0 \tag{5.99}$$

$$mg - L = 0 \tag{5.100}$$

Taking moments about the centre of mass

$$M = M_f + M_{ac} + L(h - h_0)\bar{c} - L_t l_t = 0 \tag{5.101}$$

Dividing through by $L\bar{c}$ gives

$$\frac{C_{Mf}}{C_L} + \frac{C_{Mac}}{C_L} + (h - h_0) - \frac{C_{Lt} S_t l_t}{C_L S \bar{c}} = 0 \tag{5.102}$$

The term $\frac{S_t l_t}{S\bar{c}} = \bar{V}$, the tail volume ratio, so (5.102a)

$$\frac{C_{Mf}}{C_L} + \frac{C_{Mac}}{C_L} + (h - h_0) - \frac{C_{Lt}}{C_L}\bar{V} = 0 \tag{5.103}$$

For static stability we require that $\frac{\partial M}{\partial \alpha} < 0$; that is, if the angle of incidence increases the moment generated should be negative so as to restore θ to the original state. Operating on equation (5.101)

$$\frac{\partial M}{\partial \alpha} = \frac{\partial (C_{Mf} + C_{Mac})}{\partial \alpha} \frac{1}{2} \rho U^2 S\bar{c} + \frac{\partial C_L}{\partial \alpha} \frac{1}{2} \rho U^2 S(h - h_0)\bar{c}$$

$$- \frac{\partial C_{Lt}}{\partial \alpha} \frac{1}{2} \rho U^2 S_t l_t < 0 \tag{5.104}$$

Dividing through by $\frac{1}{2}\rho U^2 S\bar{c}$ and noting that $\partial C_{Mac}/\partial \alpha = 0$ yields

$$\frac{\partial C_{Mf}}{\partial \alpha} + a_1(h - h_0) - a_t\bar{V} < 0 \tag{5.105}$$

where the symbol a stands for the gradient of the lift coefficient versus the angle of incidence.

The critical, or neutral, value of h, h_n, is that which makes $\partial M/\partial \alpha = 0$, so

$$h_n = h_0 + \frac{a_t}{a_1}\bar{V} - \frac{1}{a_1}\frac{\partial C_{Mf}}{\partial \alpha} \tag{5.106}$$

The *stick-fixed static CG margin* is defined to be

$$h_n - h = (h_0 - h) + \frac{a_t}{a_1}\bar{V} - \frac{1}{a_1}\frac{\partial C_{Mf}}{\partial \alpha} \tag{5.107}$$

and is the distance of the CG in front of the neutral point as a fraction of the *aerodynamic mean chord*. The term 'stick-fixed' signifies that the elevator is held fixed and not allowed to float.

For dynamic stability we consider that the aircraft has pitched a small angle θ to the horizontal and that there are increments in speed in the x direction of u and in the z direction of w. Referring to Fig. 5.20 we see that for small deviations in angles and speeds the variation in the angle of incidence at the wing is

$$\delta\alpha = \theta + \frac{w}{U} \tag{5.108}$$

and the variation in airspeed

$$\delta U = u \tag{5.109}$$

Therefore

$$\delta(U^2) = 2Uu \tag{5.110}$$

The effect of $\delta\alpha$ is twofold. One is to increase the magnitude of the lift and drag terms and the other is to rotate the directions of the lift and drag terms relative to the xyz axes.

In the x direction the changes in the force terms are equated to the rate of change of momentum in the x direction as given in equation (5.92)

$$-mg\theta + L\delta\alpha - \delta D = m\dot{u}$$

$$-mg\theta + L(\theta + \frac{w}{U}) - C_D\frac{1}{2}\rho 2UuS - \frac{\partial C_D}{\partial \alpha}\frac{1}{2}\rho U^2 S(\theta + \frac{w}{U}) = m\dot{u}$$

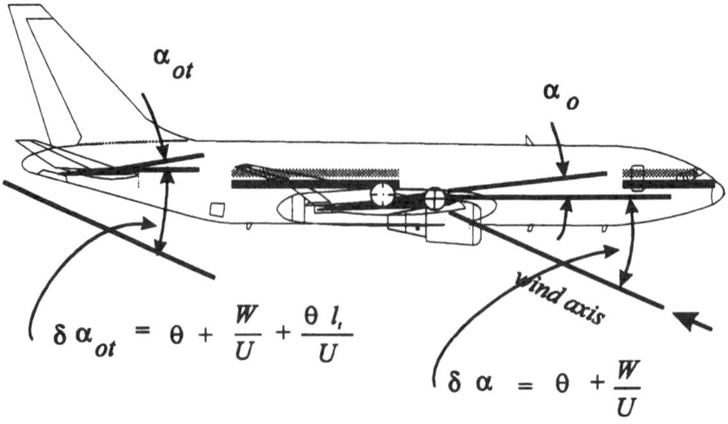

Fig. 5.20

Differentiating equation (5.96a) we obtain

$$\frac{\partial C_D}{\partial \alpha} = \frac{2}{\pi(AR)} C_L a_1$$

Substituting for $\partial C_D/\partial\alpha$ and dividing through by $mg = L$ gives

$$\left(\frac{w}{U}\right) - \frac{C_D}{C_L} 2\left(\frac{u}{U}\right) - \frac{2a_1}{\pi(AR)}\left(\theta + \frac{w}{U}\right) = \frac{1}{g}\dot{u} = \frac{U}{g}\frac{d}{dt}\left(\frac{u}{U}\right) \qquad (5.111)$$

Let $(\frac{u}{g}\frac{d}{dt}) = \bar{D}$, a non-dimensional operator, and let $u/U = \bar{u}$ and $w/U = \bar{w}$. Hence equation (5.111) becomes

$$\left(\bar{D} + \frac{2C_D}{C_L}\right)\bar{u} - \bar{w}\left(1 - \frac{2a_1}{\pi(AR)}\right) + \frac{2a_1}{\pi(AR)}\theta = 0 \qquad (5.112)$$

In the z direction, following equation (5.93),

$$-\delta L - D\,\delta\alpha = m(\dot{w} - \dot{\theta}U)$$

$$-a_1(\theta + w/U)\frac{1}{2}\rho U^2 S - C_L\frac{1}{2}\rho 2UuS - C_D\frac{1}{2}\rho U^2 S(\theta + w/U) = m(\dot{w} - \dot{\theta}U)$$

Dividing through by $L = mg$

$$-\frac{a_1}{C_L}(\theta + \bar{w}) - 2\bar{u} - \frac{C_D}{C_L}(\theta + \bar{w}) = \frac{U}{g}\frac{d}{dt}(\bar{w} - \theta)$$

or

$$2\bar{u} + \left(\bar{D} + \frac{a_1}{C_L} + \frac{C_D}{C_L}\right)\bar{w} - \left(\bar{D} - \frac{a_1}{C_L} - \frac{C_D}{C_L}\right)\theta = 0 \qquad (5.113)$$

Taking moments of the force variations about the y axis

$$\delta M_f + \delta M_{ac} + \delta L(h - h_0)\bar{c} - \delta L_t l_t = I_y\ddot{\theta}$$

Evaluating the variations and dividing through by $L\bar{c} = mg\bar{c}$ we arrive at

$$\frac{1}{C_L}\frac{\partial C_{Mf}}{\partial \alpha}\left(\theta + \frac{w}{U}\right) + \left(\frac{C_{Mac} + C_{Mf}}{C_L}\right)2\frac{u}{U} + \frac{a_1}{C_L}\left[\theta + \frac{w}{U}\right](h - h_0)$$

$$+ 2u/U(h - h_0) - \frac{a_t}{C_L}\left(\theta + \frac{w}{U} + \frac{d\theta}{dt}\frac{l_t}{U}\right)\frac{S_t l_t}{S\bar{c}} - 2\frac{u}{U}\frac{C_{Lt}S_t l_t}{C_L S\bar{c}}$$

$$= \frac{k_y^2}{g\bar{c}}\ddot{\theta} = \frac{k_y^2 g}{\bar{c}U^2}\frac{U^2}{g^2}\frac{d^2\theta}{dt^2}$$

Collecting terms and replacing d/dt by $(g/U)\bar{D}$ leads to

$$\left((h - h_0) + \frac{C_{Mac}}{C_L} + \frac{C_{Mf}}{C_L} - \frac{C_{Lt}}{C_L}\bar{V}\right)2\bar{u}$$

$$+ \left(\frac{1}{C_L}\frac{\partial C_{Mf}}{\partial \alpha} + \frac{a_1}{C_L}(h - h_0) - \frac{a_t}{C_L}\bar{V}\right)\bar{w}$$

$$+ \left(\frac{1}{C_L}\frac{\partial C_{Mf}}{\partial \alpha} + \frac{a_1}{C_L}(h - h_0) - \frac{a_t}{C_L}\bar{V}\right.$$

$$\left. - \frac{a_t}{C_L}\bar{V}\frac{l_t g}{U^2}\bar{D} - \frac{k_y^2 g}{\bar{c}U^2}\bar{D}^2\right)\theta = 0 \tag{5.114}$$

Note that from equation (5.103) the coefficient of $2u$ in the above equation is zero. Also from equation (5.107) the coefficient of $\bar{w}$ may be written as $-(\mathrm{SM})a_1/C_L$, where SM is the stick-fixed static CG margin. Therefore equation (5.114) becomes, after a change of sign,

$$\left(\frac{(\mathrm{SM})a_1}{C_L}\right)\bar{w} + \left(\frac{(\mathrm{SM})a_1}{C_L} + \frac{a_t}{C_L}\bar{V}\frac{l_t g}{U^2}\bar{D} + \frac{k_y^2 g}{\bar{c}U^2}\bar{D}^2\right)\theta = 0 \tag{5.115}$$

Equations (5.112), (5.113) and (5.115) are three simultaneous equations in the non-dimensional variables $\bar{u}$, $\bar{w}$ and θ, with $\bar{D}$ being a differential operator in non-dimensional time.

Defining the following constants

$$K_1 = k_y^2 g/(\bar{c}U^2)$$

$$K_2 = a_t l_t g\bar{V}/(C_L U^2)$$

$$K_3 = (\mathrm{SM})a_1/C_L$$

$$K_4 = (a_1 + C_D)/C_L$$

$$K_5 = 2a_1/\pi(\mathrm{AR})$$

$$K_6 = 2C_D/C_L$$

the three equations may be written in matrix form as

$$\begin{bmatrix} (\bar{D} + K_6) & -(1 - K_5) & K_5 \\ 2 & (\bar{D} + K_4) & -(\bar{D} - K_4) \\ 0 & K_3 & (K_1\bar{D}^2 + K_2\bar{D} + K_3) \end{bmatrix}\begin{bmatrix} \bar{u} \\ \bar{w} \\ \theta \end{bmatrix} = \begin{bmatrix} \\ 0 \\ \\ \end{bmatrix} \tag{5.116}$$

The differential operator $\bar{D}$ has been defined to be

$$\bar{D} = \frac{U}{g}\frac{\mathrm{d}}{\mathrm{d}t} = \frac{\mathrm{d}}{\mathrm{d}(tg/U)} = \frac{\mathrm{d}}{\mathrm{d}\tau}$$

where $\tau = tg/U$ is the non-dimensional time.

Assume that

$$\bar{u} = \tilde{u}e^{\lambda\tau}$$
$$\bar{w} = \tilde{w}e^{\lambda\tau}$$
$$\theta = \theta e^{\lambda\tau}$$

Substitution into equation (5.116) gives the following set of three simultaneous algebraic equations

$$\begin{bmatrix} (\lambda + K_6) & -(1 - K_5) & K_5 \\ 2 & (\lambda + K_4) & -(\lambda - K_4) \\ 0 & K_3 & (K_1\lambda^2 + K_2\lambda + K_3) \end{bmatrix} \begin{bmatrix} \tilde{u} \\ \tilde{w} \\ \theta \end{bmatrix} = \begin{bmatrix} \\ 0 \\ \\ \end{bmatrix} \qquad (5.117)$$

For a non-trivial solution the determinant of the 3 × 3 matrix has to be zero, yielding a quartic in λ in the form

$$A\lambda^4 + B\lambda^3 + C\lambda^2 + D\lambda + E = 0 \qquad (5.118)$$

Expansion of the determinant leads to

$$A = K_1$$
$$B = K_2 + K_4K_1 + K_1K_6$$
$$C = 2K_3 + K_4K_2 + 2(1 - K_5)K_1 + (K_2 + K_4K_1)K_6$$
$$D = 2(1 - K_5)K_2 + (2K_3 + K_4K_2)K_6$$
$$E = 2K_3$$

The overall stability of the system can be tested using the Routh–Hurwitz criteria (see Harrison and Nettleton 1994) which states that

(i) all coefficients > 0

(ii) $\begin{vmatrix} D & E \\ B & C \end{vmatrix} > 0$

(iii) $\begin{vmatrix} D & E & 0 \\ B & C & D \\ 0 & A & B \end{vmatrix} > 0$

Conditions (ii) and (iii) imply that

$$(DC - BE) > 0$$

and

$$B(DC - BE) - AD^2 > 0 \qquad (5.119)$$

Figure 5.21 shows the results of numerical integration of equations (5.112), (5.113) and (5.115) for the data given.

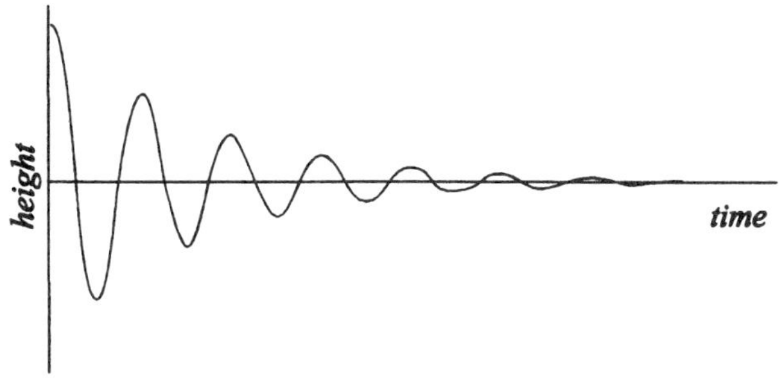

Fig. 5.21 Phugoid oscillation

In general the solution to the equations is two sinusoids. A high-frequency heavily damped mode and a low-frequency low-damped mode which is known as *phugoid motion.* Equation (5.118) may be written in the form

$$(\lambda^2 + \alpha\lambda + \beta)(\lambda^2 + \gamma\lambda + \delta) = 0$$

which expands to

$$\lambda^4 + (\alpha + \gamma)\lambda^3 + (\alpha\gamma + \beta + \delta)\lambda^2 + (\alpha\delta + \beta\gamma)\lambda + \beta\delta = 0 \qquad (5.120)$$

and comparing with equation (5.118)

$$\begin{aligned} \alpha + \gamma &= B/A \\ \alpha\gamma + \beta + \delta &= C/A \\ \alpha\delta + \beta\gamma &= D/A \\ \beta\delta &= E/A \end{aligned} \qquad (5.121)$$

Rearranging the equations as follows

$$\begin{aligned} \alpha &= B/A - \gamma \\ \beta &= C/A - \delta - \alpha\gamma \\ \delta &= (E/A)/\beta \\ \gamma &= (D/A - \alpha\delta)/\beta \end{aligned} \qquad (5.122)$$

gives an iterative procedure which converges rapidly.

From vibration theory the roots of the quadratic indicate two damped sinusoidal motions of damped natural frequency ω_{d1} and ω_{d2} with the corresponding damping ratios ζ_1 and ζ_2

$$\omega_{d1} = \sqrt{(\beta)}\sqrt{(1 - \zeta_1^2)}$$

$$\zeta_1 = \alpha/(2\sqrt{(\beta)})$$

and

$$\omega_{d2} = \sqrt{(\delta)}\sqrt{(1 - \zeta_2^2)}$$

$$\zeta_2 = \gamma/(2\sqrt{(\delta)})$$

The periods of oscillation are

$$T_1 = 2\pi/\omega_{d1}$$

$$T_2 = 2\pi/\omega_{d2}$$

For the majority of aircraft the first mode is a heavily damped short-period (a few seconds) oscillation and the second is a low-damped long-period (of the order of minutes) oscillation known as a phugoid oscillation.

A large static margin will usually lead to high stability but may make the aircraft difficult to manoeuvre.

5.12 Stability of a road vehicle

We shall consider the stability of a four-wheeled motor vehicle for small deviations from straight line motion and for a steady turn. The effects of roll on the suspension geometry and tyre characteristics will be neglected. The approach will be similar to the treatment of aircraft stability given in the previous section. In this case the two front tyres act in a similar manner to one aerofoil and the rear pair of tyres will be thought of as another aerofoil.

The tyre is a complex component with a non-linear behaviour. The characteristic which concerns us most is the relationship between the lateral force and the side-slip angle, *see* Fig. 5.22. For small lateral forces this force, F, varies linearly with side-slip angle, α, and the initial gradient, $\frac{\partial F}{\partial \alpha} = C$, is the lateral force coefficient. For any given tyre this coefficient depends on the vertical load, tyre pressure, camber angle and the type of surface on which the tyre is running. Braking and traction will also affect the coefficient. We will treat the two front tyres as a single tyre with a fixed lateral force coefficient C_f and the two rear tyres as a single tyre with coefficient C_r.

The notation for the rigid body motion shown in Fig. 5.23 is the same as that used for the aircraft in Fig. 5.18. The forces involved are shown in Fig. 5.24. The angle of steer of the front wheels is δ_f and that of the rear wheels is δ_r. The centre of mass is located a distance a from the front axle and b from the rear axle. The sum of these two is the wheelbase L. Traction has been assumed to be at the front wheels.

From Fig. 5.23 we see that the tangent of the angle of the direction of motion of the front wheel relative to the X axis is

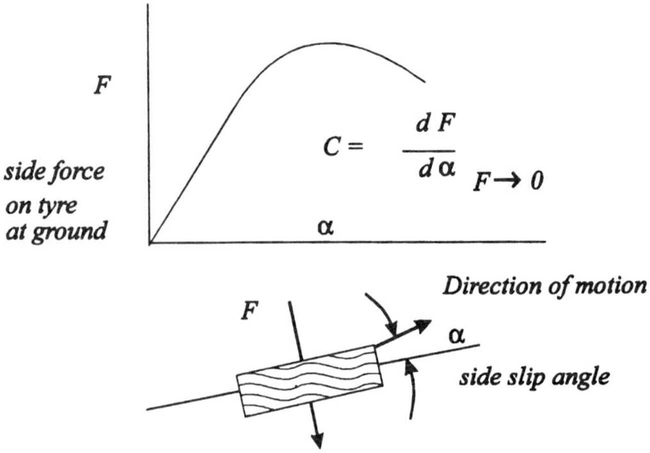

Fig. 5.22

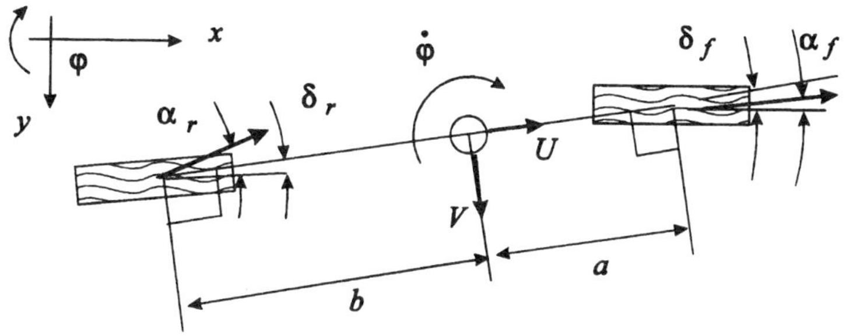

Fig. 5.23

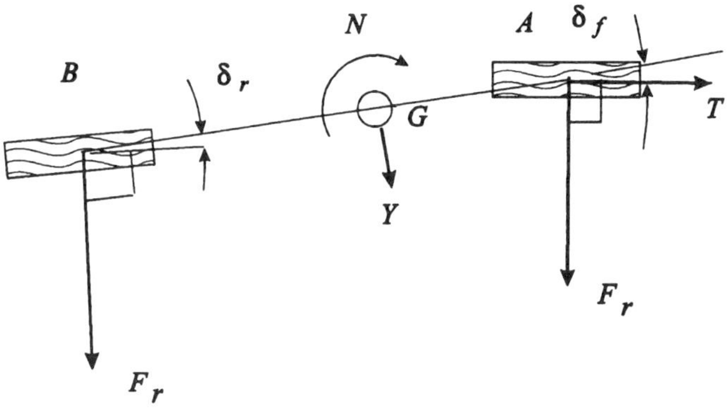

Fig. 5.24

$$\tan(\delta_f - \alpha_f) = \frac{V + \dot{\psi}a}{U} \tag{5.123}$$

and for the rear wheel

$$\tan(\alpha_r - \delta_r) = \frac{\dot{\psi}b - V}{U}$$

or

$$\tan(\delta_r - \alpha_r) = \frac{V - \dot{\psi}b}{U} \tag{5.124}$$

Resolving forces in the X direction gives

$$X + T\cos\delta_f - F_f\sin\delta_f - F_r\sin\delta_r = m(\dot{U} - V\dot{\psi}) \tag{5.125}$$

in the Y direction

$$Y + T\sin\delta_f + F_f\cos\delta_f + F_r\cos\delta_r = m(\dot{V} - U\dot{\psi}) \tag{5.126}$$

and taking moments about the Z axis through the centre of mass

$$N + T\sin(\delta_f)a + F_f\cos(\delta_f)a + F_r\cos(\delta_r)b = mk_G^2\ddot{\psi} \tag{5.127}$$

where k_G is the radius of gyration.

The lateral forces are related to the side-slip angles by the lateral force coefficients

$$F_f = C_f\alpha_f \tag{5.128}$$

$$F_r = C_r\alpha_r \tag{5.129}$$

We shall assume that the rear steering angle is proportional to the lateral force with a constant of proportionality $1/K_r$. Thus

$$\delta_r = \frac{F_r}{K_r} \tag{5.130}$$

Straight line stability is a measure of the vehicle's ability to proceed in a straight line with the steering angle fixed at zero. Substituting equations (5.128) to (5.130) into equations (5.123) and (5.124) gives

$$-\frac{F_f}{C_f} = \frac{V}{U} + \frac{\dot{\psi}a}{U} \tag{5.131}$$

and

$$-\frac{F_r}{C_r} + \frac{F_r}{K_r} = \frac{V}{U} - \frac{\dot{\psi}a}{U}$$

or

$$-\frac{F_r}{C_r'} = \frac{V}{U} - \frac{\dot{\psi}a}{U} \tag{5.132}$$

where

$$C_r' = \frac{C_r}{(1 - C_r/K_r)} \tag{5.133}$$

The term C_r' is an effective coefficient. It is seen that a positive K_r has the same effect as increasing the lateral force coefficient so in subsequent equations the prime will be dropped. It is assumed that should rear wheel steering be present then its effect will have been already incorporated.

Substituting these equations into equations (5.128) and (5.129) and assuming that the side-slip angles are small gives

$$Y - \frac{C_fV}{U} - \frac{C_fa\dot{\psi}}{U} - \frac{C_rV}{U} + \frac{C_rb\dot{\psi}}{U} = m\dot{V} + mU\dot{\psi}$$

$$N - \frac{C_fVa}{U} - \frac{C_fa^2\dot{\psi}}{U} + \frac{C_rVb}{U} - \frac{C_rb^2\dot{\psi}}{U} = mk_G^2\ddot{\psi}$$

Using the D operator the two equations above may be put into matrix form

$$\begin{bmatrix} mD + \dfrac{(C_f + C_r)}{U} & mU - \dfrac{(C_rb + C_fa)}{U} \\ -\dfrac{(C_rb + C_fa)}{U} & mk_G^2D + \dfrac{(C_rb^2 + C_fa^2)}{U} \end{bmatrix} \begin{bmatrix} V \\ \dot{\psi} \end{bmatrix} = \begin{bmatrix} Y \\ N \end{bmatrix} \tag{5.134}$$

For the case when $Y = 0$ and $N = 0$ we assume that $V = \tilde{V}e^{\lambda t}$ and $\dot{\psi} = \tilde{\dot{\psi}}e^{\lambda t}$ where λ may be complex. Substitution into equation (5.134) and dividing through by $e^{\lambda t}$ leads to

$$\begin{bmatrix} m\lambda + \dfrac{(C_f + C_r)}{U} & mU - \dfrac{(C_rb - C_fa)}{U} \\ -\dfrac{(C_rb + C_fa)}{U} & mk_G^2\lambda + \dfrac{(C_rb^2 + C_fa^2)}{U} \end{bmatrix} \begin{bmatrix} \tilde{V} \\ \tilde{\dot{\psi}} \end{bmatrix} = \begin{bmatrix} 0 \\ 0 \end{bmatrix} \tag{5.135}$$

For a non-trivial solution the determinant of the square matrix must equate to zero yielding the characteristic equation

$$\lambda^2[m^2k_G^2] + \lambda[m(C_r b^2 + C_f a^2) + mk_G^2(C_f + C_r)]/U$$
$$+ [(C_f + C_r)(C_r b^2 + C_f a^2) - (C_r b - C_f a)^2 + m(C_r b - C_f a)U^2]/U^2$$
$$= 0$$

which reduces to

$$\lambda^2[m^2k_G^2U^2] + \lambda U[C_r b^2 + C_f a^2 + k_G^2(C_f + C_r)] + [C_f C_r L^2 + mU^2(C_r b - C_f a)] = 0 \quad (5.136)$$

From the theory of differential equations it is known that if the motion is stable then all coefficients must be positive. The coefficient of λ^2 is always positive and the coefficient of λ is also positive. It should be noticed that if U is negative then the definition of the transverse force coefficient requires both C_f and C_r to be negative.

The constant term will also be positive for all speeds if $(C_r b - C_f a) > 0$. So for the case of a vehicle with four identical tyres $b > a$, that is the centre of mass must be forward of the mid-point of the wheelbase.

If the above condition is not met then the vehicle will be stable only if

$$U^2 < \frac{C_f C_r L^2}{m(C_f a - C_r b)} = U^2_{\text{critical}} \quad (5.137)$$

Therefore as the centre of mass moves towards the rear axle the critical speed, U_{critical}, becomes lower.

The concept of *static margin* is similar to that defined in the previous section on aircraft stability. Figure 5.25 shows the case of a car under the action of a steady side load Y located a distance n forward of the rear axle and such that now yaw is produced. This

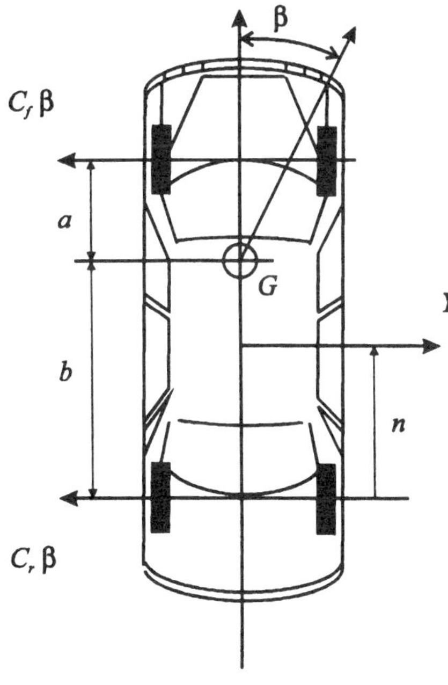

Fig. 5.25

point is known as the *neutral steer point*. Let the common side-slip angle be β so that in the Y direction

$$(C_f + C_r)\beta = Y \tag{5.138}$$

and taking moments about the rear axle gives

$$Yn - C_f\beta L = 0 \tag{5.139}$$

Thus

$$n = \frac{C_f L}{(C_f + C_r)} \tag{5.140}$$

The static margin is defined as the position of the centre of mass ahead of the neutral steer point expressed as a fraction of the wheelbase. Therefore the static margin is given by

$$\mathrm{SM} = \frac{(b - n)}{L} = \frac{(C_r b - C_f a)}{L(C_f + C_r)} \tag{5.141}$$

From this we see that if the static margin is positive a side load to the right applied at the centre of mass will give rise to positive yaw, that is the car will turn to the right, and left hand steer is required to maintain the same heading. This condition is called understeer and is seen to correspond to the condition for stability at all speeds. A negative static margin gives rise to oversteer and corresponds to the condition where there exists a critical speed. We shall discuss this further when dealing with a steady turn.

Figure 5.26 shows the geometry for a steady turn. The radius of the turn is defined to be the distance from the centre of the turning circle to the centre of the rear axle. For a low-

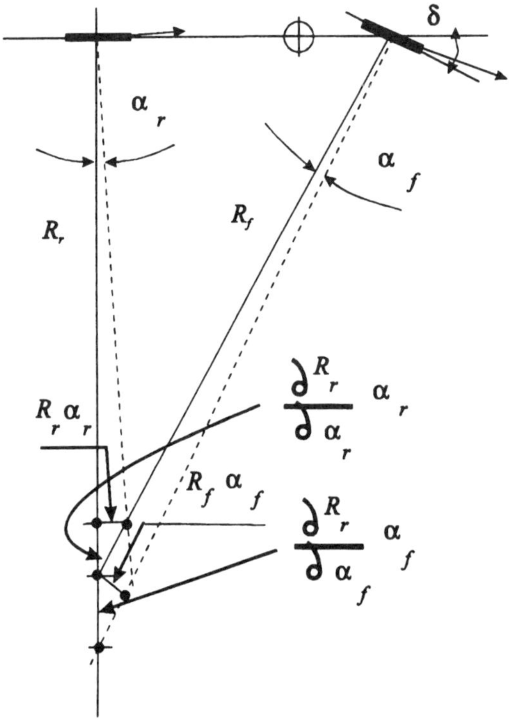

Fig. 5.26

speed turn the centre will be the intersection of the normal to the rear wheels and the normal to the front wheels. Thus

$$R_r = L/\tan(\delta) \tag{5.142}$$

Also

$$U = \dot{\psi} R_r \tag{5.143}$$

We shall now look at the variation in the radius of turn due to small changes in the side-slip angles. From the figure

$$\mathrm{d}R_r = \frac{R_f \alpha_f}{\sin\delta} - \frac{R_r \alpha_r}{\tan\delta} \tag{5.144}$$

Resolving in the Y direction

$$C_r \alpha_r + C_f \alpha_f \cos\delta = mU\dot{\psi} \tag{5.145}$$

and by moments about the centre of mass

$$C_f \alpha_f \cos(\delta) a - C_r \alpha_r b = 0 \tag{5.146}$$

Now $\tan\delta = L/R_r$ and $\sin\delta = L/R_f$, and therefore equation (5.144) becomes

$$\mathrm{d}R_r = \frac{L\alpha_f}{\sin^2\delta} - \frac{L\alpha_r}{\tan^2\delta} \tag{5.147}$$

Eliminating $C_f \alpha_f$ from equations (5.145) and (5.146) gives

$$C_r \alpha_r (1 + b/a) = mU\dot{\psi}$$

or

$$\alpha_r = \frac{amU^2}{LC_r R_r} \tag{5.148}$$

and

$$\alpha_f = \frac{bmU^2}{L\cos(\delta) C_f R_r} \tag{5.149}$$

Substituting into equation (5.147) and rearranging gives

$$\mathrm{d}R_r = \frac{mU^2}{LC_r C_f \sin\delta}\left(\frac{bC_r}{\cos\delta} - aC_f \cos^2\delta\right) \tag{5.150}$$

If the steering angle is small then the radius of turn reduces to

$$R_r = \frac{L}{\delta} + \frac{mU^2}{LC_r C_f \delta}(C_r b - C_f a)$$

$$= \frac{L}{\delta}\left(1 + \frac{mU^2}{L}\frac{(C_r b - C_f a)}{LC_r C_f}\right) \tag{5.151}$$

or

$$\delta = \frac{L}{R_r}\left(1 + \frac{mU^2}{L}\frac{(C_r b - C_f a)}{LC_r C_f}\right) \tag{5.152}$$

Figure 5.27 shows the plot of path curvature versus speed for a given steering angle and Fig. 5.28 is a plot of steering angle versus speed for a given radius of turn. It is seen that when the static margin is positive the steering angle has to be increased to maintain the same radius of turn and the vehicle is understeering. When the static margin is negative the steering angle has to be reduced in order to keep the radius constant and the vehicle is oversteering. After the critical speed has been reached the vehicle is unstable.

Looking at the situation with fixed steering angle it is seen that an understeering vehicle will run wide if the speed is increased whilst an oversteering one will tighten its turn if the speed is increased.

It must be noted that traction has an effect on the static margin: for front wheel drive cars the static margin is made more positive whilst for rear wheel drive cars it is made more negative.

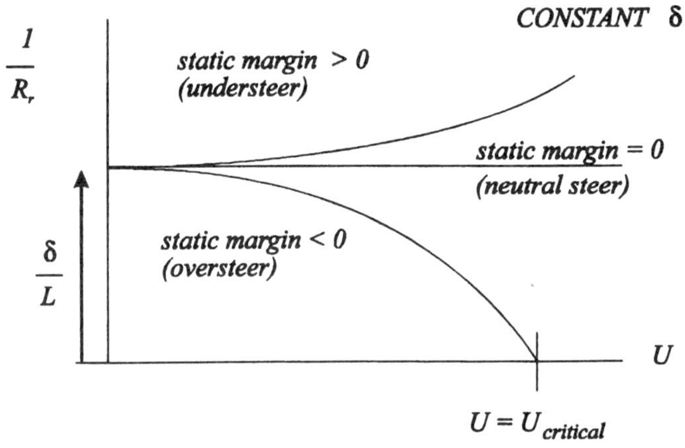

Fig. 5.27

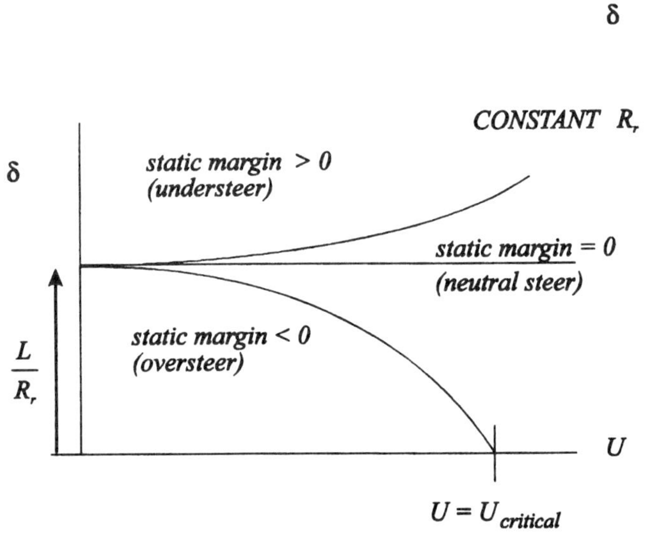

Fig. 5.28

6

Impact and One-Dimensional Wave Propagation

6.1 Introduction

This chapter deals with the propagation of waves or pulses in an elastic medium. The most fruitful application of this theory is in the study of impact. Transient phenomena can be dealt with using vibration methods but for short-duration impacts a large number of the normal modes have to be considered and in these cases a wave technique often leads to a simpler solution.

The simplest form of wave propagation is the *non-dispersive* wave. A non-dispersive wave is one which travels at a fixed speed through the medium without change in shape, for example a pulse of the form of a half sine wave will always remain a half sine wave. Physical systems which approximate to this condition are longitudinal waves in a uniform bar, torsional waves in a uniform bar and small-amplitude waves in a stretched string. However, bending, or lateral, waves in a bar are dispersive so that the shape of a transverse wave will be continually changing. This corresponds to different wavelengths travelling at different speeds so here there is no fixed speed of propagation. In the interior of an elastic medium plane waves are non-dispersive; these will be discussed in the next chapter.

6.2 The one-dimensional wave

A wave may be pictured either as the variation in time of some physical quantity, u, at a fixed location or as a variation with distance at a fixed time. Figure 6.1 shows a series of pulses at constant times and at constant positions. An arbitrary function of argument z is given in Fig. 6.2 in mathematical terms $u = f(z)$. In Fig. 6.3 this shape is used to represent a plot of u versus time t for a given position $x = 0$. If the pulse is assumed to be travelling along the positive x axis at a constant speed c then there will be a time delay of x/c. We shall change the time variable from t to ct so that both axes have the dimensions of length. Now at $x = 0$ we can represent the pulse as $u = f(ct)$ and at $x = x$ by $u = f(ct - x)$.

We may now write for an arbitrary pulse moving at a speed c along the positive x axis

$$u = f(ct - (x - \beta)) \tag{6.1}$$

where β is the location of the head of the pulse when $t = 0$. In this case the argument is

$$z = ct - (x - \beta) \tag{6.2}$$

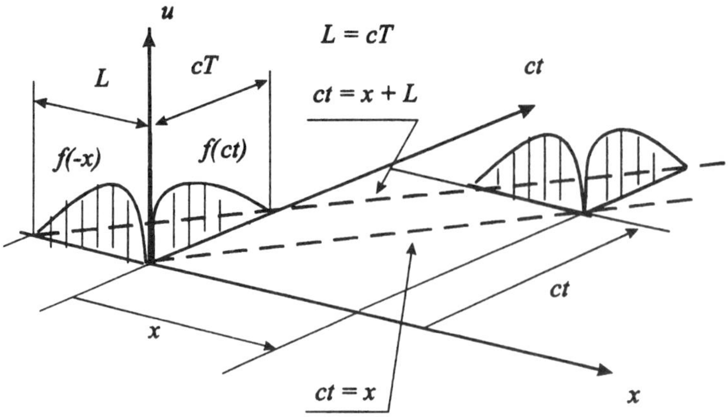

Fig. 6.1

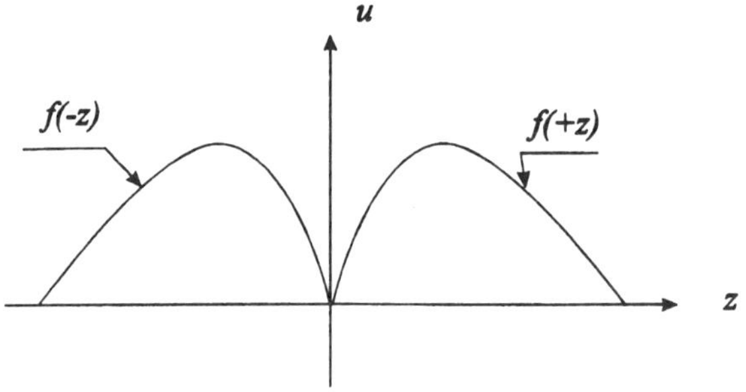

Fig. 6.2

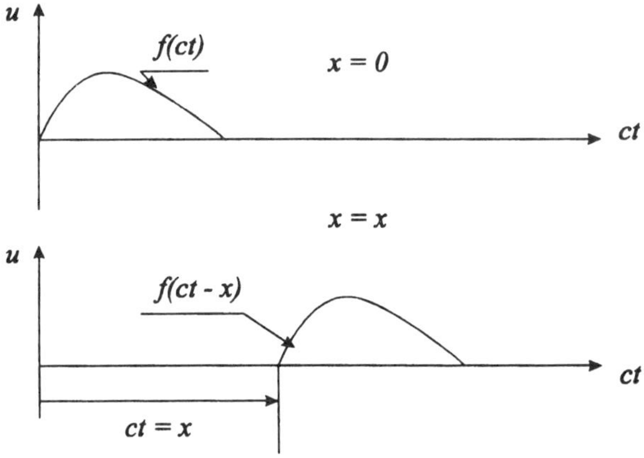

Fig. 6.3

Figure 6.4 shows the same situation but this time the pulse is shown as the variation of u against distance.

For a pulse travelling in the $-x$ direction (*see* Fig. 6.5), again at a speed c, the pulse is given by $u = g(x)$ at $t = 0$ and by $u = g(x - (-ct))$ at $t = -t$. It is convenient to express pulses travelling in the negative direction by $u = g(z) = g(ct + x)$, the reasons for which will become clear later. In the same manner as for the positive-going pulse we may generalize to

$$u = g(ct + (x - \gamma)) \tag{6.3}$$

Also

$$z = ct + (x - \gamma) \tag{6.4}$$

where γ is the position of the head of the pulse when $t = 0$.

In both cases at the head of the pulse $z = 0$. If the pulse has a finite length L in space then its duration τ will be L/c so that $c\tau = c(L/c) = L$. Thus for both waves at the tail of the pulse $z = L$.

In general waves may be travelling in both directions simultaneously and therefore

$$u = f(ct - (x - \beta)) + g(ct + (x - \gamma)) \tag{6.5}$$

Let us now evaluate the partial differentials of u with respect to x and to t

$$\frac{\partial u}{\partial x} = \frac{\mathrm{d}u}{\mathrm{d}z}\frac{\partial z}{\partial x} = \frac{\mathrm{d}f}{\mathrm{d}z}\frac{\partial z}{\partial x} + \frac{\mathrm{d}g}{\mathrm{d}z}\frac{\partial z}{\partial x}$$

$$= \frac{\mathrm{d}f}{\mathrm{d}z}(-1) + \frac{\mathrm{d}g}{\mathrm{d}z}(+1)$$

$$= -f' + g' \tag{6.6}$$

where the prime signifies differentiation with respect to the argument. Similarly

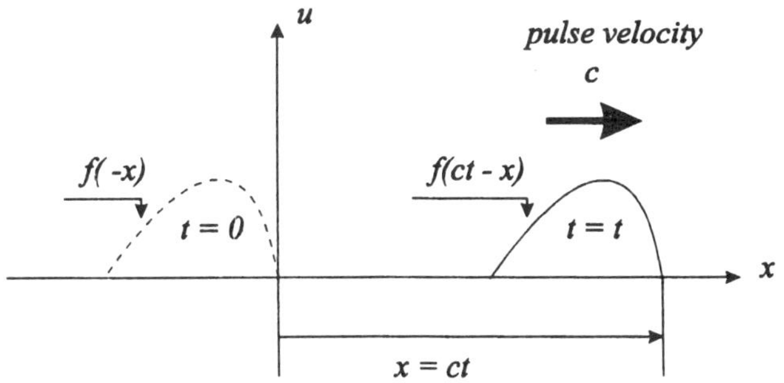

Fig. 6.4

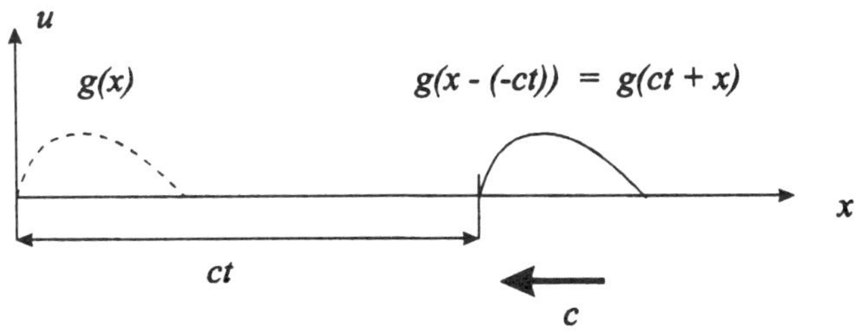

Fig. 6.5

$$\frac{\partial u}{\partial t} = \frac{\mathrm{d}u}{\mathrm{d}z}\,\frac{\partial z}{\partial t} = \frac{\mathrm{d}f}{\mathrm{d}z}\,\frac{\partial z}{\partial t} + \frac{\mathrm{d}g}{\mathrm{d}z}\,\frac{\partial z}{\partial t}$$

$$= \frac{\mathrm{d}f}{\mathrm{d}z}(c) + \frac{\mathrm{d}g}{\mathrm{d}z}(c)$$

$$= cf' + cg' \tag{6.7}$$

We see that for both f and g functions differentiation with respect to time produces a multiplication factor of c whilst differentiation with respect to x requires a factor of -1 for f functions and $+1$ for g functions. This is the reason for using different symbols for forwards and backwards travelling waves because in this way it is easy to carry out any differentiation.

Repeating the above scheme we obtain the second partial differentials

$$\frac{\partial^2 u}{\partial x^2} = f''(-1)^2 + g''(+1)^2 = f'' + g'' \tag{6.8}$$

$$\frac{\partial^2 u}{\partial t^2} = f''(c)^2 + g''(c)^2 = c^2 f'' + c^2 g'' \tag{6.9}$$

By inspection of equations (6.8) and (6.9) we see that

$$\frac{\partial^2 u}{\partial t^2} = c^2\,\frac{\partial^2 u}{\partial x^2} \tag{6.10}$$

This important equation is well known in many branches of physics and is called the wave equation. It follows immediately that any physical system which yields this equation will have non-dispersive waves travelling at a speed c as a solution.

6.3 Longitudinal waves in an elastic prismatic bar

Figure 6.6 shows a portion of a long uniform elastic bar. Young's modulus is E and the density of the bar is ρ. The cross-sectional area is A, which is also constant. The co-ordinate x is the location of a given cross-section in the quiescent state. A small movement of the particles at this location is designated as u and this movement is assumed to be constant across the cross-section, that is plane sections remain plane. In this system x is in effect the name of the group of particles at a given cross-section and u is their displace-

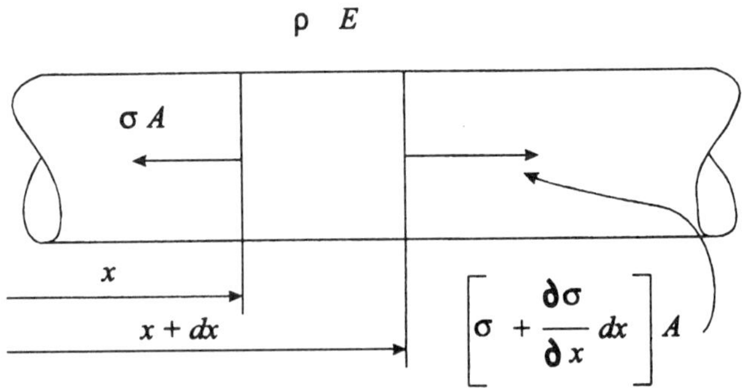

Fig. 6.6

ment. This scheme is known as *Lagrangian co-ordinates* as opposed to *Eulerian co-ordinates* where x is a fixed location in space (*see* Harrison and Nettleton (1994) or any book on fluid dynamics).

The mass of the element is $\rho A\,dx$ and this is a constant even if A and ρ vary with stress because this is the mass between two marks on the bar. Measurements made on the bar would usually be made using strain gauges or accelerometers which are attached to the bar and move with it. This is in contrast to most measurements in fluids where the measuring device, such as a pressure sensor, would be attached to the vessel containing the fluid.

The force acting on a cross-section is the product of the stress σ and the original cross-sectional area A. Equating the resultant force to rate of change of momentum gives

$$\left(\sigma + \frac{\partial \sigma}{\partial x}\,dx\right) A - \sigma A = \frac{\partial}{\partial t}\left(\rho A\,dx \frac{\partial u}{\partial t}\right)$$

or

$$\frac{\partial \sigma}{\partial x} = \rho \frac{\partial^2 u}{\partial t^2} \tag{6.11}$$

By definition the strain, ε, is the change in length per unit length. Thus

$$\varepsilon = \frac{(u + \partial u/\partial x dx) - u}{dx} = \frac{\partial u}{\partial x} \tag{6.12}$$

Hooke's law gives us

$$\sigma = E\varepsilon \tag{6.13}$$

Substituting equations (6.13) and (6.12) into equation (6.11) leads to

$$E \frac{\partial^2 u}{\partial x^2} = \rho \frac{\partial^2 u}{\partial t^2}$$

or

$$c^2 \frac{\partial^2 u}{\partial x^2} = \frac{\partial^2 u}{\partial t^2} \tag{6.14}$$

where $c = \sqrt{(E/\rho)}$. (For steel and aluminium this wave speed is of the order of 5 000 m/s or 5 mm/μs.)

Consider first the case in which a wave is moving in the positive x direction only. Here

$$u = f(ct - x)$$

The choice of $\beta = 0$ signifies that the head of the wave is at the origin when $t = 0$. The strain is

$$\varepsilon = \frac{\partial u}{\partial x} = -f' \tag{6.15a}$$

and the particle velocity

$$v = \frac{\partial u}{\partial t} = cf' \tag{6.15b}$$

From equations (6.15a) and (6.15b) we get

$$\varepsilon = -\frac{v}{c} \tag{6.16}$$

Also

$$\sigma = E\varepsilon = -\frac{Ev}{c}$$

From equation (6.14) $E = \rho c^2$ and therefore

$$\sigma = -(\rho c)v \tag{6.17}$$

The quantity ρc is called the *characteristic impedance*, Z_c, of the material. Thus

$$Z_c = \rho c = E/c = \sqrt{(E\rho)} \tag{6.18}$$

Notice that for a wave which is moving only in one direction there is a direct proportionality between stress and velocity, not acceleration. From equation (6.11) it is seen that acceleration is related to the spatial rate of change of stress. Therefore an accelerometer used in impact situations may exhibit very high values of acceleration but it is the integral of acceleration which is related to the stress in the bar. The relationship between acceleration and stress is only relevant when a body is behaving as a rigid body. This implies that the change in stress is small in the time taken for a wave to traverse the body and return. This point will be explored later. Even in cases of steady vibration it can be shown that maximum stress is related to maximum velocity, though not necessarily at the same location. In earthquake engineering a pseudo velocity is used to assess damage.

6.4 Reflection and transmission at a boundary

A boundary is a position where there is a sudden change in the material characteristics and this may be associated with a small change in the cross-sectional area. A large change in area will make the assumption of plane waves less acceptable. Figure 6.7 shows a change of properties at $x = 0$. The incident wave is $u_i = f(c_1 t - x)$, the reflected wave is $u_r = g(c_1 t + x)$ and the transmitted wave is $u_t = F(c_2 t - x)$. At $x = 0$ there must be continuity of velocity and there must also be continuity of force. Thus for continuity of velocity

$$\frac{\partial u_i}{\partial t} + \frac{\partial u_r}{\partial t} = \frac{\partial u_t}{\partial t}$$

$$c_1 f' + c_1 g' = c_2 F' \tag{6.19}$$

and for force

$$(EA)_1 \frac{\partial u_i}{\partial x} + (EA)_1 \frac{\partial u_r}{\partial x} = (EA)_2 \frac{\partial u_t}{\partial x}$$

$$-(EA)_1 f' + (EA)_1 g' = (EA)_2 F' \tag{6.20}$$

Eliminating F' from equations (6.19) and (6.20)

$$f'\left(\frac{c_1}{c_2} - \frac{(EA)_1}{(EA)_2}\right) + g'\left(\frac{c_1}{c_2} + \frac{(EA)_1}{(EA)_2}\right) = 0 \tag{6.21}$$

Impedance Z is defined to be the ratio of the force acting on a surface to the velocity of that surface in the direction of the force. Therefore

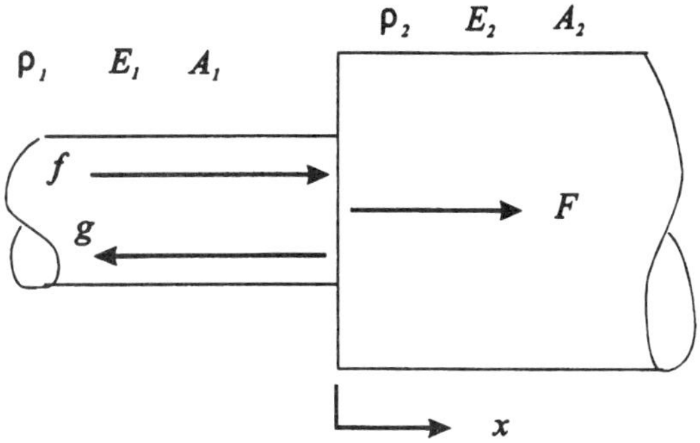

Fig. 6.7

$$Z = -E\varepsilon A/v = -\ EAf'\ (-1)/(cf') = EA/c = Z_c A \tag{6.22}$$

Multiplying equation (6.21) by c_2/c_1 and introducing the impedance gives

$$f'\left(1 - \frac{Z_1}{Z_2}\right) + g'\left(1 + \frac{Z_1}{Z_2}\right) = 0$$

or

$$g' = -f'\left(\frac{Z_2 - Z_1}{Z_2 + Z_1}\right) \tag{6.23}$$

From equation (6.19)

$$F' = \frac{c_1}{c_2}(f' + g')$$

and substituting from equation (6.23)

$$F' = \frac{c_1}{c_2} f'\left(\frac{2Z_1}{Z_2 + Z_1}\right) \tag{6.24}$$

From the above equations

$$\frac{\text{reflected strain}}{\text{incident strain}} = \frac{-g'}{-f'} = (Z_2 - Z_1)/(Z_2 + Z_1) \tag{6.25a}$$

$$\frac{\text{reflected velocity}}{\text{incident velocity}} = \frac{c_1 g'}{c_1 f'} = -(Z_2 - Z_1)/(Z_2 + Z_1) \tag{6.25b}$$

$$\frac{\text{transmitted force}}{\text{incident force}} = \frac{(EA)_2 F'}{(EA)_1 f'} = \frac{Z_2}{Z_1} 2Z_1/(Z_2 + Z_1) \tag{6.25c}$$

$$\frac{\text{transmitted velocity}}{\text{incident velocity}} = \frac{c_2 F'}{c_1 f'} = 2Z_1/(Z_2 + Z_1) \tag{6.25d}$$

For the case of a free end, that is $Z_2 = 0$, we see that the reflected strain is of the opposite sign thereby making the strain at the end zero as would be expected for a free end. The related velocity in this case is of the same sign thus doubling the velocity at the free end. A true fixed end is not possible but if $Z_2 \gg Z_1$ then the velocity at the 'fixed' end would be zero whilst the strain would be doubled. For the transmitted pulse both the strain and the velocity will be of the same sign as those of the incident pulse.

6.5 Momentum and energy in a pulse

Consider a pulse, shown in Fig. 6.8, such that the displacement is $u = f(ct - x)$ and is non-zero only between $z = 0$ and $z = L$. The momentum carried by the pulse is

$$G = \int_0^L \rho A v \, d(x) = \rho A L c \int_0^1 f' \, d(x/L) \tag{6.26}$$

The kinetic energy is

$$T = \int_0^L \frac{\rho}{2} A v^2 \, dx = \frac{\rho A L}{2} c^2 \int_0^1 (f')^2 \, d(x/L)$$

and the strain energy is

$$V = \int_0^L \frac{1}{2} E A \varepsilon^2 dx = \frac{EAL}{2} \int_0^1 (f')^2 \, d(x/L)$$

Now since $c^2 = E/\rho$ the expression for kinetic energy is identical to that for strain energy, so the energy is equally partitioned. Therefore we may write the total energy as

$$\mathfrak{E} = \rho A L c^2 \int_0^1 (f')^2 \, d\left(\frac{x}{L}\right) \tag{6.27}$$

Returning to the previous section

$$\frac{\text{transmitted energy}}{\text{incident energy}} = \frac{(\rho A L c^2)_2}{(\rho A L c^2)_1} \left(\frac{c_1}{c_2} \frac{2Z_1}{(Z_1 + Z_2)} \right)^2$$

The length of the pulse will change as it passes from one region to the other but the duration will remain constant, so

$$L_1 / c_1 = L_2 / c_2$$

or

$$L_2 / L_1 = c_2 / c_1$$

Noting that $Z_1 = \rho_1 A_1 c_1$ and $Z_2 = \rho_2 A_2 c_2$, then

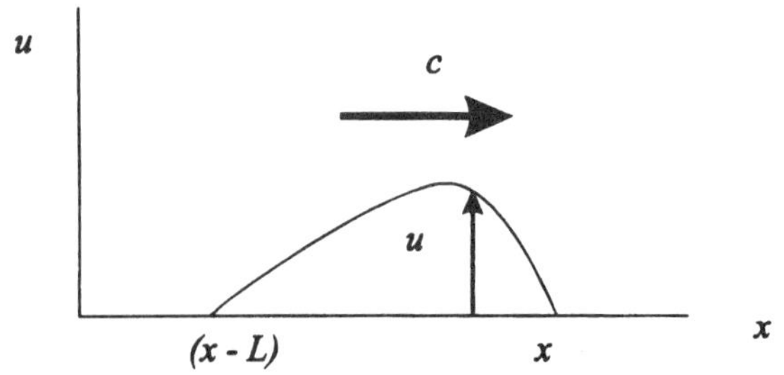

Fig. 6.8

$$\frac{\text{transmitted energy}}{\text{incident energy}} = \frac{Z_2 c_2^2}{Z_1 c_1^2}\left(\frac{c_1}{c_2}\frac{2Z_1}{(Z_1 + Z_2)}\right)^2$$

$$= \frac{4Z_2 Z_1}{(Z_1 + Z_2)^2} \tag{6.28}$$

Note that this ratio varies between 0 and 1 and is symmetrical, so the amount of transmitted energy does not depend on the direction of travel.

6.6 Impact of two bars

Two bars are shown in Fig. 6.9. The first bar has a finite length L and the second is long, such that the reflected wave from its far end will arrive after the impact has ceased. It is assumed that the impact occurs over a plane surface; in the next section we shall investigate the effect of a spherical contact surface.

The bars collide with an approach velocity of V. It is assumed that both bars are stress free before impact and the long bar is stationary. At impact a wave g_0 moves to the left in the short bar and a wave F_0 moves to the right in the long bar. The wave reaches the end of the short bar and a reflected wave f_1 is generated so that the strain at that end is zero. When this wave returns to the impacted end a new set of waves are generated. This process is shown diagrammatically in Fig. 6.10.

In general

$$f_n = f_n(c_1 t - x - n2L)$$
$$g_n = g_n(c_1 t + x - n2L)$$
$$F_n = F_n(c_2 t - x - n2Lc_2/c_1)$$

with $f_0 = 0$.

The arguments can be verified by inspection of the diagrams. The constant is the apparent position of the head of the wave at $t = 0$. Alternatively it gives the time when the waves originate either at $x = 0$ (for g_n or F_n) or $x = -L$ (for f_n).

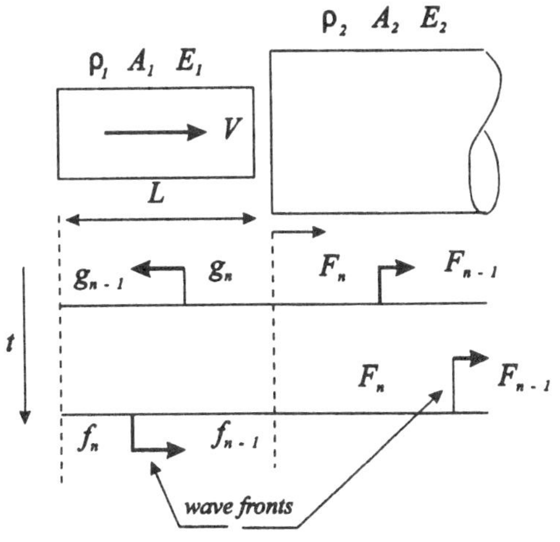

Fig. 6.9

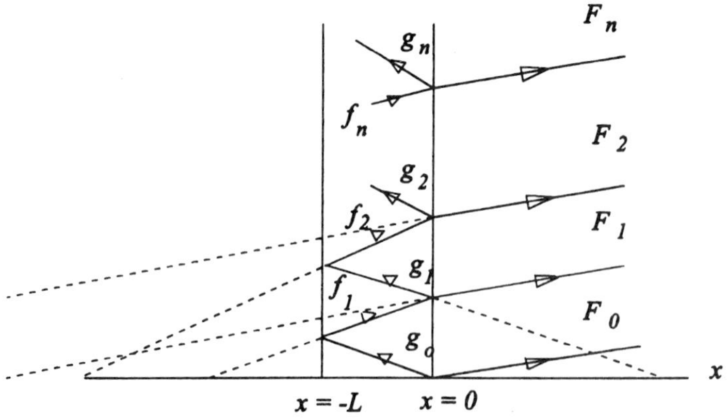

Fig. 6.10

At $x = -L$ the strain is always zero and hence

$$g'_{n-1}\ (c_1 t + (-L) - (n - 1)\,2L) - f'_n\,(c_1 t - (-L) - n2L) = 0$$

or

$$f'_n = g'_{n-1} \tag{6.29}$$

At $x = 0$ there must be continuity of velocity. For the short bar the particle velocity is superimposed on the pre-impact speed of V. Thus

$$V + c_1 f'_n + c_1 g'_n = c_2 F'_n \tag{6.30}$$

and the contact force is

$$c_1 Z_1 f'_n - c_1 Z_1 g'_n = -c_2 Z_2 F'_n \tag{6.31}$$

(note that Z = force/velocity).

From equations (6.29), (6.30) and (6.31) we obtain

$$g'_n = \frac{2V/c_1 - (1 - Z_1/Z_2)g'_{n-1}}{(1 + Z_1/Z_2)} \tag{6.32}$$

and

$$F'_n = \frac{c_1 Z_1}{c_2 Z_2}\left(\frac{V/c_1 + 2g'_{n-1}}{(1 + Z_1/Z_2)}\right) \tag{6.33}$$

Since the first waves are g_0 and F_0 it follows that $f_0 = 0$, $g_{-1} = 0$ and $F_{-1} = 0$.

Let us first examine the waves immediately after the impact, that is for $n = 0$

$$g'_0 = \frac{-V/c_1}{(1 + Z_1/Z_2)} \tag{6.34}$$

$$F'_0 = \frac{c_1 Z_1}{c_2 Z_2}\ \frac{V/c_1}{(1 + Z_1/Z_2)} \tag{6.35}$$

and for $n = 1$

$$F_1' = \frac{V}{c_2} \frac{Z_1/Z_2}{(1 + Z_1/Z_2)} \frac{(Z_1/Z_2 - 1)}{(Z_1/Z_2 + 1)} \tag{6.36}$$

$$f_1' = g_0' = \frac{-2V/c_1}{(1 + Z_1/Z_2)} \tag{6.37}$$

If Z_1 is less than or equal to Z_2 then F_1' is zero or negative. This means that the strain is zero or positive, that is tensile. Because a tensile strain is not possible at the interface the contact is terminated, the contact time being $2L/c_1$.

The velocity at the interface is

$$v = V + c_1 g_0' = c_2 F_0'$$
$$= V \frac{Z_1/Z_2}{(1 + Z_1/Z_2)} \tag{6.38}$$

In the special case when $Z_1 = Z_2$, $v = V/2$. Figure 6.11 shows the progress of the wave.

If Z_1 is greater than Z_2 then equations (6.32) and (6.33) can be used repeatedly to determine the wave functions. With a little algebra it can be shown that

$$g_n' = \frac{V}{2c}\left[1 - \left(\frac{Z_1 - Z_2}{Z_1 + Z_2}\right)^{n+1}\right] \tag{6.39}$$

and

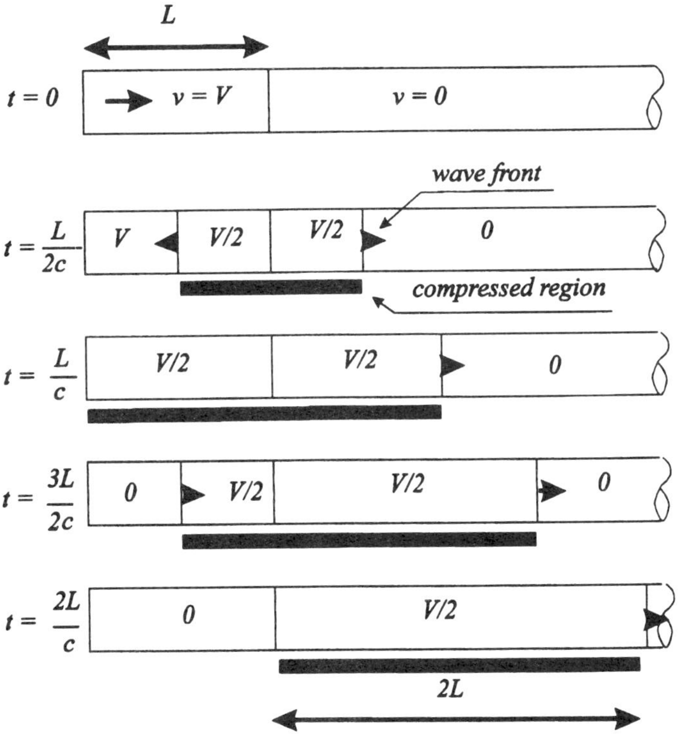

Fig. 6.11

$$F'_n = \frac{Z_1 V/c_2}{Z_1 + Z_2}\left(\frac{Z_1 - Z_2}{Z_1 + Z_2}\right)^n \tag{6.40}$$

so that the force transmitted is

$$(EA)_2 F'_n = \frac{Z_1 V Z_2}{Z_1 + Z_2}\left(\frac{Z_I - Z_2}{Z_1 + Z_2}\right)^n \tag{6.41}$$

from which we see that force decays exponentially. Also the velocities decay to zero so the coefficient of restitution, defined in the usual rigid body way, is zero. In the case where the two bars have the same properties and the second bar is the same length as the first the coefficient of restitution is unity, showing that this quantity can range from 0 to 1 even though the process is elastic.

6.7 Constant force applied to a long bar

We shall now consider a long bar under the action of a constant force X applied to the face at $x = 0$ as shown in Fig. 6.12. If we assume that a wave travels into the bar with a speed c then we may use

force = rate of change of momentum

$$X = \frac{\mathrm{d}}{\mathrm{d}t}(\rho A v\,(ct)) = \rho A v c$$

so

$$-\varepsilon = X/(AE) = \rho v c/E$$

By definition

$$-\varepsilon = vt\,/\,(ct) = v/c$$

Equating the two expressions for ε gives

$$\rho v c/E = v/c$$

or $c^2 = E/\rho$ as before.

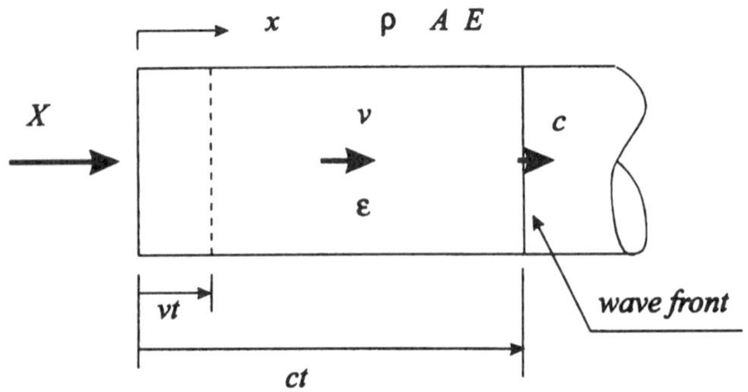

Fig. 6.12

Now let the bar be of finite length L, as shown in Fig. 6.13. At $x = L$ the strain has to be zero. Therefore at any time

$$\varepsilon = -f'_n + g'_{n-1} = 0$$

or

$$g'_{n-1} = f'_n$$

At $x = 0$ the force, X, is constant and therefore

$$X = -EA\ (-f'_n + g'_n)$$
$$= EA\ (f'_n - f'_{n-1}) \tag{6.42}$$

and

$$v = c\ (f'_n + g'_n)$$
$$= c\ (f'_n + f'_{n-1}) \tag{6.43}$$

From equation (6.42)

$$f'_n = \frac{X}{EA} + f'_{n-1}$$

Thus

$$f'_0 = \frac{X}{EA}$$

$$f'_1 = \frac{X}{EA} + f'_0 = \frac{2X}{EA}$$

Hence

$$f'_n = \frac{(n+1)X}{EA}$$

Substituting into equation (6.43)

$$v = c\left(\frac{(n+1)X}{EA} + \frac{nX}{EA}\right)$$

$$= \frac{cX}{EA}(2n+1)$$

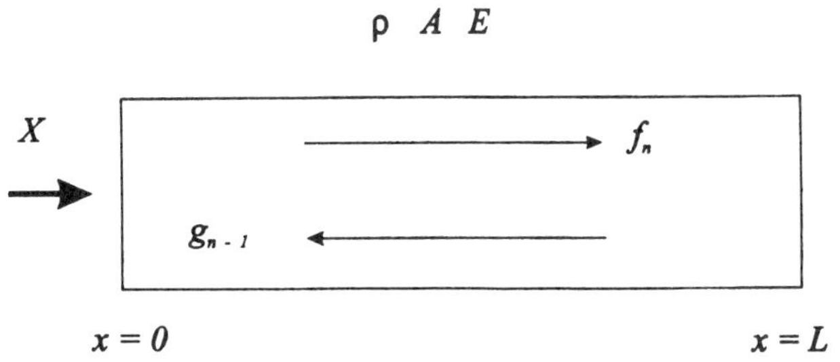

Fig. 6.13

now time $t = n2L/c$ so the average acceleration is

$$\frac{v}{t} = \frac{cX}{EA}(2n+1)\frac{c}{2nl}$$

$$= \frac{X}{\rho AL}(1 + 1/2n)$$

As n tends to infinity

$$\frac{v}{t} = \frac{X}{\rho AL}$$

So we see that the result is that which would have been given by elementary means. From this we learn the very important lesson that rigid body behaviour may be assumed when the variation of force is small compared with the time taken for the wave to traverse the body and return. After a few reflections the body behaves like a body with vibratory modes superimposed on the rigid body modes.

The wave method is most suitable when dealing with the initial stages which, in the case of impacting solids, may well be when the maximum strains occur. As mentioned earlier a vibration approach will require a large number of principal modes to be included.

6.8 The effect of local deformation on pulse shape

In the previous analysis for which impact occurred between plane surfaces it is seen that the leading edge is sharp leading to instantaneous changes in strain and velocity. Although these are not precluded in continuum mechanics, in practice some rounding of the leading edge occurs largely due to the impacting surfaces not being plane. We shall assume that in the immediate vicinity of the impact point the material behaves as an elastic spring with linear or non-linear characteristics.

Referring to Fig. 6.14 we see that the impacting surfaces are convex and the separation of the two reference planes is denoted by $(s_0 - \alpha)$, α being the compression. It is assumed that the compressive force deflection law is of the form $X = k\alpha^m$.

The rate of approach of the two reference planes is

$$\dot{\alpha} = V + c_1 g' - c_1 f' \tag{6.44}$$

and the contact force

$$X = -(EA)_1 g' = +(EA)_2 f' \tag{6.45}$$

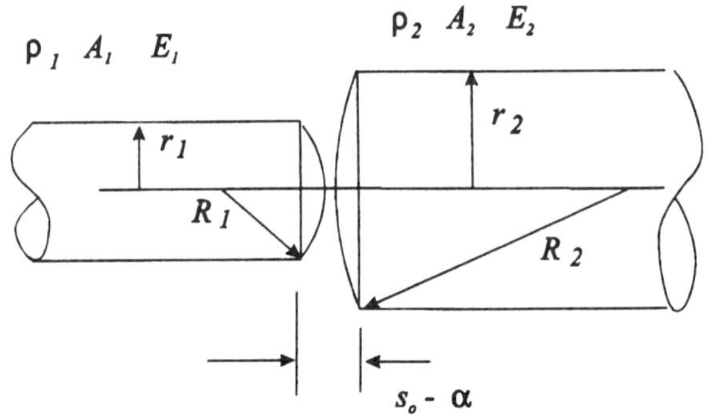

Fig. 6.14

Eliminating g' and f' we get

$$\dot{\alpha} = V - \frac{Xc_1}{(EA)_1} - \frac{Xc_2}{(EA)_2} = V - k\alpha^m(1/Z_1 + 1/Z_2) \tag{6.46}$$

Let $\lambda = k(1/Z_1 + 1/Z_2)$ so that equation (6.46) becomes

$$\dot{\alpha} + \lambda\alpha^m = V \tag{6.47}$$

If $m = 1$ then the interface behaves like a linear spring and the solution is, with $\alpha = 0$ at $t = 0$,

$$\alpha = \frac{V}{\lambda}(1 - e^{-\lambda t})$$

and

$$X = k\alpha$$

$$= V\frac{Z_1 Z_2}{Z_1 + Z_2}(1 - e^{\lambda t}) \tag{6.48}$$

from which we see that the maximum force is as given by equation (6.41) with $n = 0$.

The Hertz theory of contact for two hemispherical bodies in contact states that

$$\alpha = X^{2/3}\left[\frac{9\pi}{16}(\mu_1 + \mu_2)^2\left(\frac{1}{R_1} + \frac{1}{R_2}\right)\right]^{1/3} \tag{6.49}$$

where R is the radius and $\mu = (1 - \upsilon)/(\pi E)$. ($\upsilon$ = Poisson's ratio). We may write

$$X = k\alpha^{3/2}$$

where

$$k = \left[\frac{3\pi}{4}(\mu_1 + \mu_2)\sqrt{\left(\frac{1}{R_1} + \frac{1}{R_2}\right)}\right]^{-1}$$

Equation (6.47) now becomes

$$\dot{\alpha} + \lambda\alpha^{3/2} = V \tag{6.50}$$

or

$$\int_0^{\alpha}\frac{d\alpha}{V - \lambda\alpha^{3/2}} = \int_0^t dt$$

Using the substitution

$$\beta^3 = \frac{\lambda\alpha^{3/2}}{V}$$

leads eventually to

$$t = \frac{2}{3}\frac{1}{V}\left(\frac{V}{\lambda}\right)^{2/3}\left[\frac{1}{2}\ln\left(\frac{\beta^2 + \beta + 1}{(1 - \beta)^2}\right) - \sqrt{3}\arctan\left(\frac{2\beta + 1}{\sqrt{3}}\right) + \frac{\pi\sqrt{3}}{6}\right] \tag{6.51}$$

Now

$$X = k\alpha^{3/2} = \frac{kV}{\lambda}\beta^3$$

and because as $\beta \to \infty$, $t \to 1$,

$$X_{\max} = kV/\lambda = \frac{V}{(1/Z_1 + 1/Z_2)} \tag{6.52}$$

Thus

$$\frac{X}{X_{\max}} = \beta^3$$

Introducing a non-dimensional time

$$\tilde{t} = t\lambda^{2/3}V^{1/3} \tag{6.53}$$

leads to a plot of $X/X_{\max}$ versus $\tilde{t}$ being made. Figure 6.15 shows the plot. Equation (6.53) can be rearranged as

$$\tilde{t} = \left(\frac{2\sqrt{2}}{3\pi(1-\upsilon^2)}\right)^{2/3}\left(\frac{VR}{rc}\right)^{1/3}\left(\frac{ct}{r}\right) \tag{6.54}$$

and taking $\upsilon = 0.3$ the constant evaluates to 0.478. Note that from equation (6.52) $X_{\max}$ is proportional to V.

Also shown on Fig. 6.15 is a plot of

$$X/X_{\max} = (1 - e^{-\tilde{t}}) \tag{6.55}$$

and this shows a reasonably close resemblance to the plot of (6.54). Equation (6.55) is of the same form as equation (6.48) which was obtained from the linear spring model. Thus by equating the exponents an equivalent linear spring may be obtained.

Therefore

$$\lambda_1 t = \tilde{t} = \lambda^{2/3}V^{1/3}t \tag{6.56}$$

or

$$k_1(1/Z_1 + 1/Z_2) = (k(1/Z_1 + 1/Z_2))^{2/3}\, V^{1/3} \tag{6.57}$$

where k_1 and λ_1 refer to the linear spring model.

Figure 6.16 shows a plot of the rise time to three different fractions of the maximum versus the product of impact velocity and nose radius. It is seen that as the nose radius tends to infinity the rise time tends to zero as was predicted for a plane-ended impact. Also as the impact velocity (or the maximum force) increases then the rise time decreases.

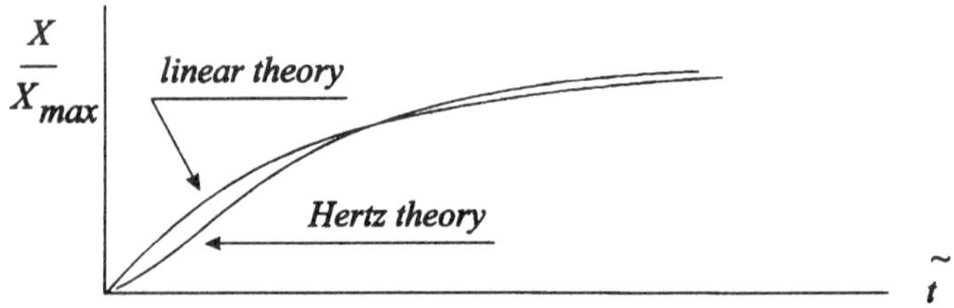

Fig. 6.15

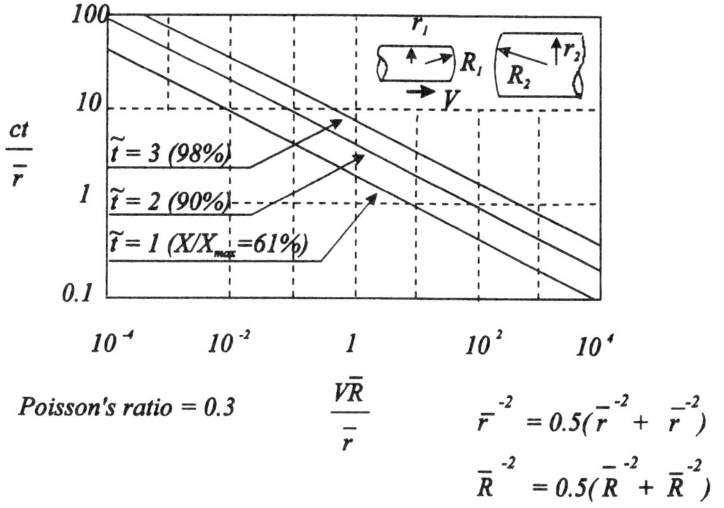

Fig. 6.16 Rise time based on Hertz theory of contact

6.9 Prediction of pulse shape during impact of two bars

We shall consider the impact of two bars having equal properties. One bar is of length L whilst the other is sufficiently long so that no reflection occurs in that bar during the time of contact. If we assume a plane-ended impact then the contact will cease after the wave has returned from the far end of the short bar, that is the duration of impact is $2L/c$.

Because the rise time, in practice, is finite several reflections will occur before the contact force reduces to zero and remains zero in the long bar. The leading edge profile has been predicted in the previous section using the Hertz theory of contact where it was also shown that this could be approximately represented by an exponential expression. To simplify the computation we shall adopt the exponential form.

Figure 6.17 shows the x,t diagram (which is similar to Fig. 6.10).

At $x = -L$, $\varepsilon = 0$ and thus

$$f'_n - g'_{n-1} = 0$$

or

$$f'_n = g'_{n-1} \quad (g'_{-1} = 0) \tag{6.58}$$

At $x = 0$ the difference in the velocity of the reference faces is $\dot{\alpha}$,

$$(V + cf'_n + cg'_n) - cF'_n = \dot{\alpha}_n$$

or

$$V/c + f'_n + g'_n - F'_n = \dot{\alpha}_n/c \tag{6.59}$$

Also, by continuity of force,

$$-(EAg'_n - EAf'_n) = -(-EAF'_n) = k\alpha_n$$

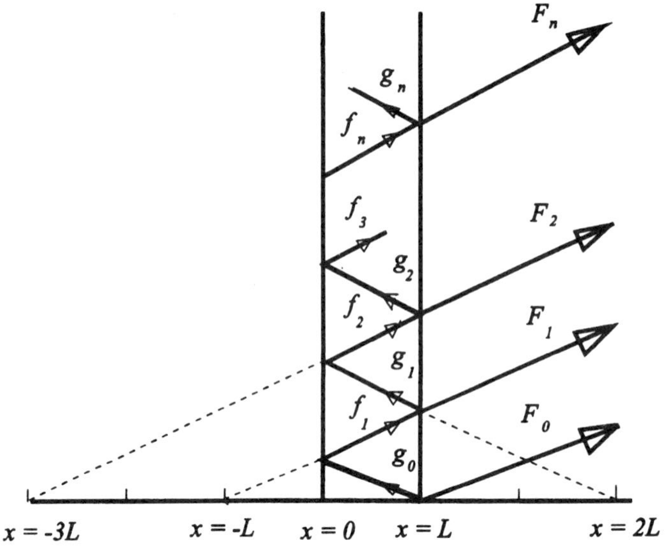

Fig. 6.17

or

$$f'_n - g'_n = \frac{k}{EA}\alpha_n \tag{6.60}$$

and

$$F'_n = \frac{k}{EA}\alpha_n \tag{6.61}$$

Adding equations (6.59), (6.60) and (6.61) gives

$$\frac{V}{c} + 2f'_n = \frac{\dot{\alpha}_n}{c} + \frac{2k}{EA}\alpha_n \tag{6.62}$$

From equations (6.58) and (6.60)

$$f'_n = g'_{n-1} = f'_{n-1} - \frac{k}{EA}\alpha_{n-1}$$

As $f'_0 = 0$

$$f'_1 = -\frac{k}{EA}\alpha_0$$

and

$$f'_2 = f'_1 - \frac{k}{EA}\alpha_1 = -\frac{k}{EA}(\alpha_0 + \alpha_1)$$

so

$$f'_n = -\frac{k}{EA}\sum_0^{n-1}\alpha_i \tag{6.63}$$

Substituting equation (6.63) into equation (6.62) gives

$$\frac{V}{c} - \frac{2k}{EA}\sum_{0}^{n-1}\alpha_i = \frac{1}{c}\frac{d\alpha_n}{dt} + \frac{2k}{EA}\alpha_n \tag{6.64}$$

We define the non-dimensional quantities

$$\tilde{\alpha} = \alpha\,\frac{2kc}{EAV} \tag{6.65}$$

and

$$\tilde{t} = t\,\frac{2kc}{EA} \tag{6.66}$$

Thereby equation (6.64) can be written as

$$1 + \sum_{0}^{n-1}\tilde{\alpha}_i = \frac{d\tilde{\alpha}_n}{d\tilde{t}} + \tilde{\alpha}_n \tag{6.67}$$

As $\tilde{\alpha}$ has to be continuous

$$\tilde{\alpha}_n(0) = \tilde{\alpha}_{n-1}(p) \tag{6.68}$$

The parameter p is the value of $\tilde{t}$ when $ct = 2L$, that is the time at which the wave in the short bar returns to the impact point.
From equation (6.66)

$$p = \frac{4Lk}{EA} \tag{6.69}$$

Multiplying equation (6.67) by $e^{\tilde{t}}$ gives

$$e^{\tilde{t}}\left(1 + \sum_{0}^{n-1}\tilde{\alpha}_i\right) = e^{\tilde{t}}\left(\frac{d\tilde{\alpha}_n}{d\tilde{t}} + \tilde{\alpha}_n\right) = \frac{d}{d\tilde{t}}\left(e^{\tilde{t}}\,\tilde{\alpha}_n\right)$$

and integrating produces

$$\tilde{\alpha}_n = e^{-\tilde{t}}\;[\int e^{\tilde{t}}\,(1 - \sum_{0}^{n-1}\tilde{\alpha}_i)\,d\tilde{t} + \text{constant}] \tag{6.70}$$

Carrying out the integration

$$\tilde{\alpha}_0 = e^{-\tilde{t}}\,[\int e^{\tilde{t}}\,(1 - 0)\,d\tilde{t} + \text{constant}]$$
$$= e^{-\tilde{t}}\,[e^{\tilde{t}} - 1] = (1 - e^{-\tilde{t}}) \tag{6.71}$$

(the constant = 1 as $\tilde{\alpha}_0 = 0$ when $\tilde{t} = 0$)

$$\tilde{\alpha}_1 = e^{-\tilde{t}}\;[\int e^{\tilde{t}}\,[1 - (1 - e^{-\tilde{t}})]\,d\tilde{t} + \text{constant}]$$
$$= e^{-\tilde{t}}\,[\tilde{t} + \tilde{\alpha}_0(p)] \tag{6.72}$$

the constant being determined by the fact that $\tilde{\alpha}_1(0) = \tilde{\alpha}_0(p)$. Continuing the process

$$\tilde{\alpha}_2 = e^{-\tilde{t}}\,[\int e^{\tilde{t}}\,\{1 - e^{-\tilde{t}}\,[e^{\tilde{t}} - 1 + \tilde{t} + \tilde{\alpha}_0(p)]\}\,d\tilde{t} + \text{constant}]$$
$$= e^{-\tilde{t}}\,[e^{\tilde{t}} - e^{-\tilde{t}} + \tilde{t} - \frac{\tilde{t}^2}{2} - \tilde{t} - \tilde{\alpha}(p) + \text{constant}]$$

$$= e^{-\tilde{t}}\left[-\frac{\tilde{t}^2}{2} + \tilde{t}\,[1 - \tilde{\alpha}_0(p)] + \tilde{\alpha}_1(p)]\right] \quad (6.73)$$

In the same way the next two functions may obtained; they are

$$\tilde{\alpha}_3 = e^{-\tilde{t}}\left[\frac{\tilde{t}^3}{6} - \frac{\tilde{t}^2}{2}[2 - \tilde{\alpha}_0(p)] + \tilde{t}\,[1 - \tilde{\alpha}_0(p) - \tilde{\alpha}_1(p)] + \tilde{\alpha}_2(p)\right] \quad (6.74)$$

and

$$\tilde{\alpha}_4 = e^{-\tilde{t}}\left[-\frac{\tilde{t}^4}{24} + \frac{\tilde{t}^3}{6}[3 - \tilde{\alpha}_0(p)] + \frac{\tilde{t}^2}{2}[3 - 2\tilde{\alpha}_0(p) - \tilde{\alpha}_1(p)]\right.$$

$$\left. + \tilde{t}\,[1 - \tilde{\alpha}_0(p) - \tilde{\alpha}_1(p)] - \tilde{\alpha}_2(p) + \tilde{\alpha}_3(p)\right] \quad (6.75)$$

Figure 6.18 shows the results of the above analysis. This figure should be compared with Fig. 6.19(a) and (b) which are copies of actual measurements made on a Hopkinson bar.

The *Hopkinson bar* is a similar arrangement to that described above. In the above case the contact force can be deduced either by measuring the strain (ε) by means of strain gauges

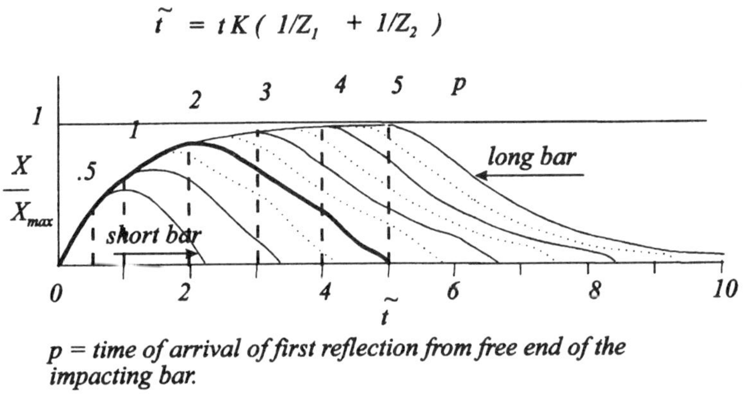

Fig. 6.18 Pulse shapes for varying lengths of impacting bar

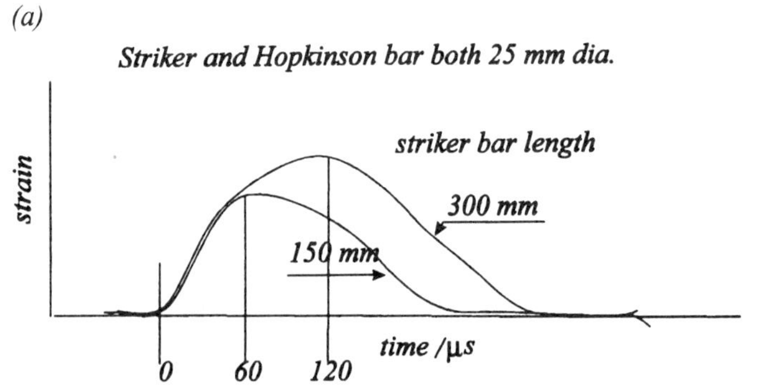

Fig. 6.19(a) Measured pulse shapes showing variation with bar length

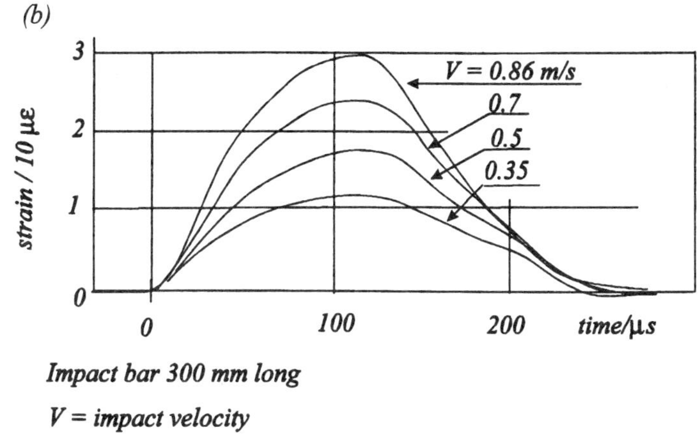

Fig. 6.19(b) Measured pulse shapes showing variation with V

sited away from the reflecting surfaces of the long bar, or by measuring the acceleration at the far end of the long bar and integrating to obtain the velocity (v). The contact force is given either by $X = -\varepsilon EA$ or by $X = (v/c)EA$.

By measurements on the impacting bar it is possible to deduce the compression across the contact region and thereby obtain the dynamic characteristics of that region or of a specimen of other material cemented there. This is particularly useful for tests at high strain rate where the bars can be sufficiently long for the reflected waves to arrive after the period of interest.

6.10 Impact of a rigid mass on an elastic bar

In this example the impacting body is assumed to be short compared with the bar but of comparable mass. The reflections of the strain waves in the body are assumed to be of such short duration that a rigid body approximation is practicable. Figure 6.20 shows the relevant details; for this exercise the far end of the bar is taken to be fixed.

At the far end the particle velocity is zero and thus when $x = L$

$$v = cf'_n + cg'_{n+1} = 0$$

or

$$g'_n = f'_{n-1} \tag{6.76}$$

At $x = 0$ the contact force is

$$X = -(-EAf'_n + EAg'_n) = -M\frac{\partial^2 u}{\partial t^2} = -M(c^2 f''_n + c^2 g''_n)$$

Let

$$\frac{EA}{Mc^2} = \frac{\rho AL}{ML} = \frac{\mu}{L}$$

Therefore

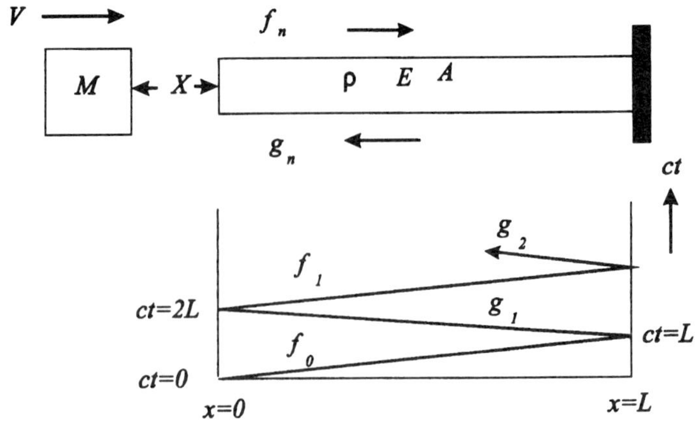

Fig. 6.20

$$f''_n + \frac{\mu}{L} f'_n = f''_{n-1} - \frac{\mu}{L} f'_{n-1}$$

Multiplying by $e^{\mu z/L}$ we get

$$\frac{d}{dz}\left(e^{\mu z/L} f'_n \right) = e^{\mu z/L} \left(- f''_{n-1} - \frac{\mu}{L} f'_{n-1} \right)$$

Integrating gives

$$f'_n = e^{-\mu z/L} \left[\int e^{\mu z/L} \left(f''_{n-1} - \frac{\mu}{L} f'_{n-1} \right) dz + B_n \right] \tag{6.77}$$

where B_n is a constant of integration.
As $f_{-1} = 0$ the first function is

$$f'_0 = e^{-\mu z/L} \, (0 + B_0)$$

When $z = 0$, $v = V$ and therefore

$$v = c f'_0 \, (0) = c e^0 \, (B_0) = V$$

Thus $B_0 = V/c$ and

$$f'_0 = \frac{V}{c} \, e^{-\mu z/L} \tag{6.78}$$

This function is valid until the wave returns from the far end, that is, when $z = 2L$ or $t = 2L/c$ at $x = 0$.

For $t > 2L/c$

$$f'_1 = e^{-\mu z/L} \left[\int e^{\mu z/L} \left(-\frac{\mu V}{Lc} e^{-\mu z/L} - \frac{\mu V}{Lc} e^{-\mu z/L} \right) dz + B_1 \right]$$

$$= e^{-\mu z/L} \left[- \frac{2\mu V}{Lc} z + B_1 \right]$$

Now the velocity must be continuous at $x = 0$ so that

$$cf'_1(0) + cg'_1(0) = cf'_0(2L) + cg'_0(2L)$$

Using equation (6.76) we obtain

$$f'_1(0) = f'_0(2L) + f'_0(0) - f'_{-1}(2L)$$

Now $f'_n(0) = B_1$ and $f'_{-1} = 0$. Therefore

$$B_1 = \frac{V}{c} e^{-2\mu} + \frac{V}{c} \tag{6.79}$$

and hence

$$f'_1 = \frac{V}{c} e^{-\mu z/L} \left(- \frac{2\mu}{L} z + e^{-2\mu} + 1 \right) \tag{6.80}$$

The same procedure can be employed to generate the subsequent functions. The results for the next two are given below

$$f'_2 = \frac{V}{c} e^{-\mu z/L} \left[- \frac{\mu}{L} (4 + 2e^{-2\mu}) z + 2 \left(\frac{\mu}{L} \right)^2 z^2 + B_2 \right] \tag{6.81}$$

where

$$B_2 = 1 + e^{-2\mu} (1 - 4\mu) + e^{-4\mu}$$

and

$$f'_3 = \frac{V}{c} e^{-\mu z/L} \left[- \frac{2\mu}{L} [e^{-4\mu} + e^{-2\mu} (2 - 4\mu) + 3]z \right.$$

$$\left. + 2\left(\frac{\mu}{L} \right)^2 (3 + e^{-2\mu}) z^2 - \frac{4}{3} \left(\frac{\mu}{L} \right)^3 z^3 + B_3 \right] \tag{6.82}$$

where

$$B_3 = e^{-6\mu} + e^{-4\mu} (1 - 8\mu) + e^{-2\mu} (1 - 8\mu + 8\mu^2) + 1$$

At the impact point, $x = 0$, the velocity and the strain are given by

$$v_0 = cf'_n + cg'_n = c(f'_n - f'_{n-1}) \tag{6.83}$$

and

$$\varepsilon_0 = -f'_n + g'_n = -(f'_n + f'_{n-1}) \tag{6.84}$$

At the fixed end the velocity is zero but the strain is

$$\varepsilon_L = -f'_n + g'_{n+1} = -2f'_n \tag{6.85}$$

Figure 6.21 gives plots of these functions versus time. It is apparent from the graphs that the highest strains occur at the fixed end at the beginning of the periods, that is for $z = 0$ and $x = L$. Figure 6.22 gives a plot of the maxima versus $1/\mu$.

As the ratio of the impacting mass to that of the rod increases it is possible to use an approximate method to determine the maximum strain. It is well known from vibration theory that a first approximation in this type of problem is to add one-third of the mass of the rod to that of the end mass. For this case we equate the initial kinetic energy with the final strain energy

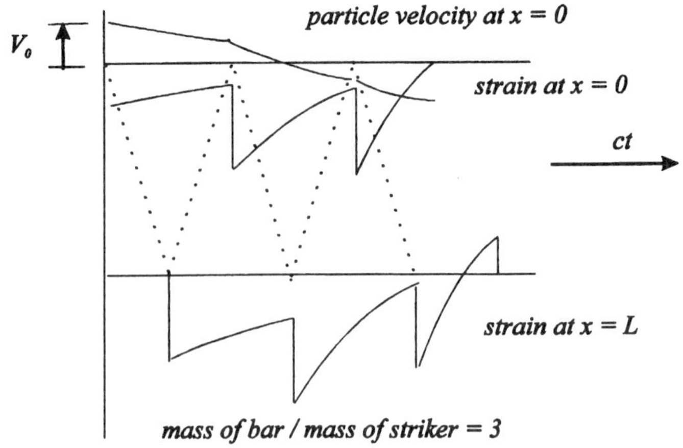

Fig. 6.21

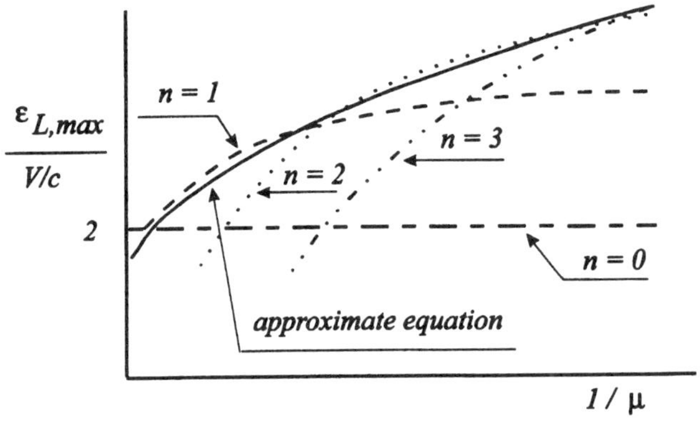

Fig. 6.22 Maximum strain at $x = L$ for various period numbers, n

$$\frac{1}{2}(M + \rho AL/3)\,V^2 = \frac{1}{2}\frac{X^2}{k} = \frac{1}{2}\frac{X^2 L}{EA}$$

$$= \frac{1}{2}EAL\varepsilon^2 = \frac{1}{2}c^2\rho AL\varepsilon^2$$

Hence

$$\varepsilon = \frac{V}{c}\left(\frac{M + \rho AL/3}{\rho AL}\right)^{1/2}$$

$$= \frac{V}{c}\left(\frac{1}{\mu} + \frac{1}{3}\right)^{1/2}$$

where $\mu = \rho AL/M$.

This is adequate for large values of $1/\mu$ but not for small. A better approximation can be made by adding the result obtained in the previous analysis, which was that the initial strain is V/c. Therefore

$$\varepsilon = \frac{V}{c}\left[1 + \left(\frac{1}{\mu} + \frac{1}{3}\right)^{1/2}\right] \tag{6.86}$$

A plot of this aproximation is included on Fig. 6.22.

6.11 Dispersive waves

Let us first discuss the sinusoidal travelling wave. The argument for a sinusoidal function is required to be non-dimensional so we shall adopt for a wave along the positive axis

$$\begin{aligned} z &= k(ct - x) \\ &= (ckt - kx) \end{aligned}$$

where k is a parameter with dimensions 1/length.

A typical wave would be

$$u = U\cos(ckt - kx) \tag{6.87}$$

Figure 6.23 shows a plot of u against t for $x = 0$. If the argument increases by 2π, the time is the periodic time T. Thus

$$ckT = 2\pi$$

or

$$ck = 2\pi/T = 2\pi\upsilon = \omega$$

where υ is the frequency and ω is the circular frequency.

Figure 6.24 is a similar plot but this time versus x. An increase of 2π in the argument corresponds to a change in x of one wavelength λ. Thus

$$k\lambda = 2\pi \tag{6.88}$$

or

$$k = 2\pi/\lambda \tag{6.89}$$

k is known as the *wavenumber*.

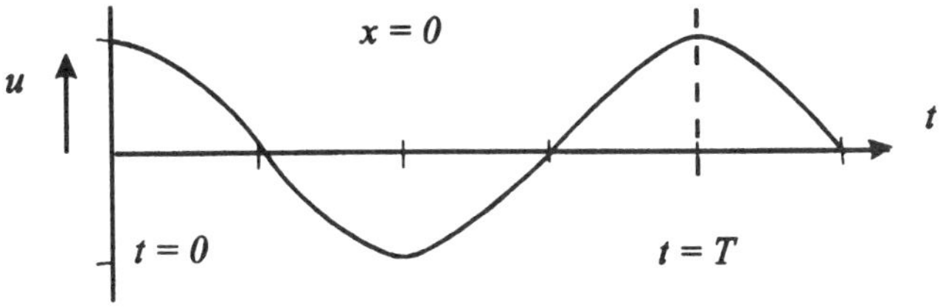

Fig. 6.23

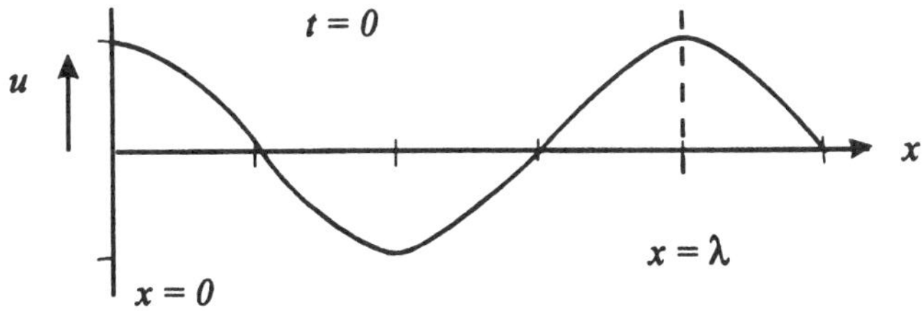

Fig. 6.24

Hence

$$z = ckt - kx = \omega t - kx \tag{6.90}$$

and

$$c = \frac{\omega}{k} = c_{\text{p}} \tag{6.91}$$

This quantity is called the *phase velocity* as it is the speed at which points of constant phase move through the medium. In a dispersive medium c_{p} varies with frequency (or wavelength) and it will be shown that this is not the speed at which energy is propagated.

Consider now two waves of equal amplitude moving to the right as shown in Fig. 6.25. Note that both arguments are zero for $t = 0$ and $x = 0$. The displacement is

$$u = U\cos(\omega_1 t - k_1 x) + U\cos(\omega_2 t - k_1 x)$$

and using the formula for the addition of two cosines

$$u = U\cos\left(\frac{\omega_1 + \omega_2}{2}t - \frac{k_1 + k_2}{2}x\right)\cos\left(\frac{\omega_1 - \omega_2}{2}t - \frac{k_1 - k_2}{2}x\right)$$

$$= U\cos(\omega_0 t - k_0 x)\cos\left(\frac{\Delta\omega}{2}t - \frac{\Delta\omega}{2}k\right) \tag{6.91a}$$

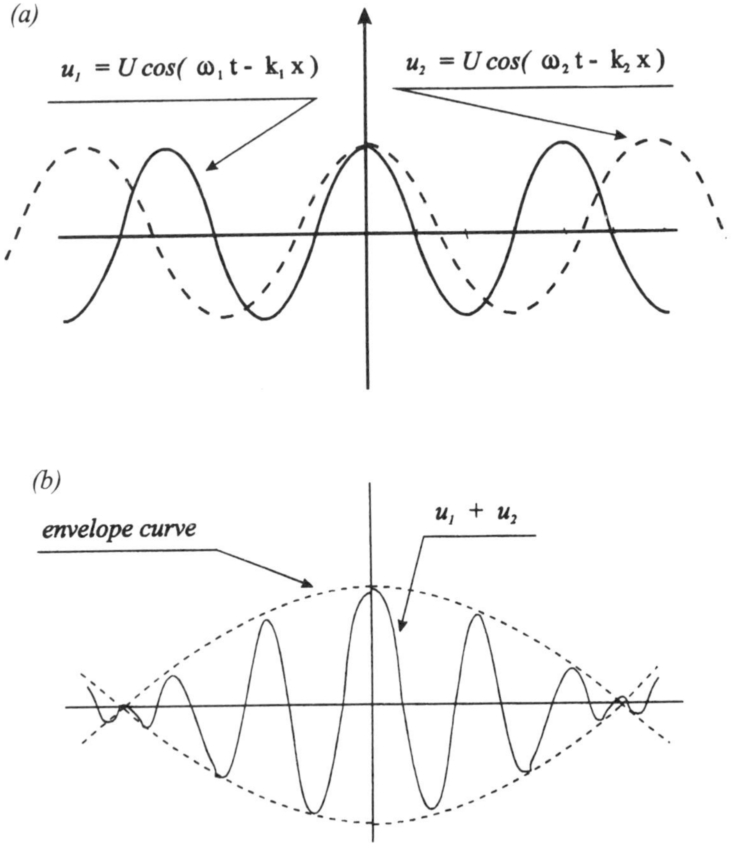

Fig. 6.25(a) and **(b)**

where the suffix 0 refers to a mean value and Δ signifies the difference. Figure 6.25b shows the plot of equation (6.91a). The individual crests are still moving with a phase velocity ω_0/k_0 but the envelope curve is travelling at a speed $\Delta\omega_0/\Delta k_0$ which is known as the *group velocity*, c_g. If the two frequencies are very close together then, in the limit,

$$\text{phase velocity } c_p = \frac{\omega_0}{k_0} \tag{6.92}$$

and

$$\text{group velocity } c_g = \frac{d\omega}{dk} \tag{6.93}$$

A graph of ω versus k is called the dispersion diagram. For a non-dispersive wave ω is proportional to k so that c_p and c_g are identical constants. For the dispersive wave there is a functional relationship between ω and k and it is found that the gradient can be negative as well as positive; also the curve need not pass through the origin. Figure 6.26 shows a typical curve.

To reinforce the concept of group velocity consider a packet of waves with frequencies in a narrow bandwidth on the dispersion diagram. The bandwidth is narrow enough for the gradient to be constant. Thus in this region the gradient is

$$c_g = \frac{\omega_i - \omega_0}{k_i - k_0}$$

or

$$\omega_i = c_g(k_i - k_0) + \omega_0 \tag{6.94}$$

Now a typical wave in this region is

$$u = U_i \cos(\omega_i t - k_i x)$$

Substituting from equation (6.94) gives

$$\begin{aligned} u &= U_i \cos\{[c_g(k_i - k_0) + \omega_0]t - k_i x\} \\ &= U_i \cos[(\omega_0 - c_g k_0)t + k_i(c_g t - x)] \end{aligned} \tag{6.95}$$

This equation represents a wave moving to the right with a phase velocity $c_p = \omega_i/k_i$, which will only vary slightly over the narrow frequency band.

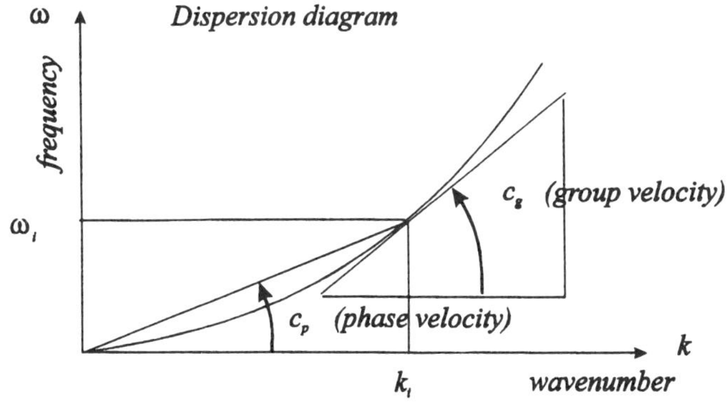

Fig. 6.26

If we change our origin so that $x = c_g t$, which is equivalent to moving along the x axis at the group velocity, then equation (6.95) becomes independent of the wavenumber k_i (and of the frequency ω_i) and we can then sum for any number of waves within the frequency band

$$\begin{aligned} u &= \sum_i U_i \cos[(\omega_0 - c_g k_0)t] \\ &= (\sum_i U_i)\cos[(\omega_0 - c_g k_0)t] \\ &= (\sum_i U_i)\cos[(c_p - c_g)(\omega_0/c_p)t] \end{aligned} \tag{6.96}$$

Thus if we move along the x axis at the group velocity we see a constant amplitude, which is the envelope curve, with the displacement varying at a frequency which depends on the difference between the phase and group velocities.

In this example the group velocity is shown as greater than the phase velocity so the individual peaks will be seen to retreat within the envelope curve. This is displayed in Fig. 6.27.

If there is a short-duration pulse then the frequency band is wide; a simple rule is that the product of bandwidth (in hertz) and pulse duration (in seconds) is approximately unity. Let us assume that the pulse is represented by a sum of sinusoids

$$u = \sum U_i \cos(\omega_i t - k_i x) \tag{6.97}$$

At $t = 0$ and $x = 0$ all displacements are additive, and the pulse is symmetrical about both the time and the space origins. The amplitudes U_i are functions of the frequency (and wavenumber). The functions depend on the shape of the pulse and are given by standard Fourier transform techniques. We seek the peak of the pulse therefore at the peak

$$\frac{\partial u}{\partial x} = \sum U_i k_i \sin(\omega_i t - k_i x) = 0 \tag{6.98}$$

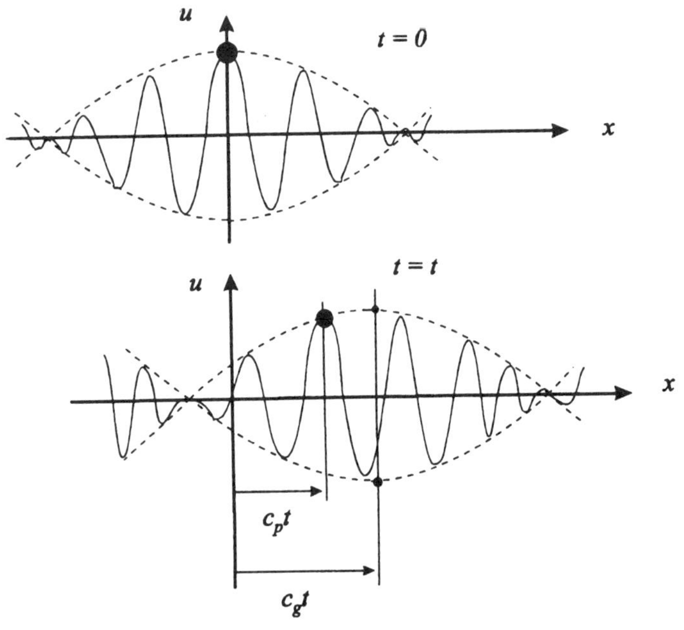

Fig. 6.27

In the neighbourhood of the peak and for small dispersion the argument of the function will be small so that sin z may be replaced by z. This gives

$$t \Sigma U_i k_i \omega_i = x \Sigma U_i k_i^2$$

Hence the velocity of the peak will be

$$c_{pp} = \frac{x}{t} = \frac{\Sigma U_i k_i \omega_i}{\Sigma U_i k_i^2} \tag{6.99}$$

If U is a continuous function of k then

$$c_{pp} = \frac{\int_0^{k_m} U(k)\, k\omega(k)\, \mathrm{d}k}{\int_0^{k_m} U(k)\, k^2 \mathrm{d}k} \tag{6.100}$$

where k_m is the value of k when ω = the bandwidth, that is $\omega\tau = 2\pi$.

The quantity c_{pp} will be known in this book as the *pulse-peak velocity*. The expression for pulse-peak velocity is only meaningful if the dispersion is moderate, otherwise the pulse will be changing rapidly and no distinct peak will be observed. Since the amplitude appears in both numerator and denominator small variations will have little effect. This is borne out by Figs 6.28 and 6.29 where the curves for a square pulse and triangular pulse are seen to be very close together.

Figure 6.28 shows the three wave velocities versus k for a concave downwards dispersion curve. Here the group velocity drops quickest and eventually becomes negative, the curve for phase velocity dropping less quickly. The plots of pulse-peak velocity for a square pulse and for a triangular pulse are shown but on the scale of the diagram the difference is not measurable. Figure 6.29 shows a similar plot but in this case the dispersion curve is concave

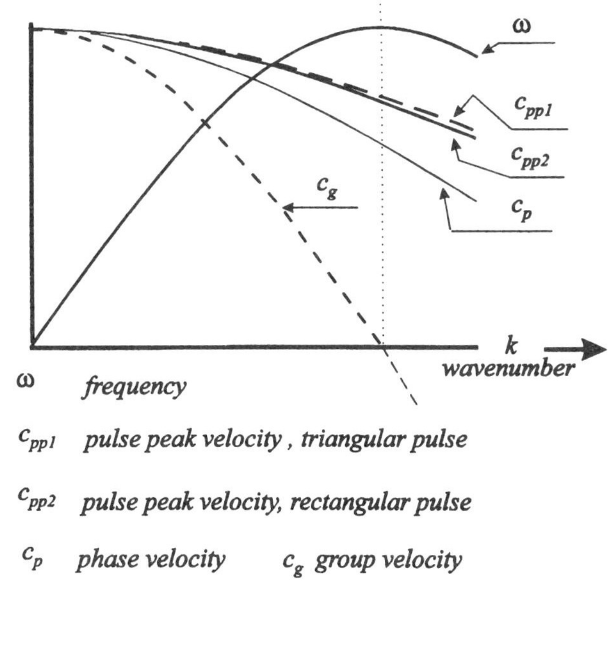

Fig. 6.28 Wave velocities for negative curvature dispersion curve

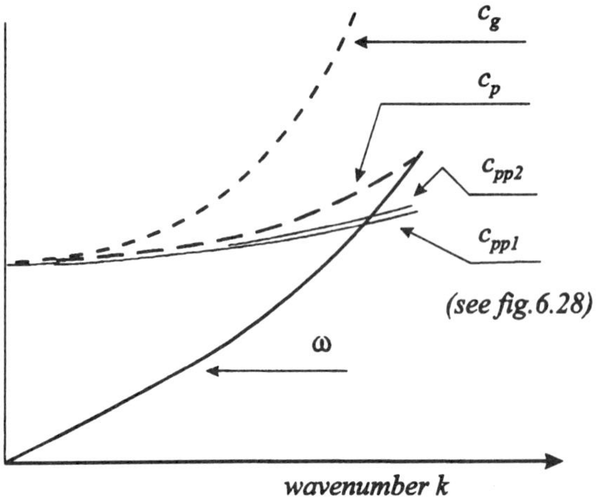

Fig. 6.29 Wave velocities for positive curvature dispersion curve

upwards. It is conventional to plot the curves with k as the independent variable because ω is univalued for a given k, but not vice versa.

Figure 6.30 shows four symmetrical pulse shapes and their respective Fourier transforms. The ordinate is the square of the amplitude as this gives a measure of the energy and it is

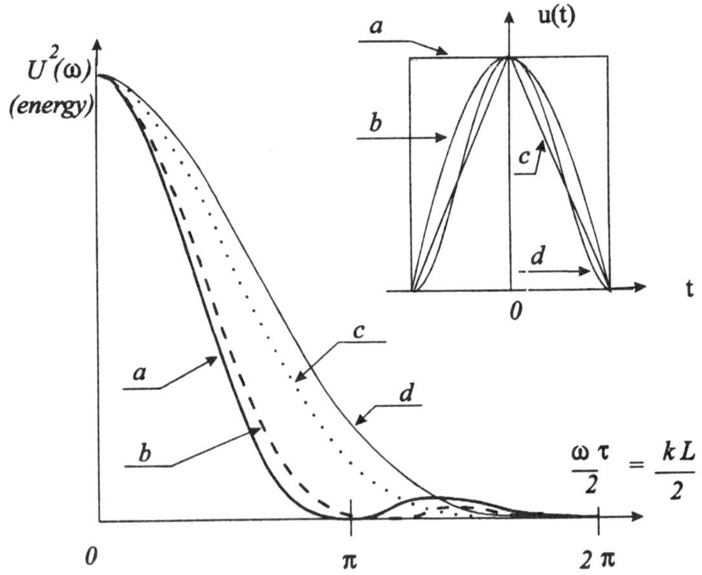

Fig. 6.30 Fourier transform pairs

seen that most of the energy is accounted for before the abscissa value reaches π. This gives credence to the approximations $\tau/T_m = 1$ and $L/\lambda_m = 1$.

The practical difficulty is the determination of the width of the peak and hence the appropriate value of k_m. Measured values tend to be towards the phase velocity but are far from the group velocity. Although the pulse-peak velocity is as precisely defined as the other two it does emphasize the fact that the group velocity is valid only for a narrow frequency band. In a highly dispersive situation the group velocity gives the arrival time of specific wavelengths which were generated by an impact.

Dispersion indicates that short pulses will spread out but the total energy remains constant. The term is not to be confused with dissipation in which some of the mechanical energy is converted to thermal energy.

6.12 Waves in a uniform beam

In this section we shall be examining lateral waves in a long uniform beam, shown in Fig. 6.31, with a cross-sectional area A and a second moment of area I about the z axis through the centroid. The xy plane is a plane of symmetry. The material has a Young's modulus E, a shear modulus G and the density is ρ. We are going to use Hamilton's principle to obtain the equations of motion because it is easier to modify the model. The exact equations are very involved and therefore approximations are required.

A first approximation is to consider only kinetic energy due to lateral motion and strain energy due to bending strains. Later we shall include rotary inertia and shear strain energy. The simple case can be obtained by free-body diagrams and Newton's laws but we shall use the variational method and subsequently modify the Lagrangian to take into account the extra terms.

Referring to Fig. 6.32 we can write an expression for the kinetic energy (note that in this section v is the deflection in the y direction)

$$T_1 = \int_0^L \frac{\rho A}{2} \left(\frac{\partial v}{\partial t} \right)^2 \mathrm{d}x \tag{6.101}$$

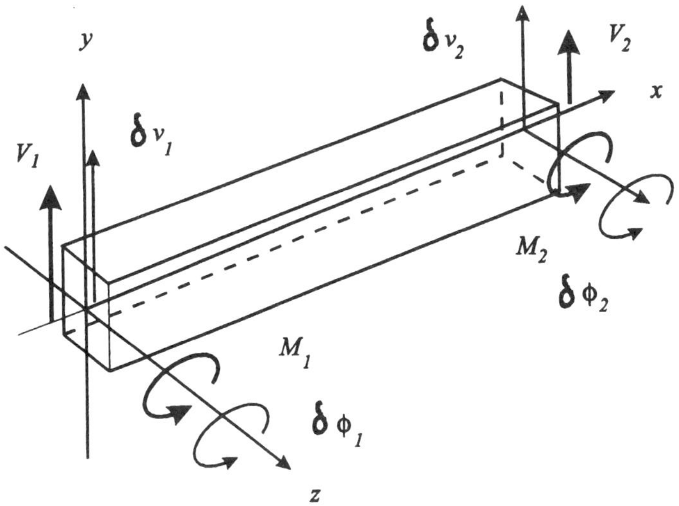

Fig. 6.31

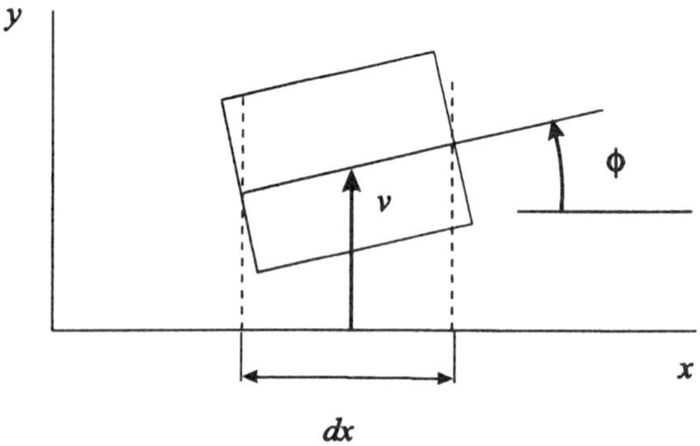

Fig. 6.32

and for the strain energy

$$V_1 = \int_0^L \frac{EI}{2}\left(\frac{\partial ø}{\partial x}\right)^2 \mathrm{d}x \tag{6.102}$$

If we assume that there are no external forces Hamilton's principle states that

$$\delta \int_{t_1}^{t_2} (T_1 - V_1)\,\mathrm{d}t = 0$$

or

$$\delta \int_{t_1}^{t_2}\int_0^L \left[\frac{\rho A}{2}\left(\frac{\partial v}{\partial t}\right)^2 - \frac{EI}{2}\left(\frac{\partial ø}{\partial x}\right)^2 \right] \mathrm{d}x\ \mathrm{d}t = 0 \tag{6.103}$$

Carrying out the variation first

$$\int_0^L \int_{t_1}^{t_2} \left[\rho A \frac{\partial v}{\partial t}\, \sigma \left(\frac{\partial v}{\partial t}\right) - EI\frac{\partial ø}{\partial x}\, \sigma \left(\frac{\partial ø}{\partial t}\right) \right. \mathrm{d}x\,\mathrm{d}t = 0 \tag{6.104}$$

For the first term we reverse the order of integration and integrate by parts to obtain

$$\int_0^L \left[\int_{t_1}^{t_2} \rho A \frac{\partial v}{\partial t}\, \delta \left(\frac{\partial v}{\partial t}\right) \mathrm{d}t \right] \mathrm{d}x = \int_0^L \left[\rho A \frac{\partial v}{\partial t}\, \delta v \Big|_{t_1}^{t_2} - \int_{t_1}^{t_2} \frac{\partial^2 v}{\partial t^2}\, \delta v\ \mathrm{d}t \right] \mathrm{d}x \tag{6.105}$$

Because δV vanishes, by definition, at t_1 and t_2 the first term in the square brackets is zero.
For the second term we integrate by parts with respect to x to obtain

$$\int_{t_1}^{t_2} \int_0^L \left[EI \frac{\partial ø}{\partial x}\, \delta \left(\frac{\partial ø}{\partial t}\right)\right] \mathrm{d}x\ \mathrm{d}t = \int_0^{t_2} \left[EI\frac{\partial ø}{\partial x}\, \delta ø \Big|_0^L - \int_0^L EI\, \frac{\partial^2 ø}{\partial x^2}\, \delta ø\ \mathrm{d}x \right] \mathrm{d}t \tag{6.106}$$

We have already assumed for this exercise that there are no external active forces. Therefore the end constraints must be workless and this implies that either $\delta ø$ or $\partial ø/\partial x$ must be zero at the ends. Hence the first term is zero.

Combining the two results, equations (6.105) and (6.106), gives

$$\int_{t_1}^{t_2}\int_0^L \left(EI \frac{\partial^2 ø}{\partial x^2}\, \delta ø \; - \; \rho A \frac{\partial^2 v}{\partial t^2}\, \delta v \right) \mathrm{d}x\, \mathrm{d}t = 0 \tag{6.107}$$

v and $ø$ are not independent but are related by geometry

$$\frac{\partial v}{\partial x} = ø \tag{6.108}$$

so that

$$\delta ø = \delta \left(\frac{\partial v}{\partial x} \right) \tag{6.109}$$

Integrating the first term once more by parts and noting that the end forces are workless leads to

$$\int_{t_1}^{t_2}\int_0^L \left(-EI \frac{\partial^3 ø}{\partial x^3}\, \delta v \; - \; \rho A \frac{\partial^2 v}{\partial t^2}\, \delta v \right) \mathrm{d}x\, \mathrm{d}t = 0$$

and finally since $ø = \partial v/\partial x$

$$\int_{t_1}^{t_2}\int_0^L \left(-EI \frac{\partial^4 v}{\partial x^4} - \rho A \frac{\partial^2 v}{\partial t^2} \right) \delta v \; \mathrm{d}x\, \mathrm{d}t = 0 \tag{6.110}$$

Because δv is arbitrary (except at t_1 and t_2 where it is zero) the expression in the large parentheses must be zero. Thus

$$EI \frac{\partial^4 v}{\partial x^4} + \rho A \frac{\partial^2 v}{\partial t^2} = 0 \tag{6.111}$$

This equation is known as *Euler's equation* for beam vibration and is widely used. This form can readily be deduced from free-body diagrams in a similar method to that used for longitudinal waves in a bar. The reason for using Hamilton's principle here is to expose the details of the method and to form the basis for development of a more refined model.

We now add an extra kinetic energy term to take into account rotary inertia. For a thin element of beam the moment of inertia about a z axis through the centroid is $\rho I\, \mathrm{d}x$, where I is the second moment of area. Therefore the kinetic energy is

$$T_2 = \int_0^L \frac{\rho I}{2} \left(\frac{\partial ø}{\partial t} \right)^2 \mathrm{d}x \tag{6.112}$$

Thus

$$\delta \int_{t_1}^{t_2}\int_0^L \frac{\rho I}{2} \left(\frac{\partial ø}{\partial t} \right)^2 \mathrm{d}x\, \mathrm{d}t$$

$$= \int_0^L \int_{t_1}^{t_2} \rho I \frac{\partial ø}{\partial t} \delta \frac{\partial ø}{\partial t} \, dt \, dx$$

$$= \int_0^L \int_{t_1}^{t_2} \left(-\rho I \frac{\partial^2 ø}{\partial t^2} \delta ø \right) dt \, dx \tag{6.113}$$

The rate at which the lateral deflection changes with x is now augmented by γ, the shear strain, giving

$$\frac{\partial v}{\partial x} = ø + \gamma \tag{6.114}$$

and the additional strain energy is

$$V_2 = \int_0^L \frac{\kappa GA}{2} \gamma^2 \, dx \tag{6.115}$$

The constant κ (kappa), which is greater than unity, corrects for the fact that the shear stress is not uniformly distributed across the cross-sectional area. For a rectangular cross-section $\kappa = 6/5$; this is based on a parabolic shear stress distribution. Thus

$$\delta \int_{t_1}^{t_2} V_2 \, dt = \int_{t_1}^{t_2} \int_0^L \kappa GA \left(\frac{\partial v}{\partial x} - ø \right) \left(\delta \frac{\partial v}{\partial x} - \delta ø \right) dx \, dt$$

$$= \int_{t_1}^{t_2} \left[\kappa GA \left(\frac{\partial v}{\partial x} - ø \right) \delta v \Big|_0^L - \kappa GA \int_0^L \left(\frac{\partial^2 v}{\partial x^2} - \frac{\partial ø}{\partial x} \right) dv \, dx \right] dt \tag{6.116}$$

This time we shall not make the first term zero as we are now admitting external forces.

From Fig. 6.31 the virtual work done by the external forces is (note that a vector sign convention is used)

$$\delta W = M_1 \delta ø_1 + M_2 \delta ø_2 + S_1 \delta v_1 + S_2 \delta v_2 \tag{6.117}$$

Therefore Hamilton's principle for the modified model including external forces is

$$\delta \int_{t_1}^{t_2} (T_1 + T_2 - V_1 - V_2) \, dt + \int_{t_1}^{t_2} \delta W \, dt = 0 \tag{6.118}$$

Using equations (6.105), (6.106), (6.113), (6.116) and (6.117), equation (6.118) may be written

$$\int_{t_1}^{t_2} \int_0^L \{ [-\rho A V_{,tt} + \kappa GA(v_{,xx} - ø_{,x}] \, dv \ldots$$

$$+ [-\rho I \, ø_{,tt} + EI \, ø_{,xx} + \kappa GA(v_{,x} - ø)] \, \delta ø \} \, dx \, dt \ldots$$

$$+ \int_{t_1}^{t_2} \{ | -EI \, ø_{,x} \delta ø - \kappa GA(v_{,x} - ø) \, \delta v \, |_0^L \quad \ldots$$

$$+ [M_1 \delta ø_1 + M_2 \delta ø_2 + V_1 \delta v_1 + V_2 \delta v_2] \} \, dt \tag{6.119}$$

Here the notation $\partial^2 v / \partial x^2 = v_{,xx}$ etc. is used.

Because δv and $\delta ø$ are arbitrary between t_1 and t_2 the factors of δv and of $\delta ø$ under the double integrals must each equate to zero. Thus the two equations of motion are

$$\frac{\partial^2 v}{\partial t^2} - \kappa c_t^2 \left(\frac{\partial^2 v}{\partial x^2} - \frac{\partial ø}{\partial x} \right) = 0 \qquad (6.120)$$

and

$$\frac{\partial ø^2}{\partial t^2} - c_0^2 \frac{\partial ø^2}{\partial x^2} - \frac{\kappa c_t^2}{(I/A)} \left(\frac{\partial v}{\partial x} - ø \right) = 0 \qquad (6.121)$$

Summing each of the coefficients of $\delta ø_1$, $\delta ø_2$, δv_1 and δv_2 to zero gives the boundary conditions

$$\left(EI \frac{\partial ø}{\partial x} \right)_1 = -M_1 \qquad (6.122)$$

$$\left(EI \frac{\partial ø}{\partial x} \right)_2 = M_2 \qquad (6.123)$$

$$\kappa GA \left(\frac{\partial \upsilon}{\partial x} - ø \right)_1 = -V_1 \qquad (6.124)$$

$$\kappa GA \left(\frac{\partial \upsilon}{\partial x} - ø \right)_2 = -V_2 \qquad (6.125)$$

It is now possible to eliminate ø between equations (6.120) and (6.121). Equation (6.120) can be written as

$$v_{,tt} - \kappa c_t^2 \, v_{,xx} + \kappa c_t^2 \, ø_{,x} = 0$$

Therefore

$$ø_{,x} = v_{,xx} - \frac{1}{\kappa c_t^2} v_{,tt} \qquad (6.126)$$

Equation (6.121) is written as

$$ø_{,tt} - c_0^2 \, ø_{,xx} - \frac{\kappa c_t^2}{(I/A)} v_{,x} + \frac{\kappa c_t^2}{(I/A)} ø = 0$$

and differentiating partially with respect to x gives

$$ø_{,ttx} - c_0^2 \, ø_{,xxx} - \frac{\kappa c_t^2}{(I/A)} v_{,xx} + \frac{\kappa c_t^2}{(I/A)} ø_{,x} = 0 \qquad (6.127)$$

Substituting from equation (6.126) we obtain

$$v_{,ttxx} - \frac{1}{\kappa c_t^2} v_{,tttt} - c_0^2 \, v_{,xxxx} + \frac{c_0^2}{\kappa c_t^2} v_{,ttxx} \ldots$$

$$- \frac{\kappa c_t^2}{(I/A)} v_{,xx} + \frac{\kappa c_t^2}{(I/A)} v_{,xx} - \frac{1}{(I/A)} v_{,tt} = 0$$

or

$$c_0^2 \frac{\partial^4 v}{\partial x^4} + \frac{1}{(I/A)} \frac{\partial^2 v}{\partial t^2} - \left(1 + \frac{c_0^2}{\kappa c_t^2} \right) \frac{\partial^4 v}{\partial x^2 \, \partial t^2} + \frac{1}{(I/A)} \frac{\partial^4 v}{\partial t^4} = 0 \qquad (6.128)$$

This is known as the *Timoshenko beam equation.*

For a running wave solution

$$v = \tilde{v}\, e^{j(\omega t - kx)}$$

so substituting into equation (6.128) and dividing through by the common factor gives

$$c_0^2 k^4 - \frac{1}{(I/A)}\omega^2 - \left(1 + \frac{c_0^2}{\kappa c_t^2}\right)\omega^2 k^2 + \frac{1}{(I/A)}\omega^4 = 0 \qquad (6.129)$$

which is the dispersion equation for bending waves in a uniform beam. This equation is a quadratic in ω^2 and therefore yields two values of ω for any value of k. The lower of the two, the first mode, approximates to Euler's equation for small values of k ($k < 0.2$, i.e. wavelengths longer than about five times the beam depth).

Plots of ω, c_p and c_g are shown in Fig. 6.33. The phase velocity tends to a maximum value which is close to the velocity of pure shear waves (*see* section 7.5). The group velocity also tends to the same value but passes through a maximum for wavelengths of the order of the depth of the beam. It follows that after a short-duration impact these wavelengths are the first to arrive at a distant point but most of the energy will follow at longer wavelengths. The model becomes invalid when the wavelengths are very short compared with the depth of the beam in which case the wave speed will tend to that of surface waves which have a speed a little less than the shear wave speed.

Figure 6.34 is a similar plot but for the higher, or second, mode. It is seen that there is a minimum frequency; below this no travelling wave is possible in this mode. The consequence of this is that the phase velocity tends to infinity at very low wavenumbers but the group velocity remains finite and less than c_0. The validity of this mode is not as good as the first mode and is probably only witnessed in I-section beams where the end load is carried mainly by the flanges and the shear is carried mainly by the web.

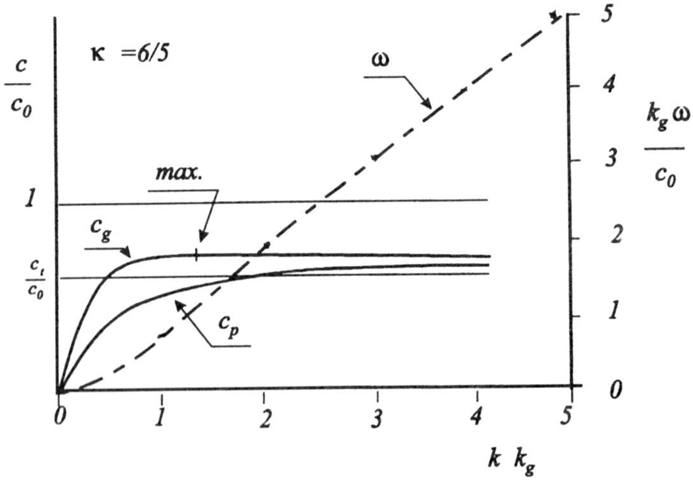

Fig. 6.33 Uniform beam mode 1

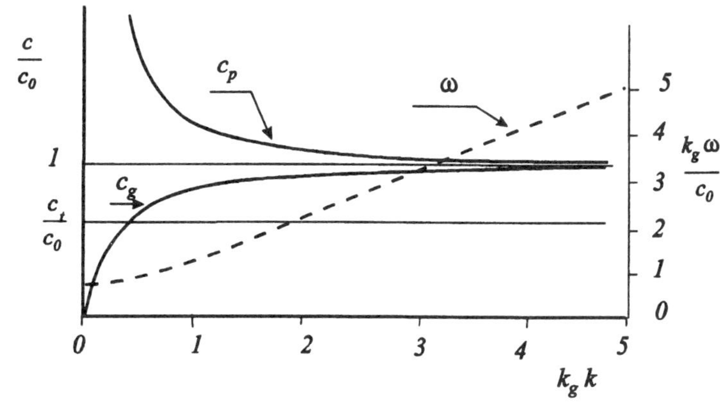

Fig. 6.34 Uniform beam mode 2

6.13 Waves in periodic structures

The type of structure envisaged here is the continuous mass–spring system shown in Fig. 6.35. Away from any boundary we can use the same form of expression for displacement as used in the earlier sections but in place of the continuous location x we have a discrete number of locations n. The mass of each body in the system is m and the stiffness of each spring is s. For the nth body the equation of motion is

$$s\,(u_{n+1} - u_n) - s\,(u_n - n_{n-1}) = m\ddot{u}_n \qquad (6.130)$$

Let us assume

$$u_n = Ue^{\mathrm{j}(\omega t - kn)} = Ue^{\mathrm{j}\omega t}\,e^{-\mathrm{j}kn} \qquad (6.131)$$

Substituting equation (6.131) into equation (6.130) and dividing through by the common factor $Ue^{\mathrm{j}\omega t}$ we obtain

$$s\,(e^{-\mathrm{j}k(n+1)} - e^{-\mathrm{j}kn}) - s\,(e^{-\mathrm{j}kn} - e^{-\mathrm{j}k(n-1)}) = -m\omega^2 e^{-\mathrm{j}kn}$$

Dividing further by $se^{-\mathrm{j}kn}$ gives

$$(e^{-\mathrm{j}k} - 1) - (1 - e^{\mathrm{j}k}) = -\,m\omega^2/s$$

and as

$$e^{\pm\mathrm{j}k} = \cos k \pm \mathrm{j}\,\sin k$$

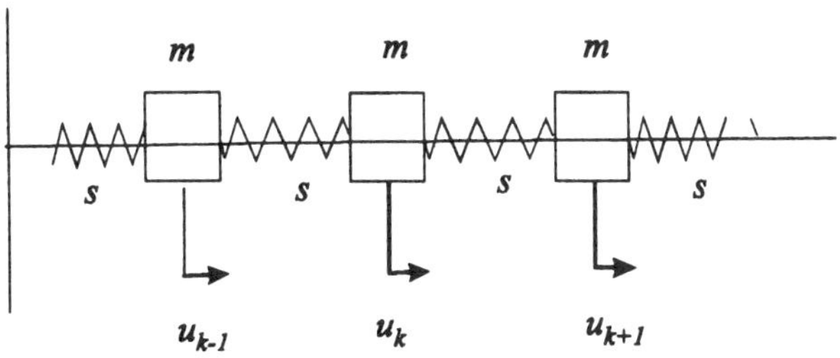

Fig. 6.35

we get

$$2\cos(k) - 2 = -\frac{m\omega^2}{s}$$

or

$$2(-2\sin^2(k/2)) = -\frac{m\omega^2}{s}$$

giving

$$\omega = 2\sqrt{(s/m)}\sin(k/2) \tag{6.132}$$

Figure 6.36 is a plot of the dispersion diagram. From this we see that there is a cut-off frequency, $\omega_{co} = 2\sqrt{(s/m)}$, above which no continuous wave will propagate. At this point $k = \pi$ so

$$u = Ue^{j\omega t}e^{-j\pi n} = Ue^{j\omega t}(-1)^n$$

that is, each body is in phase opposition with its neighbours.

The phase velocity is

$$c_p = \frac{\omega}{k} = \sqrt{(s/m)}\frac{\sin(k/2)}{k/2} \tag{6.133}$$

and the group velocity is

$$c_g = \frac{d\omega}{dk} = \sqrt{(s/m)}\cos(k/2) \tag{6.134}$$

If we impose a vibration above the cut-off frequency then one solution is to assume that each body has the opposite phase to its neighbour and that the amplitude decays exponentially,

$$u_n = U(-1)^n e^{j\omega t} e^{-k'n} \tag{6.135}$$

Substituting equation (6.135) into equation (6.130) leads to

$$\omega = 2\sqrt{(s/m)}\cosh(k'/2) \tag{6.136}$$

for $\omega > \omega_{co}$. Here the disturbance remains local to the point of initial excitation and does not propagate; such a mode is said to be *evanescent*.

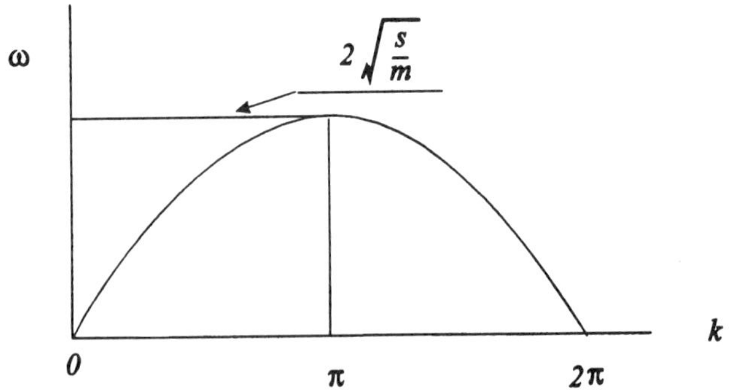

Fig. 6.36 Dispersion diagram for mass-spring system

6.14 Waves in a helical spring

The helical spring will be treated as a thin wire, that is plane cross-sections remain plane. This assumption has been shown to be acceptable by mechanical testing. An element of the wire has six degrees of freedom, three displacements and three rotations, so the possibility exists for six modes of propagation. If the helix angle is small these modes separate into two groups each having three degrees of freedom, one set consisting of the in-plane motion and the other the out-of-plane motion. The effect of helix angle will be discussed later but here we shall develop the theory for out-of-plane motion for a spring with zero helix angle. This, as we shall see, is associated with axial motion of the spring, that is with the spring being in its compression or tensile mode.

Figure 6.37 defines the co-ordinate system to be used. The unit vector $\boldsymbol{i}$ is tangent to the axis of the wire, $\boldsymbol{j}$ is along the radius of curvature directed towards the centre and $\boldsymbol{k}$ completes the right-handed triad. For zero helix angle $\boldsymbol{k}$ is parallel to the axis of the spring. The radius of curvature is R and s is the distance measured along the wire. θ is the angle through which the radius turns. Thus

$$\mathrm{d}s = R\ \mathrm{d}\theta \tag{6.137}$$

First we shall consider the differentiation of an arbitrary vector $\boldsymbol{V}$ with respect to s

$$\frac{\mathrm{d}\boldsymbol{V}}{\mathrm{d}s} = \frac{\mathrm{d}'\boldsymbol{V}}{\mathrm{d}s} + \boldsymbol{\Omega} \times \boldsymbol{V} \tag{6.138}$$

where the prime signifies differentiation with respect to non-rotating axes and Ω is the rate of rotation of the axes with distance s. Thus

$$\boldsymbol{\Omega} = \frac{\mathrm{d}\boldsymbol{\theta}}{\mathrm{d}s} = \frac{\mathrm{d}\theta}{\mathrm{d}s}\ \boldsymbol{k}$$

and using equation (6.137)

$$\boldsymbol{\Omega} = \frac{1}{R}\ \boldsymbol{k} \tag{6.139}$$

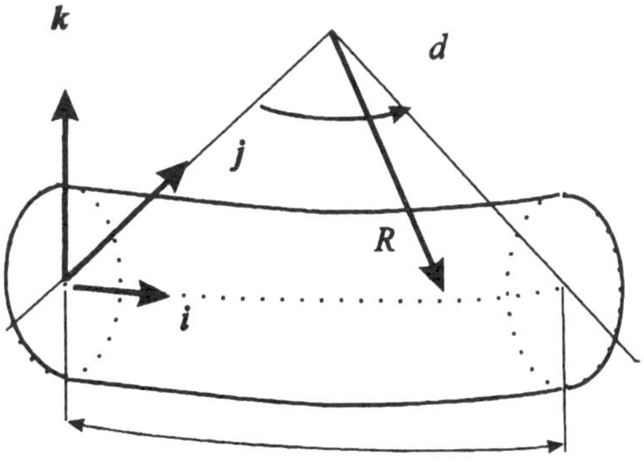

Fig. 6.37

From equation (6.138) the components of the derivative are

$$\frac{dV_i}{ds} = \frac{d'V_i}{ds} - \frac{1}{R} V_j \tag{6.140}$$

$$\frac{dV_j}{ds} = \frac{d'V_j}{ds} + \frac{1}{R} V_i \tag{6.141}$$

$$\frac{dV_k}{ds} = \frac{d'V_k}{ds} \tag{6.142}$$

or in matrix form

$$\frac{d(V)}{ds} = [T_1](V) \tag{6.143}$$

where

$$[T1] \equiv \begin{bmatrix} p & -1/R & 0 \\ 1/R & p & 0 \\ 0 & 0 & p \end{bmatrix} \tag{6.144}$$

and

$$p \equiv \frac{d'}{ds} \tag{6.145}$$

A cross-section has a displacement u and a rotation ø from its equilibrium position. The spatial rate of change of u is due to stretching and shearing of the element of wire and also to rigid body rotation, so the strain

$$\begin{aligned}(\varepsilon) &= [T_1]\,(u) - [ø]^{\times}\,(ds) / ds \\ &= [T_1]\,(u) + [ds/ds]^{\times}\,(ø)\end{aligned} \tag{6.146}$$

Now (d**s**) = (ds 0 0) and therefore

$$\frac{[ds]^{\times}}{ds} = \begin{bmatrix} 0 & 0 & 0 \\ 0 & 0 & -1 \\ 0 & 1 & 0 \end{bmatrix} \equiv [T_2] \tag{6.147}$$

(see appendix 1)

The components of strain are, therefore,

axial strain $\varepsilon_i = pu_i - u_j/R$ (6.148)

shear strain $\varepsilon_j = pu_j + u_i/R - ø_k$ (6.149)

shear strain $\varepsilon_k = pu_k + ø_j$ (6.150)

These are related to the elastic constants by

$$\varepsilon_i = \frac{P_i}{EA} \tag{6.151}$$

$$\varepsilon_j = \frac{P_j}{GqA} \tag{6.152}$$

$$\varepsilon_k = \frac{P_k}{GqA} \tag{6.153}$$

where q is a factor to allow for the shear stress distribution not being uniform. A typical value for circular cross-sections is 0.9.

The relationship between the elastic constants, Young's modulus E, shear modulus G and Poisson's ratio υ is $E = 2G(1 + \upsilon)$, so let

$$m \equiv \frac{E}{2G} = (1 + \upsilon) \tag{6.154}$$

Combining equations (6.148) to (6.154)

$$P_i = GA\,(2mpu_i - 2mu_j/R) \tag{6.155}$$

$$P_j = GA\,(qpu_j + qu_i/R - q\o_k) \tag{6.156}$$

$$P_k = GA\,(qpu_k + q\o_j) \tag{6.157}$$

For bending we use the usual engineering relationships for bending and torsion of shafts. If the shape of the cross-section has point symmetry then with J being the polar second moment of area

$$I_j = I_k = I_i/2 = J/2$$

Also $E = G2m$ so that $EI_j = EI_k = mGJ$, and therefore

$$M_i = GI_i \frac{\mathrm{d}\o_i}{\mathrm{d}s} = GJ\,(p\o_i - \o_j/R) \tag{6.158}$$

$$M_j = EI_j \frac{\mathrm{d}\o_j}{\mathrm{d}s} = GJm\,(p\o_j - \o_i/R) \tag{6.159}$$

$$M_k = EI_k \frac{\mathrm{d}\o_k}{\mathrm{d}s} = GJm\,(p\o_k) \tag{6.160}$$

The equations of motion can be derived with reference to Fig. 6.38. Resolving forces acting on the element, neglecting any external forces,

$$\left((P) + \frac{\mathrm{d}(P)}{\mathrm{d}s_{t\,=\,\text{constant}}}\,\mathrm{d}s \right) - (P) = \rho A\,\mathrm{d}s\,\frac{\partial^2(u)}{\partial t^2}$$

or, letting $D = \dfrac{\partial}{\partial t}$

$$\frac{\mathrm{d}(P)}{\mathrm{d}s_{t\,=\,\text{constant}}} = \rho A D^2(u)$$

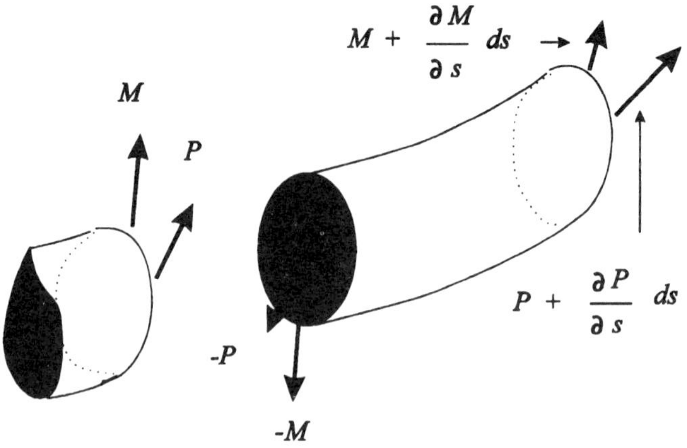

Fig. 6.38

or

$$[T_1]\,(P) = \rho A D^2(u) \tag{6.161}$$

The component equations are

$$pP_i - P_j/R = \rho A D^2 u_i \tag{6.162}$$

$$pP_j + P_i/R = \rho A D^2 u_k \tag{6.163}$$

$$pP_k = \rho A D^2 u_k \tag{6.164}$$

Now considering moments about the centre of mass of the element

$$\left((M) + \frac{\mathrm{d}(M)}{\mathrm{d}s_{t\ =\ \mathrm{constant}}}\,\mathrm{d}s \right) - (M) + (\mathrm{d}\boldsymbol{s})^{\times}(P) = \frac{\partial(L)}{\partial t} \tag{6.165}$$

where (L) is the moment of momentum. For small rotations

$$(L) = \begin{bmatrix} \rho I_i \mathrm{d}s\, D\text{ø}_i \\ \rho I_j \mathrm{d}s\, D\text{ø}_j \\ \rho I_k \mathrm{d}s\, D\text{ø}_k \end{bmatrix} \tag{6.166}$$

and

$$\frac{\partial(L)}{\partial t} = \rho J D^2 \begin{bmatrix} \text{ø}_i \\ \text{ø}_j/2 \\ \text{ø}_k/2 \end{bmatrix} \mathrm{d}s = \frac{\mathrm{d}(M)}{\mathrm{d}s_{t=\mathrm{constant}}}\,\mathrm{d}s + (\mathrm{d}\boldsymbol{s})^{\times}(P) \tag{6.167}$$

The three component equations are, after dividing by $\mathrm{d}s$,

$$pM_i - M_j/R = \rho J D^2 \text{ø}_i \tag{6.168}$$

$$pM_j + M_i/R - P_k = \rho J D^2 \text{ø}_j/2 \tag{6.169}$$

$$pM_k + P_j = \rho J D \text{ø}_k/2 \tag{6.170}$$

Substituting the six equations of state ((6.155) to (6.160)) into the equations of motion ((6.162–6.164) and (6.168–6.170)) will yield six equations in the six co-ordinates and these will separate into two groups of three. So, substituting equations (6.157), (6.158) and (6.159) into equations (6.164), (6.168) and (6.169) leads to

$$GJ\,(p^2ø_i - pø_j/R) - GJ\frac{m}{R}\,(pø_j + ø_i/R) = \rho JD^2ø_i \tag{6.171}$$

$$GJm\,(p^2ø_j + pø_i/R) + \frac{GJ}{R}\,(pø_i - ø_j/R) - qGA\,(pu_k + ø_j) = \rho JD^2ø_j/2 \tag{6.172}$$

$$qGA\,(p^2u_k + pø_j) = \rho AD^2u_k \tag{6.173}$$

which contain only the three out-of-plane co-ordinates.

It is convenient for discussion purposes to put the above equations into non-dimensional form. To this end we define the following terms

$$\tilde{p} \equiv \frac{\partial}{\partial\theta} = R\frac{\partial}{\partial s} = Rp \tag{6.174}$$

$$U_k \equiv u_k/R \tag{6.175}$$

$$\tilde{D} \equiv \frac{RD}{c_2} \tag{6.176}$$

Thus equations (6.171) to (6.173) may be written in matrix form as

$$\begin{bmatrix} (\tilde{p}^2 - m) & -\tilde{p}(1 + m) & 0 \\ \tilde{p}\,(1 + m) & (m\tilde{p}^2 - 1 - \alpha^2 q) & -q\alpha^2\tilde{p} \\ 0 & q\tilde{p} & q\tilde{p}^2 \end{bmatrix} \begin{bmatrix} ø_i \\ ø_j \\ U_k \end{bmatrix} = \tilde{D}^2 \begin{bmatrix} ø_i \\ ø_j/2 \\ U_k \end{bmatrix} \tag{6.177}$$

(Note that this matrix equation can be written in symmetrical form; however, we can discuss the manner of wave propagation just as well in the current form.)

In a manner similar to previous cases we shall assume a wave travelling along the axis of the wire. Thus

$$ø_i = \tilde{ø}_i\,e^{j(\omega t - ks)} \tag{6.178}$$

$$ø_j = \tilde{ø}_j\,e^{j(\omega t - ks)} \tag{6.179}$$

$$U_k = \tilde{U}_k\,e^{j(\omega t - ks)} \tag{6.180}$$

Now

$$\tilde{p}ø_i = R\,(-jk)ø_i \tag{6.181}$$

and

$$\tilde{D}ø_i = \frac{R}{c_2}\,(i\omega)\,ø_i \tag{6.182}$$

with similar expressions for the other two co-ordinates.

Let us define the non-dimensional wavenumber

$$K \equiv Rk \tag{6.183}$$

and the non-dimensional frequency

$$W \equiv \omega\frac{R}{c_2} \tag{6.184}$$

from which we have

$$\frac{W}{K} = \frac{c_p}{c_2} \tag{6.185}$$

the non-dimensional phase velocity.
From equations (6.181) to (6.184) we write

$$\tilde{p} = jK \tag{6.186}$$

$$\tilde{D} = jW \tag{6.187}$$

so equation (6.177) can now be written as

$$\begin{bmatrix} (W^2 - m - K^2) & jK(1 + m) & 0 \\ -jK(1 + m) & (\frac{1}{2}W^2 - 1 - \alpha^2 q) & -qjK\alpha^2 \\ 0 & -qjK & (W^2 - qK^2) \end{bmatrix} \begin{bmatrix} \text{ø}_i \\ \text{ø}_j \\ U_k \end{bmatrix} = \begin{bmatrix} 0 \\ 0 \\ 0 \end{bmatrix} \tag{6.188}$$

For a non-trivial solution the determinant of the square matrix must equate to zero. Thus

$$\begin{vmatrix} (W^2 - m - K^2) & jK(1 + m) & 0 \\ -jK(1 + m) & (\tfrac{1}{2}W^2 - 1 - \alpha^2 q) & -qjK\alpha^2 \\ 0 & -qjK & (W^2 - qK^2) \end{vmatrix} = 0 \tag{6.189}$$

This leads to a cubic in W^2 for any given value of K. There are, therefore, three branches to the dispersion diagram and these are shown in Fig. 6.39. The fourth root of ω is plotted as ordinate in order to compress the scale. The two highest modes have cut-off frequencies and therefore sinusoidal wave propagation only exists in these modes at high frequencies. The important lowest mode is shown in detail in Fig. 6.40.

The W–K diagram for the lowest mode exhibits a zero frequency when $K = 0$ and also when $K = 1$ or $\lambda = 2\pi R$. At this wavelength particles having maximum positive velocity are one turn apart and the maximum negative velocity particles are diametrically opposite.

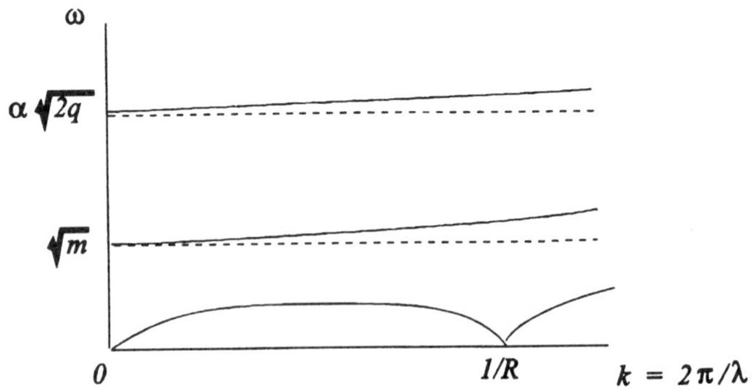

Fig. 6.39 Dispersion diagram for helical spring

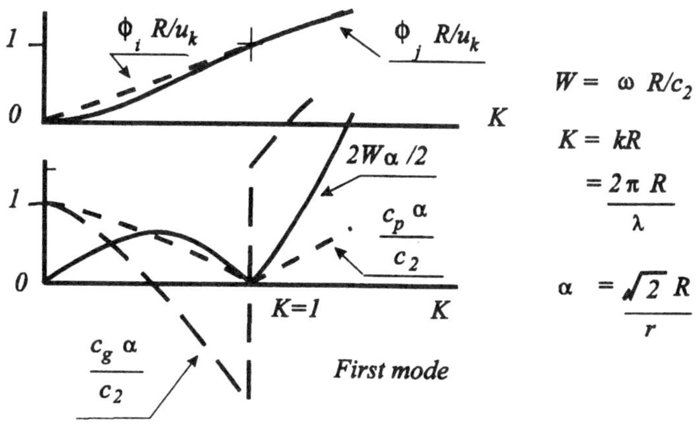

Fig. 6.40

For low values of wavenumber equation (6.189) reduces to

$$W^2 = \frac{K^2 q}{1 + q\alpha^2} \tag{6.190}$$

Now $\alpha^2 = 2R^2/r^2$ for solid circular cross-section wire. The spring index (R/r) is unlikely to be less than 3 so the minimum value of α^2 is about 18; a more typical index of 5 gives $\alpha^2 = 50$. Since q is of the order unity equation (6.190) is, to a close approximation,

$$W^2 = (K/\alpha)^2 \tag{6.191}$$

Returning to the dimensional form

$$\omega = \frac{c_2}{\alpha} k = \frac{rc_2}{R\sqrt{2}} k \tag{6.192}$$

from which the phase velocity and the group velocity are given by

$$c_p = c_g = c_2/\alpha \tag{6.193}$$

This approximation is quite reasonable for wavelengths longer than five turns. Also shown on Fig. 6.40 are the amplitude ratios and it is interesting to note that although the strain associated with long wavelengths is torsional in nature there is very little rotation about the wire axis ($ø_i \to 0$). This is true for the static case, represented here by zero frequency and infinitely long wavelength.

The dispersion diagram shown is a plot of $W\alpha$ versus K for $\alpha = 10$ but on the scale used no difference is seen for α ranging from 3 to 30.

The effect of the helix angle being greater than zero is to couple the in-plane and out-of-plane co-ordinates, but for small helix angle and low wavenumber the essential nature of the curves does not change. The more noticeable effect is around $K = 1$ where the curve is more rounded for the lowest longitudinal mode and the curve for the torsional mode no longer goes to zero. The two lowest dispersion curves are shown in Fig. 6.41 which also shows the results of mechanical steady-state vibration tests. Impact tests were also carried out from which the arrival times of various frequency components were measured and compared with the theory; some results are shown in Fig. 6.42.

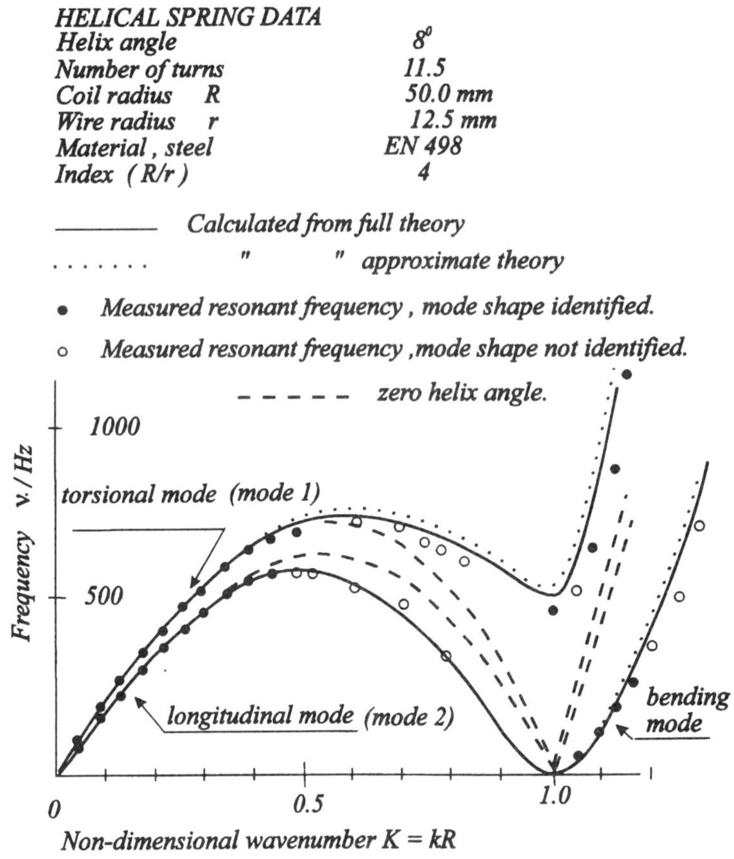

Fig. 6.41 Dispersion curve for helical spring (data from Ph.D thesis, H.R. Harrison 1971)

In section 6.12 the dispersion diagram for a periodic mass–spring system was developed and shown in Fig. 6.36. The similarity with the lowest mode for the spring as shown on Fig. 6.40 is quite noticeable. The numerical similarity is strong if in the lumped parameter model the mass and the stiffness of the components are those of a single turn of the spring. This model gives good agreement for wavelengths as short as one turn of the helix.

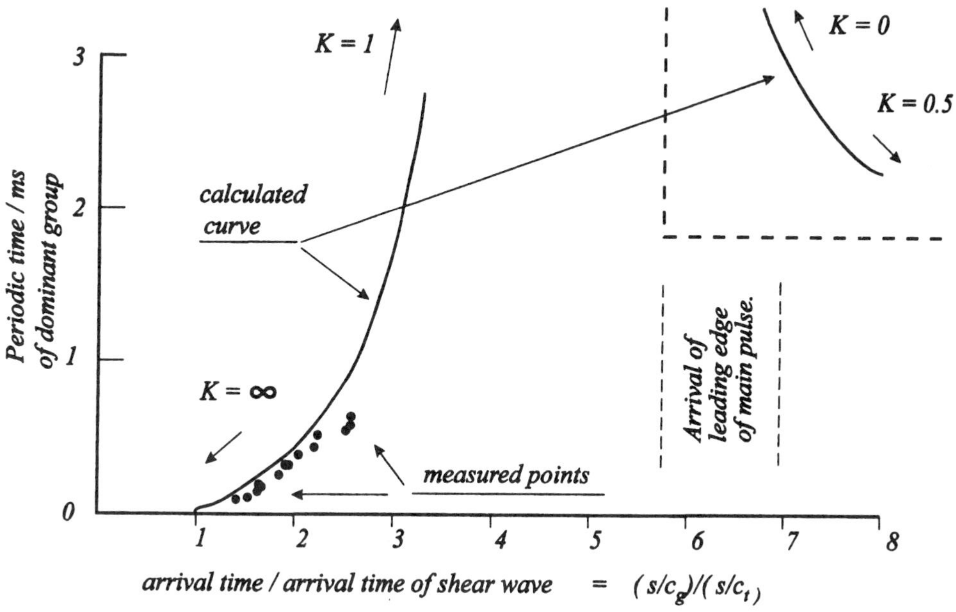
K = 1
K = 0
K = 0.5
calculated curve
Periodic time / ms of dominant group
K = ∞
measured points
Arrival of leading edge of main pulse.
3
2
1
0
1
2
3
4
5
6
7
8
arrival time / arrival time of shear wave = $(s/c_g)/(s/c_t)$

Fig. 6.42

7

Waves In a Three-Dimensional Elastic Solid

7.1 Introduction

All the examples of wave motion considered in Chapter 6 have been one dimensional, that is only one spatial dimension is required to define the direction of wave propagation. In the case of the helical spring the path of propagation is curvilinear, namely that along the wire axis. We now consider a homogeneous, isotropic, linearly elastic solid. The dynamics of such a solid are completely defined by three constants: the density and two elastic moduli. There are six elastic constants in general use: Young's modulus E, the shear modulus or modulus of rigidity G, Poisson's ratio υ, the bulk modulus K and the Lamé constants λ and μ. Any two will do but we shall find the Lamé constants the most convenient for this topic; these will be defined below.

The methods used for the bars and beams were approximations but this is justified by the fact that the boundary value problem to the exact equations, which we shall develop, has only been solved for a limited number of cases. However, a knowledge of the propagation of plane waves in an infinite, and semi-infinite, solid provides much insight into the physical nature of the phenomena previously studied.

We shall develop the required equations in a compact, though complete, notation but the reader not familiar with three-dimensional elasticity should consult the appropriate texts.

7.2 Strain

Referring to Fig. 7.1 the point P is located at a position $\boldsymbol{r}$ and a nearby point P′ is located at $\boldsymbol{r} + \mathrm{d}\boldsymbol{r}$. The displacement of point P is $\boldsymbol{u}$ and that of P′ is $\boldsymbol{u} + \mathrm{d}\boldsymbol{u}$. In terms of Cartesian co-ordinates

$$\begin{aligned} \mathrm{d}\boldsymbol{r} &= \mathrm{d}x\boldsymbol{i} + \mathrm{d}y\boldsymbol{j} + \mathrm{d}z\boldsymbol{k} \\ &= (\boldsymbol{i}\,\boldsymbol{j}\,\boldsymbol{k})\begin{bmatrix}\mathrm{d}x\\ \mathrm{d}y\\ \mathrm{d}z\end{bmatrix} \\ &= (e)^{\mathrm{T}}(\mathrm{d}r) \end{aligned} \tag{7.1}$$

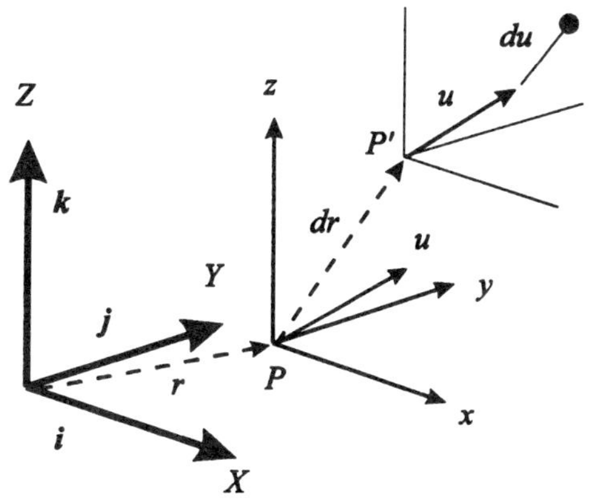

Fig. 7.1

where (e) is the column of unit vectors.

Similarly

$$\boldsymbol{u} = u_x\boldsymbol{i} + u_y\boldsymbol{j} + u_z\boldsymbol{k}$$

$$= (\boldsymbol{i}\,\boldsymbol{j}\,\boldsymbol{k})\begin{bmatrix} u_x \\ u_y \\ u_z \end{bmatrix}$$

$$= (e)^{\mathrm{T}}(u) \tag{7.2}$$

For small variations

$$\mathrm{d}u_x = \frac{\partial u_x}{\partial x}\mathrm{d}x + \frac{\partial u_x}{\partial y}\mathrm{d}y + \frac{\partial u_x}{\partial z}\mathrm{d}z$$

Therefore

$$\begin{bmatrix} \mathrm{d}u_x \\ \mathrm{d}u_y \\ \mathrm{d}u_z \end{bmatrix} = \begin{bmatrix} \dfrac{\partial u_x}{\partial x} & \dfrac{\partial u_x}{\partial y} & \dfrac{\partial u_x}{\partial z} \\ \dfrac{\partial u_y}{\partial x} & \dfrac{\partial u_y}{\partial y} & \dfrac{\partial u_y}{\partial z} \\ \dfrac{\partial u_z}{\partial x} & \dfrac{\partial u_z}{\partial y} & \dfrac{\partial u_z}{\partial z} \end{bmatrix} \begin{bmatrix} \mathrm{d}x \\ \mathrm{d}y \\ \mathrm{d}z \end{bmatrix} \tag{7.3}$$

Let us define

$$(\nabla) = \left(\frac{\partial}{\partial x}\ \frac{\partial}{\partial y}\ \frac{\partial}{\partial z}\right)^{\mathrm{T}} \tag{7.4}$$

Then equation (7.3) may be written

$$(\mathrm{d}u) = [(\nabla)(u)^{\mathrm{T}}]^{\mathrm{T}}(\mathrm{d}r) \tag{7.5}$$

The total displacement $(\mathrm{d}u)$ is due not only to a change in size and shape but also to a rotation as a rigid body. We confine our attention to small displacements so that the rotation may be regarded as a vector quantity

$$\boldsymbol{\Omega} = (\mathrm{e})^{\mathrm{T}}(\Omega_x\ \Omega_y\ \Omega_z)^{\mathrm{T}} \tag{7.6}$$

The rigid body displacement of P′ relative to P is

$$\mathrm{d}\boldsymbol{u}_{\mathrm{rot}} = \boldsymbol{\Omega}\times\mathrm{d}\boldsymbol{r} \tag{7.7}$$

or in vector–matrix terms (*see* Appendix 1), assuming the same basis on both sides of the equation

$$(\mathrm{d}u)_{\mathrm{rot}} = [\Omega]^{\times}(\mathrm{d}r) \tag{7.8}$$

where

$$[\Omega]^{\times} = \begin{bmatrix} 0 & -\Omega_z & \Omega_y \\ \Omega_z & 0 & -\Omega_x \\ -\Omega_y & \Omega_x & 0 \end{bmatrix} \tag{7.9}$$

From Fig. 7.2 it is seen that the rotation about the z axis is

$$\Omega_z = \frac{1}{2}\left(\frac{\partial u_y}{\partial x} - \frac{\partial u_x}{\partial y}\right) \tag{7.10}$$

Thus, the total rotation Ω is given by

$$\begin{bmatrix} \Omega_x \\ \Omega_y \\ \Omega_z \end{bmatrix} = \frac{1}{2}\begin{bmatrix} -\dfrac{\partial u_y}{\partial z} + \dfrac{\partial u_z}{\partial y} \\ \dfrac{\partial u_x}{\partial z} - \dfrac{\partial u_z}{\partial x} \\ -\dfrac{\partial u_x}{\partial y} + \dfrac{\partial u_y}{\partial x} \end{bmatrix}$$

$$= \frac{1}{2}\begin{bmatrix} 0 & -\dfrac{\partial}{\partial z} & \dfrac{\partial}{\partial y} \\ \dfrac{\partial}{\partial z} & 0 & -\dfrac{\partial}{\partial x} \\ -\dfrac{\partial}{\partial y} & \dfrac{\partial}{\partial x} & 0 \end{bmatrix}\begin{bmatrix} u_x \\ u_y \\ u_z \end{bmatrix} \tag{7.11}$$

So, in short matrix notation,

$$(\Omega) = \frac{1}{2}[\nabla]^{\times}(u) \tag{7.12}$$

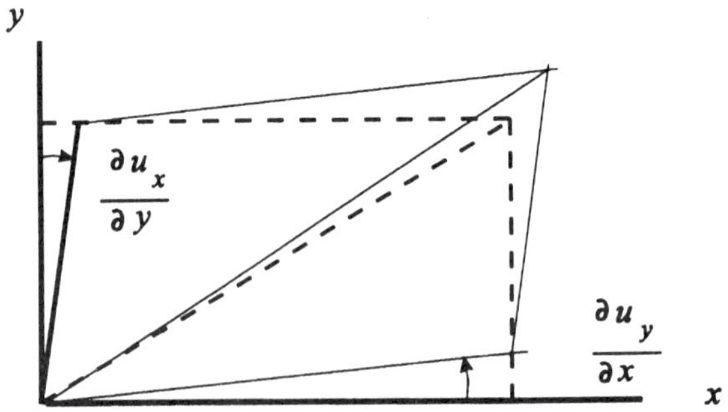

Fig. 7.2

or, in indicial notation,

$$\Omega_i = \frac{1}{2}(-u_{j,k} + u_{k,j}) \tag{7.13}$$

The elastic displacement will be the total displacement less that due to rotation, so

$$(\mathrm{d}u)_{\text{elastic}} = (\mathrm{d}u)_{\text{total}} - (\mathrm{d}u)_{\text{rot}} \tag{7.14}$$

$$(\mathrm{d}u)_{\text{elastic}} = \left\{\left[(\nabla)(u)^{\mathrm{T}}\right]^{\mathrm{T}} - \frac{1}{2}\left[[\nabla]^{\times}(u)\right]^{\times}\right\}(\mathrm{dr}) \tag{7.15}$$

From Appendix 1 or by direct multiplication we have that

$$[(\nabla)^{\times}(u)]^{\times} = (u)(\nabla)^{\mathrm{T}} - (\nabla)(u)^{\mathrm{T}} \tag{7.16}$$

Therefore

$$(\mathrm{d}u)_{\text{strain}} = \frac{1}{2}\left\{\left[(\nabla)(u)^{\mathrm{T}}\right]^{\mathrm{T}} + \left[(\nabla)(u)^{\mathrm{T}}\right]\right\}(\mathrm{d}r) \tag{7.17}$$

The first term in the braces is given in equations (7.3) and (7.5) and the second term is its transpose, so half the sum of the two is a symmetric matrix which is the strain matrix $[\varepsilon]$. Hence

$$(\mathrm{d}u)_{\text{strain}} = [\varepsilon](\mathrm{d}r) \tag{7.18}$$

where

$$[\varepsilon] = \frac{1}{2}\left\{\left[(\nabla)(u)^{\mathrm{T}}\right]^{\mathrm{T}} + \left[(\nabla)(u)^{\mathrm{T}}\right]\right\} \tag{7.19}$$

$$= \begin{bmatrix} \dfrac{\partial u_x}{\partial x} & \dfrac{1}{2}\left(\dfrac{\partial u_y}{\partial x} + \dfrac{\partial u_x}{\partial y}\right) & \dfrac{1}{2}\left(\dfrac{\partial u_z}{\partial x} + \dfrac{\partial u_x}{\partial z}\right) \\ & \dfrac{\partial u_y}{\partial y} & \dfrac{1}{2}\left(\dfrac{\partial u_z}{\partial y} + \dfrac{\partial u_y}{\partial z}\right) \\ \text{sym.} & & \dfrac{\partial u_z}{\partial z} \end{bmatrix}$$

$$= \begin{bmatrix} \varepsilon_x & \frac{1}{2}\gamma_{xy} & \frac{1}{2}\gamma_{xz} \\ & \varepsilon_y & \frac{1}{2}\gamma_{yz} \\ \text{sym.} & & \varepsilon_z \end{bmatrix} \tag{7.20}$$

Figure 7.2 shows the geometric definition of shear strain.

In indicial notation

$$\varepsilon_{ij} = \frac{1}{2}(u_{i,j} + u_{j,i}) \tag{7.21}$$

Note that for $i \neq j$, $\varepsilon_{ij} = \frac{1}{2}\gamma_{ij}$, that is half the conventional shear strain, whilst ε_{ii} is the usual tensile strain.

7.3 Stress

Figure 7.3 shows a tetrahedron with forces acting on all faces. The oblique face has an area A_n and is acted on by a force

$$\boldsymbol{F}_n = F_{nx}\boldsymbol{i} + F_{ny}\boldsymbol{j} + F_{nz}\boldsymbol{k} = (e)^T(F_n) \tag{7.22}$$

The area on which it acts is

$$\boldsymbol{A}_n = (e)^T(A_n) \tag{7.23}$$

where

$$(A_n) = (A_{nx}\ A_{ny}\ A_{nz})^T = (A_x\ A_y\ A_z)^T$$

The force on the face whose outward normal is in the x direction is

$$(F_x) = \begin{bmatrix} F_{xx} \\ F_{xy} \\ F_{xz} \end{bmatrix} = \begin{bmatrix} \sigma_{xx} \\ \tau_{xy} \\ \tau_{xz} \end{bmatrix} A_x \tag{7.24}$$

where σ_{xx} is the conventional tensile stress and τ_{xy} and τ_{xz} are the conventional shear stresses. Similarly for the forces in the other two directions.

The resultant force on the element is

$$(F_x) + (F_y) + (F_z) - (F_n) = \rho(\ddot{u})\text{volume}$$

Using equation (7.24) we have

$$\begin{bmatrix} \sigma_{xx} & \tau_{yx} & \tau_{zx} \\ \tau_{xy} & \sigma_{yy} & \tau_{zy} \\ \tau_{xz} & \tau_{yz} & \sigma_{zz} \end{bmatrix} \begin{bmatrix} A_x \\ A_y \\ A_z \end{bmatrix} - (F_n) = \rho(\ddot{u})\text{volume} \tag{7.25}$$

The right hand side is proportional to length^3 whilst the left hand side is proportional to length^2, so as $A_n \to 0$ the right hand side becomes negligible. Hence, for small volume,

$$(F_n) = \begin{bmatrix} \sigma_{xx} & \tau_{yx} & \tau_{zx} \\ \tau_{xy} & \sigma_{yy} & \tau_{zy} \\ \tau_{xz} & \tau_{yz} & \sigma_{zz} \end{bmatrix} \begin{bmatrix} A_x \\ A_y \\ A_z \end{bmatrix}$$

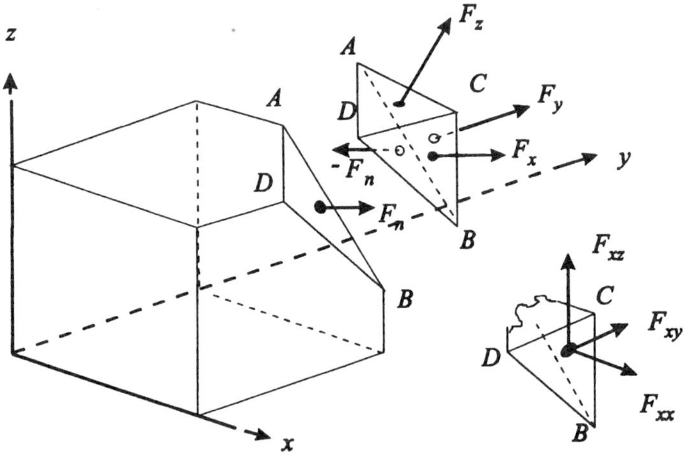

Fig. 7.3

or

$$(F_n) = [\sigma](A_n) \tag{7.26}$$

where A_x is a component of the area vector $\boldsymbol{A}_n$ and

$$[\sigma] = \begin{bmatrix} \sigma_{xx} & \tau_{yx} & \tau_{zx} \\ \tau_{xy} & \sigma_{yy} & \tau_{zy} \\ \tau_{xz} & \tau_{yz} & \sigma_{zz} \end{bmatrix} = \begin{bmatrix} \sigma_{xx} & \sigma_{yx} & \sigma_{zx} \\ \sigma_{xy} & \sigma_{yy} & \sigma_{zy} \\ \sigma_{xz} & \sigma_{yz} & \sigma_{zz} \end{bmatrix} \tag{7.27}$$

is the stress matrix.

The force vector is

$$\boldsymbol{F}_n = (e)^T[\sigma](A_n) = (e)^T[\sigma](e)(e)^T(A_n) \tag{7.28}$$

Note that $(e)(e)^T = [I]$ the unit matrix.

The term $(e)^T[\sigma](e)$ is the second-order stress tensor or dyadic. That is, it is the physical quantity which, when it premultiplies the area vector, gives the force vector.

To show that the stress matrix is symmetrical consider an elemental rectangular volume as shown in Fig. 7.4. By taking moments about an axis through the centroid parallel to the x axis

$$(\sigma_{xy}\mathrm{d}x\,\mathrm{d}z)\mathrm{d}y - (\sigma_{yx}\mathrm{d}y\,\mathrm{d}z)\mathrm{d}x = \rho\,\mathrm{d}x\,\mathrm{d}y\,\mathrm{d}z\,\frac{(\mathrm{d}x^2 + \mathrm{d}y^2)}{12}\,\ddot{\Omega}_z$$

In the limit as $\mathrm{d}x \to 0$

$$\sigma_{xy} - \sigma_{yx} = 0 \tag{7.29}$$

which demonstrates that $[\sigma]$ is symmetrical.

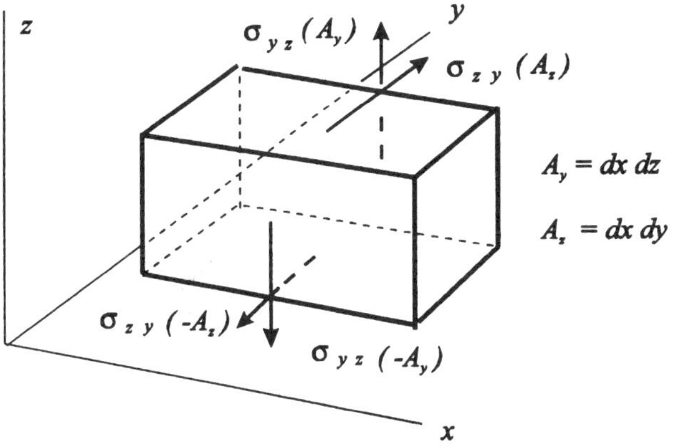

Fig. 7.4

7.4 Elastic constants

From the definitions of Young's modulus and Poisson's ratio

$$\varepsilon_{xx} = \frac{1}{E}\sigma_{xx} - \frac{\upsilon}{E}\sigma_{yy} - \frac{\upsilon}{E}\sigma_{zz}$$

$$= \left[\frac{1}{E} + \frac{\upsilon}{E}\right]\sigma_{xx} + \frac{\upsilon}{E}3p \tag{7.30}$$

where $p = -(\sigma_{xx} + \sigma_{yy} + \sigma_{zz})/3$, the mean pressure.

By definition of the bulk modulus, K,

$$p = -K\Delta \tag{7.31}$$

where $\Delta = (\varepsilon_{xx} + \varepsilon_{yy} + \varepsilon_{zz})$, the dilatation. Therefore equation (7.30) becomes

$$\varepsilon_{xx} = \left(\frac{1}{E} + \frac{\upsilon}{E}\right)\sigma_{xx} + \frac{\upsilon}{E}3K\Delta \tag{7.32}$$

or

$$\sigma_{xx} = \frac{E}{(1 + \upsilon)}\varepsilon_{xx} + \frac{3K\upsilon}{(1 + \upsilon)}\Delta \tag{7.33}$$

From elementary elasticity theory

$$E = 2G(1 + \upsilon) \tag{7.34}$$

and

$$K = \frac{E}{3(1 - 2\upsilon)} = \frac{2G(1 + \upsilon)}{3(1 - 2\upsilon)} \tag{7.35}$$

which means that equation (7.33) can be rewritten as

$$\sigma_{xx} = 2G\varepsilon_{xx} + \frac{2G\upsilon}{(1 - 2\upsilon)}\Delta \tag{7.36}$$

The Lamé constants λ and μ are defined by

$$\sigma_{xx} = 2\mu\varepsilon_{xx} + \lambda\Delta \tag{7.37}$$

from which it follows that by comparison with equation (7.36)

$$\mu = G \tag{7.38}$$

and

$$\lambda = \frac{2G\upsilon}{(1 - 2\upsilon)} \tag{7.39}$$

We also have

$$\tau_{xy} = G\gamma_{xy}$$

$$\sigma_{xy} = 2G\gamma_{xy}/2 = 2G\varepsilon_{xy}$$

or

$$\sigma_{xy} = 2\mu\varepsilon_{xy} \tag{7.40}$$

It is now possible to write equation (7.37) in matrix form

$$\begin{bmatrix} \sigma_{xx} & \sigma_{xy} & \sigma_{xz} \\ \sigma_{yx} & \sigma_{yy} & \sigma_{yz} \\ \sigma_{zx} & \sigma_{zy} & \sigma_{zz} \end{bmatrix} = 2\mu \begin{bmatrix} \varepsilon_{xx} & \varepsilon_{xy} & \varepsilon_{xz} \\ \varepsilon_{yx} & \varepsilon_{yy} & \varepsilon_{yz} \\ \varepsilon_{zx} & \varepsilon_{zy} & \varepsilon_{zz} \end{bmatrix} + \lambda\Delta[I]$$

or

$$[\sigma] = 2\mu[\varepsilon] + \lambda\Delta[I] \tag{7.41}$$

7.5 Equations of motion

A small elemental volume is shown in Fig. 7.5. The resultant force due to the stresses acting on the faces with normals in the x direction is

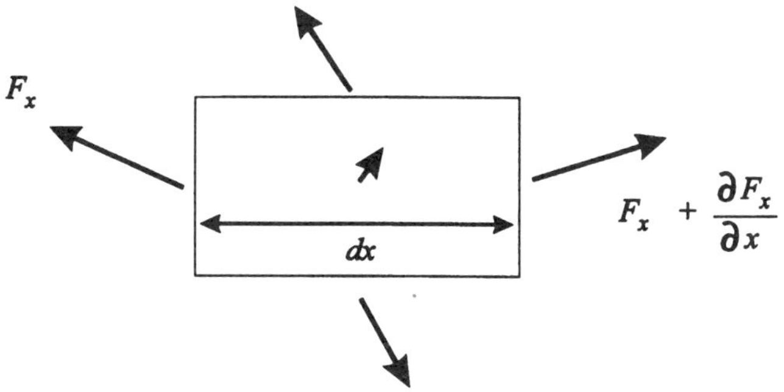

Fig. 7.5

$$\frac{\partial}{\partial x}(F_x)\mathrm{d}x = \frac{\partial}{\partial x}\begin{bmatrix}\sigma_{xx}\\ \sigma_{yx}\\ \sigma_{zx}\end{bmatrix}(\mathrm{d}y\,\mathrm{d}z)\mathrm{d}x \tag{7.42}$$

Summing for all three pairs of faces gives the total resultant force

$$\mathrm{d}(F) = \left[\frac{\partial}{\partial x}\begin{bmatrix}\sigma_{xx}\\ \sigma_{yx}\\ \sigma_{zx}\end{bmatrix} + \frac{\partial}{\partial y}\begin{bmatrix}\sigma_{xy}\\ \sigma_{yy}\\ \sigma_{zy}\end{bmatrix} + \frac{\partial}{\partial z}\begin{bmatrix}\sigma_{xz}\\ \sigma_{yz}\\ \sigma_{zz}\end{bmatrix}\right]\text{volume}$$

$$= \left[\begin{bmatrix}\sigma_{xx} & \sigma_{xy} & \sigma_{xz}\\ \sigma_{yx} & \sigma_{yy} & \sigma_{yz}\\ \sigma_{zx} & \sigma_{zy} & \sigma_{zz}\end{bmatrix}\right]\begin{bmatrix}\frac{\partial}{\partial x}\\ \frac{\partial}{\partial y}\\ \frac{\partial}{\partial z}\end{bmatrix}\text{volume}$$

$$= [\sigma](\nabla)\text{volume} \tag{7.43}$$

Equating this force to the rate of change of momentum leads to

$$[\sigma](\nabla)\text{volume} = \rho\ \text{volume}\ (\ddot{u})$$

or

$$[\sigma](\nabla) = \rho(\ddot{u}) \tag{7.44}$$

where the double dots over the u signify the second partial derivative with respect to time and it being understood that

$$\sigma\frac{\partial}{\partial x} = \frac{\partial\sigma}{\partial x} \quad \text{etc.}$$

Equation 7.44 could be written as

$$\{(\nabla)^{\mathrm{T}}[\sigma]\}^{\mathrm{T}} = \rho\frac{\partial^2(u)}{\partial t^2} \tag{7.44a}$$

7.6 Wave equation for an elastic solid

In this section we shall derive the wave equations; that is, we need to eliminate stress and strain from the equation of motion already derived. We first collect together the equations needed.

Equation of motion

$$[\sigma](\nabla) = \rho(\ddot{u}) \tag{7.44}$$

kinematics

$$\text{strain} \quad (\varepsilon) = \frac{1}{2}\left\{\left[(\nabla)(u)^{\mathrm{T}}\right]^{\mathrm{T}} + \left[(\nabla)(u)^{\mathrm{T}}\right]\right\} \tag{7.19}$$

$$\text{dilatation} \quad \Delta = (\nabla)^{\mathrm{T}}(u) = (u)^{\mathrm{T}}(\nabla) \tag{7.32}$$

$$\text{rotation} \quad (\Omega) = \frac{1}{2}[\nabla]^{\times}(u) \tag{7.12}$$

and the elastic relationships

$$[\sigma] = 2\mu[\varepsilon] + \lambda\Delta[I] \tag{7.41}$$

Substituting equations (7.41) and (7.19) into (7.44) gives

$$\mu\,((u)(\nabla)^{\mathrm{T}} + (\nabla)(u)^{\mathrm{T}}))(\nabla) + \lambda\Delta[I](\nabla) = \rho(\ddot{u}) \tag{7.45}$$

Now we premultiply by $(\nabla)^{\mathrm{T}}$ so that

$$\mu\,(((\nabla)^{\mathrm{T}}(u)(\nabla)^{\mathrm{T}}(\nabla) + (\nabla)^{\mathrm{T}}(\nabla)(u)^{\mathrm{T}}(\nabla))) + \lambda\Delta(\nabla)^{\mathrm{T}}(\nabla) = \rho(\nabla)^{\mathrm{T}}(\ddot{u})$$

and using (7.32)

$$\mu\,((\Delta(\nabla)^{\mathrm{T}}(\nabla) + (\nabla)^{\mathrm{T}}(\nabla)\Delta)) + \lambda\nabla^2\Delta = \rho\ddot{\Delta}$$

or

$$\mu\,(\Delta\nabla^2 + \nabla^2\Delta) + \lambda\nabla^2\Delta = \rho\ddot{\Delta}$$

Hence

$$(2\mu + \lambda)\nabla^2\Delta = \rho\ddot{\Delta} \tag{7.46}$$

In full

$$(2\mu + \lambda)\left(\frac{\partial^2\Delta}{\partial x^2} + \frac{\partial^2\Delta}{\partial y^2} + \frac{\partial^2\Delta}{\partial z^2}\right) = \rho\frac{\partial^2\Delta}{\partial t^2} \tag{7.46a}$$

which has the form of a classic wave equation in three dimensions.

This time we premultiply equation (7.45) by $[\nabla]^{\times}$ so that

$$\mu\,(([\nabla]^{\times}(u)(\nabla)^{\mathrm{T}} + [\nabla]^{\times}(\nabla)(u)^{\mathrm{T}}))(\nabla) + \lambda[\nabla]^{\times}\Delta(\nabla) = \rho[\nabla]^{\times}(\ddot{u})$$

Using equation (7.12) and noting that $[\nabla]^{\times}(\nabla) = (0)$ we get

$$\mu(2\Omega)\nabla^2 + (0) + (0) = \rho(2\ddot{\Omega})$$

or

$$\mu\nabla^2(\Omega) = \rho(\ddot{\Omega}) \tag{7.47}$$

Expanding we get three equations of the form

$$\mu\frac{\partial^2\Omega_x}{\partial x^2} = \rho\frac{\partial^2\Omega_x}{\partial t^2} \tag{7.47a}$$

Again this is the form of a classic wave equation in three dimensions.

The nature of the waves is best explored by the introduction of potential functions, one scalar function of position and one vector function of position. The functions are assumed to be defined such that

$$(u) = (\nabla)\text{ø} + [\nabla]^{x}(\psi) \tag{7.48}$$

The dilatation is then

$$\begin{aligned}\Delta = (\nabla)^{T}(u) &= \nabla^{2}(\text{ø}) + (\nabla)^{T}[\nabla]^{x}(\psi) \\ &= \nabla^{2}(\text{ø})\end{aligned} \tag{7.49}$$

and twice the rotation

$$\begin{aligned}2(\Omega) = [\nabla]^{x}(u) &= [\nabla]^{x}(\nabla)\text{ø} + [\nabla]^{x}[\nabla]^{x}(\psi) \\ &= [\nabla]^{x}[\nabla]^{x}(\psi)\end{aligned} \tag{7.50}$$

So we see that the dilatation is a function of the scalar function only and the rotation is a function of the vector function only.

The wave equations can be written in terms of the potential functions. Equation (7.46) becomes

$$(2\mu + \lambda)\nabla^{4}\text{ø} = \rho\nabla^{2}\ddot{\text{ø}}$$

so

$$(2\mu + \lambda)\nabla^{2}\text{ø} = \rho\ddot{\text{ø}} \tag{7.51}$$

Similarly equation (7.47) becomes

$$\mu\nabla^{2}[\nabla]^{x}[\nabla]^{x}(\psi) = \rho[\nabla]^{x}[\nabla]^{x}(\ddot{\psi})$$

or

$$[\nabla]^{x}[\nabla]^{x}\mu\nabla^{2}(\psi) = \rho[\nabla]^{x}[\nabla]^{x}(\ddot{\psi})$$

Therefore

$$\mu\nabla^{2}(\psi) = \rho(\ddot{\psi}) \tag{7.52}$$

Expanding equation (7.51) we get

$$(2\mu + \lambda)\left(\frac{\partial^{2}\text{ø}}{\partial x^{2}} + \frac{\partial^{2}\text{ø}}{\partial y^{2}} + \frac{\partial^{2}\text{ø}}{\partial z^{2}}\right) = \rho\frac{\partial^{2}\text{ø}}{\partial t^{2}} \tag{7.51a}$$

A solution to this wave equation is

$$\text{ø} = f(ct - s)$$

where s is a line with components x, y and z. It follows that since

$$s^{2} = x^{2} + y^{2} + z^{2}$$

$$\frac{\partial\text{ø}}{\partial x} = f'\frac{\partial s}{\partial x} = f'l$$

where l is the direction cosine between s and x.

Similarly

$$\frac{\partial\text{ø}}{\partial y} = f'm \quad\text{and}\quad \frac{\partial\text{ø}}{\partial z} = f'n$$

Substitution into equation (7.51a) yields

$$(2\mu + \lambda)(l^2 + m^2 + n^2)f'' = \rho(c^2)f''$$

As $l^2 + m^2 + n^2 = 1$ the phase velocity of a wave travelling in the direction of s is

$$c = \sqrt{[(2\mu + \lambda)/\rho]} \tag{7.53}$$

The displacement, from equation (7.48), is

$$(u) = (\nabla)^{\mathrm{T}}\phi$$

or

$$u_x = \frac{\partial \phi}{\partial x}$$

$$u_y = \frac{\partial \phi}{\partial y}$$

$$u_z = \frac{\partial \phi}{\partial z}$$

so without loss of generality we may take s to be in the x direction, in which case $\phi = \phi(x)$. It is clear that the particle motion is then in the direction of propagation, that is it is longitudinal. The wave is often referred to as dilatational because of the nature of equation (7.46) but, referring to Fig. 7.6, as there is no movement normal to the direction of propagation an elemental volume will change shape as the wave passes, and therefore some shear distortion occurs. The most accurate description of the wave is that it is irrotational because, as we have shown in equation (7.50), rotation is a function of ψ only.

For a disturbance which is represented by the vector function (ψ), let us consider a wave propagating in the x direction. In this case ∇^2 is a function of x only so equation (7.52) becomes

$$\mu \frac{\partial^2 \psi_x}{\partial x^2} = \rho \frac{\partial^2 \psi_x}{\partial t^2}$$

$$\mu \frac{\partial^2 \psi_y}{\partial x^2} = \rho \frac{\partial^2 \psi_y}{\partial t^2} \tag{7.54}$$

$$\mu \frac{\partial^2 \psi_z}{\partial x^2} = \rho \frac{\partial^2 \psi_z}{\partial t^2}$$

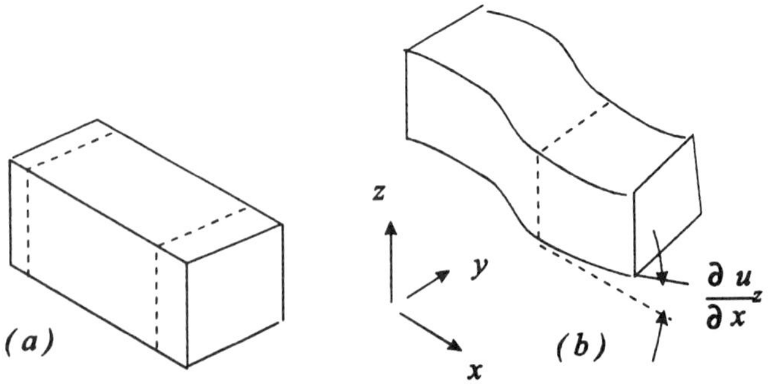

Fig. 7.6 (a) and **(b)**

The displacement is

$$(u) = [\nabla]^{\times}(\psi) = \begin{bmatrix} 0 & -\dfrac{\partial}{\partial z} & \dfrac{\partial}{\partial y} \\ \dfrac{\partial}{\partial z} & 0 & -\dfrac{\partial}{\partial x} \\ -\dfrac{\partial}{\partial y} & \dfrac{\partial}{\partial x} & 0 \end{bmatrix} \begin{bmatrix} \psi_x \\ \psi_y \\ \psi_z \end{bmatrix}$$

or

$$\begin{aligned} u_x &= -\psi_{y,z} + \psi_{z,y} \\ u_y &= \psi_{x,z} - \psi_{z,x} \\ u_z &= -\psi_{x,y} + \psi_{y,x} \end{aligned}$$

If (ψ) is a function of x only and (ψ) is constant in the yz plane then the displacement is

$$\begin{aligned} u_x &= 0 \\ u_y &= -\psi_{z,x} \\ u_z &= \psi_{y,x} \end{aligned}$$

which shows that the displacement is wholly normal to the direction of propagation. It is possible to choose the orientation of the yz axes so that $\psi_z = 0$ and the displacement is in the z direction only.

So, if

$$u_z = \frac{\partial \psi_y}{\partial x}$$

the rotation (*see* equations (7.11) and (7.12))

$$(\Omega) = \frac{1}{2}[\nabla]^{\times}(u)$$

yields

$$\Omega_y = -\frac{1}{2}\frac{\partial u_z}{\partial x} = -\frac{1}{2}\frac{\partial^2 \psi_y}{\partial x^2}$$

$$\Omega_x = \Omega_z = 0$$

which shows that the rotation is about the y axis. Figure 7.6(b) shows the deformation. From the figure it is clear that there is also shear deformation. As a result of this the wave is often called a shear wave but a more accurate description is equivoluminal because the dilatation is zero. The particle motion is transverse to the direction of propagation and is polarized because the direction of motion can be in any direction which is normal to the direction of propagation. The wave speed, from equation (7.52), is

$$\sqrt{(\mu/\rho)} \tag{7.55}$$

Summarizing the above results

Correct name	Irrotational	Equivoluminal
Common name	Dilatational	Shear
In seismology	Primary	Secondary(Horizontal or Vertical),polarization
Displacement	Longitudinal	Transverse
Wave speed	$\sqrt{[(2\mu + \lambda)/\rho]}$	$\sqrt{(\mu/\rho)}$
Symbols	c_1, c_d, c_p	c_2, c_s, c_{SH} or c_{SV}

7.7 Plane strain

In the previous section we studied a three-dimensional wave and found that two types of wave could be propagated, one with motion in the direction of travel and the other with motion normal to the direction of travel. Both types of wave arose as a solution to the three-dimensional wave equation in displacement $\boldsymbol{u}$, rotation $\boldsymbol{\Omega}$, the scalar function ø or the vector function ψ. If the direction of propagation is parallel to a vector $\boldsymbol{s} = s\boldsymbol{e}$, $\boldsymbol{e}$ being the unit vector, then we would expect a solution of the form used in the one-dimensional case to be applicable. So for any component of displacement we can write

$$u = f(ct - s)$$

for a wave travelling in the s direction.

Now, choosing a suitable origin,

$$\boldsymbol{s} = x\boldsymbol{i} + y\boldsymbol{j} + z\boldsymbol{k}$$

and

$$\boldsymbol{e} = l\boldsymbol{i} + m\boldsymbol{j} + n\boldsymbol{k}$$

where l, m and n are the direction cosines of the vector $\boldsymbol{s}$. So

$$s = \boldsymbol{e}\cdot\boldsymbol{s} = lx + my + nz$$

and therefore

$$u = f(ct - s) = f[ct - (lx + my + nz)] \tag{7.56}$$

The argument of f is,

$$(\arg f) = ct - (lx + my + nz) \tag{7.56a}$$

(we shall not use z in this chapter for the argument).

If we choose to consider a sinusoidal wave of the form

$$u = \mathrm{e}^{\mathrm{j}(\arg f)}$$

then we introduce a quantity k, as before, where k has the dimension (1/length) and is the wavenumber. If we make k the magnitude of a vector $\boldsymbol{K}$ then the argument could be written

$$(\arg f) = ckt - ks = ckt - \boldsymbol{K}\cdot\boldsymbol{s}$$

where

$$\boldsymbol{K} = k_x\boldsymbol{i} + k_y\boldsymbol{j} + k_z\boldsymbol{k} \tag{7.57}$$

As before ck is identified as the circular frequency ω, so

$$\begin{aligned}(\arg f) &= \omega t - \boldsymbol{K}\cdot\boldsymbol{s}\\ &= \omega t - (k_x x + k_y y + k_z z)\end{aligned} \tag{7.58}$$

If we wish to retain the use of an arbitrary function f then we need to modify the argument so that all wave functions have the same time component. This can be achieved by writing

$$(\arg f) = t - \left(\frac{l}{c}x + \frac{m}{c}y + \frac{n}{c}z\right) \tag{7.59}$$

This is now necessary as we can have two waves in the same medium travelling at different speeds. In the one-dimensional case, although we encountered different speeds of propagation, they were not superimposed. It was only required that the boundary conditions were satisfied.

We now restrict our attention to waves whose direction of propagation is in the xz plane and the particle motion is also confined to the xz plane. This implies that $u_y = 0$ and no quantity varies in the y direction.

Referring to Fig. 7.7

$$l = \sin\theta \quad m = 0 \quad \text{and} \quad n = \cos\theta$$

Therefore the argument is

$$(\arg f) = \left[t - \left(\frac{x}{c}\sin\theta + \frac{z}{c}\cos\theta\right)\right] \tag{7.60}$$

In terms of potentials the displacement is

$$(u) = (\nabla)^{\mathrm{T}}\phi + [\nabla]^{\times}(\psi)$$

The conditions specified are met if $\partial/\partial y = 0$ and $\psi_x = \psi_z = 0$. Thus

$$(u) = \begin{bmatrix} \dfrac{\partial\phi}{\partial x} \\ 0 \\ \dfrac{\partial\phi}{\partial z} \end{bmatrix} + \begin{bmatrix} 0 & -\dfrac{\partial}{\partial z} & 0 \\ \dfrac{\partial}{\partial z} & 0 & -\dfrac{\partial}{\partial x} \\ 0 & \dfrac{\partial}{\partial x} & 0 \end{bmatrix} \begin{bmatrix} 0 \\ \psi_y \\ 0 \end{bmatrix}$$

$$u_x = \frac{\partial\phi}{\partial x} - \frac{\partial\psi_y}{\partial z} = \phi_{,x} - \psi_{y,z} \tag{7.61}$$

$$u_z = \frac{\partial\phi}{\partial z} + \frac{\partial\psi_y}{\partial x} = \phi_{,z} + \psi_{y,x} \tag{7.62}$$

From equation (7.17)

$$(\varepsilon) = \frac{1}{2}[[(\nabla)(u)^{\mathrm{T}}]^{\mathrm{T}} + [(\nabla)(u)]]$$

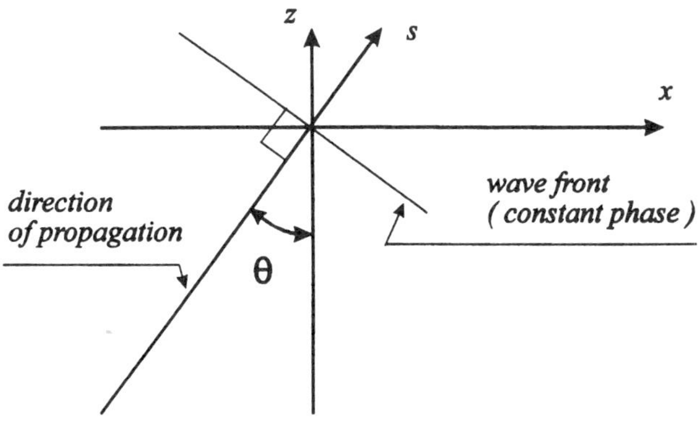

Fig. 7.7

Because the strain is planar we will need only the first and third rows and first and third columns. Thus

$$(\nabla)(u)^{\mathrm{T}} = \begin{bmatrix} \dfrac{\partial}{\partial x} \\ \dfrac{\partial}{\partial z} \end{bmatrix} (\varnothing_{,x} - \psi_{y,z} \quad \varnothing_{,z} + \psi_{y,x})$$

$$= \begin{bmatrix} \varnothing_{,xx} - \psi_{y,zx} & \varnothing_{,zx} + \psi_{y,xx} \\ \varnothing_{,xz} - \psi_{y,zz} & \varnothing_{,zz} + \psi_{y,xz} \end{bmatrix}$$

Hence the strain is

$$[\varepsilon] = \begin{bmatrix} \varnothing_{,xx} - \psi_{y,zx} & \varnothing_{,xz} - \frac{1}{2}(\psi_{y,zz} - \psi_{y,xx}) \\ \varnothing_{,xz} - \frac{1}{2}(\psi_{y,zz} - \psi_{y,xx}) & \varnothing_{,zz} + \psi_{y,xz} \end{bmatrix} \tag{7.63}$$

and the dilatation

$$\Delta = \nabla^2\varnothing = \varnothing_{,xx} + \varnothing_{,zz} \tag{7.64}$$

From equation (7.41) the stress is

$$[\sigma] = 2\mu[\varepsilon] + \lambda\Delta[I]$$

which in two dimensions is

$$[\sigma] = 2\mu \begin{bmatrix} \varnothing_{,xx} - \psi_{y,zx} & \varnothing_{,xz} - \frac{1}{2}(\psi_{y,zz} - \psi_{y,xx}) \\ \varnothing_{,xz} - \frac{1}{2}(\psi_{y,zz} - \psi_{y,xx}) & \varnothing_{,zz} + \psi_{y,xz} \end{bmatrix}$$

$$+ \lambda \begin{bmatrix} \varnothing_{,xx} + \varnothing_{,zz} & 0 \\ 0 & \varnothing_{,xx} + \varnothing_{,zz} \end{bmatrix} \tag{7.65}$$

Note that although the strain is planar there will be a σ_{yy} component of stress but this is not relevant in the problems to be discussed.

7.8 Reflection at a plane surface

We shall now use the equations developed in the last section to study the reflection of a wave incident on a surface given by $z = 0$. First we consider a dilatational wave approaching the surface such that the direction of propagation makes an angle θ_i with the normal to the surface, *see* Fig. 7.8. The potential function for this wave will be

$$\varnothing_i = \varnothing_i\left[t - \left(\frac{x}{c}\sin(\theta_i) + \frac{z}{c}\cos(\theta_i)\right)\right] \tag{7.66}$$

We now assume that both a dilatational and a shear wave will be reflected; the functions will be

$$\varnothing_r = \varnothing_r\left[t - \left(\frac{x}{c_1}\sin(\theta_r) - \frac{z}{c_1}\cos(\theta_r)\right)\right] \tag{7.67}$$

and

$$\psi_r = \psi_r\left[t - \left(\frac{x}{c_2}\sin(\alpha_r) - \frac{z}{c_2}\cos(\alpha_r)\right)\right] \tag{7.68}$$

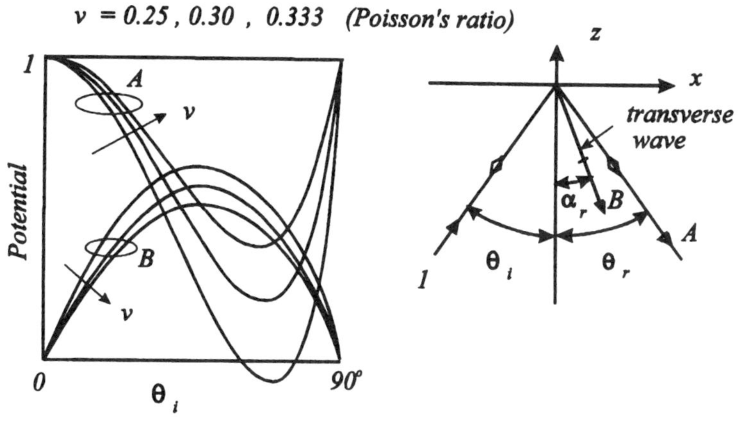

Fig. 7.8 Reflection of dilational wave at free surface

where ψ_r is the reflected ψ_y function. Note the change in sign of the z terms because the reflected waves are travelling in the negative z direction but still in the positive x direction.

For $z = 0$ the functions represent a wave moving along the surface. Therefore the phase velocity given by each function must be identical

$$c_p = \frac{x}{t} = \frac{c_1}{\sin(\theta_i)} = \frac{c_1}{\sin(\theta_r)} = \frac{c_2}{\sin(\alpha_r)} \tag{7.69}$$

Hence

$$\theta_r = \theta_i$$

and

$$\sin(\alpha_r) = \frac{c_2}{c_1}\sin(\theta_i) \tag{7.70}$$

This relationship is, of course, Snell's law.

The angle of reflection is now determined but we need to calculate the relative amplitudes of the reflected waves. These will be determined by the boundary conditions at the free surface. The conditions are that the direct stress and the shear stress at the surface shall at all times be zero. From equation (7.65)

$$\begin{aligned}\sigma_{zz} &= 2\mu(\varnothing_{i,zz} + \varnothing_{r,zz} + \psi_{r,xz}) \\ &\quad + \lambda(\varnothing_{i,xx} + \varnothing_{r,xx} + \varnothing_{i,zz} + \varnothing_{r,xx})\end{aligned} \tag{7.71}$$

and

$$\sigma_{xz} = 2\mu\left[\varnothing_{i,xz} + \varnothing_{r,xz} - \frac{1}{2}(\psi_{r,zz} - \psi_{r,xx})\right] \tag{7.72}$$

As both of these equations equate to zero we can substitute from equations (7.66) to (7.68) and divide through by the common factor to give

$$\begin{aligned}0 &= \overset{\prime\prime}{\varnothing}i\left[\frac{\cos^2\theta}{c_1^2} + \frac{\lambda}{2\mu}\left(\frac{\sin^2\theta}{c_1^2} + \frac{\cos^2\theta}{c_1^2}\right)\right] \\ &\quad + \overset{\prime\prime}{\varnothing}_r\left[\frac{\cos^2\theta}{c_1^2} + \frac{\lambda}{2\mu}\left(\frac{\sin^2\theta}{c_1^2} + \frac{\cos^2\theta}{c_1^2}\right)\right]\end{aligned}$$

$$- \psi_r''\left(\frac{\cos\alpha\ \sin\alpha}{c_2^2}\right)$$

and

$$0 = \o_i''\left(\frac{\cos\theta\ \sin\theta}{c_1^2}\right) - \o_r''\left(\frac{\sin\theta\ \cos\theta}{c_1^2}\right) - \frac{1}{2}\psi_r''\left(\frac{\cos^2\alpha\ \sin^2\alpha}{c_2^2}\right)$$

where $\theta = \theta_i = \theta_r$ and $\alpha = \alpha_r$. The primes signify differentiation with respect to the argument. Multiplying through by $2c_1^2$ and taking $\o_i''$ to be unity gives

$$\begin{bmatrix} (2\cos^2\theta + \frac{\lambda}{\mu}) & -(c_1/c_2)^2\sin 2\alpha \\ -\sin 2\theta & -(c_1/c_2)^2\cos 2\alpha \end{bmatrix} \begin{bmatrix} \o_r'' \\ \psi_r'' \end{bmatrix} = \begin{bmatrix} -(2\cos^2\theta + \frac{\lambda}{\mu}) \\ -\sin 2\theta \end{bmatrix} \tag{7.73}$$

From equations (7.38) and (7.39)

$$\frac{\lambda}{\mu} = \frac{2\upsilon}{(1 - 2\upsilon)} \tag{7.74}$$

and from equations (7.53) and (7.55)

$$\frac{c_1}{c_2} = \left(\frac{\lambda + 2\mu}{\mu}\right)^{1/2} = \left(\frac{2 - 2\upsilon}{1 - 2\upsilon}\right)^{1/2} \tag{7.75}$$

(Note that if $\upsilon = 1/3$ then $\lambda/\mu = 2$ and $c_1/c_2 = 2$.)

Given the angle of incidence equation (7.73) can be solved numerically for the relative amplitudes. The results of such calculations are given on Fig. 7.8 for three values of Poisson's ratio. Notice the sensitivity to Poisson's ratio.

The above analysis gives the ratio of the second derivatives of the arbitrary potential functions, which will be the same as the ratios of the functions themselves. From equations (7.61) and (7.62) it is seen that for the dilatational waves

$$u_x = \o_{,x} \quad \text{and} \quad u_z = \o_{,z}$$

so the displacement amplitude of a dilatational wave is

$$(u_x^2 + u_z^2)^{1/2} = \left(\frac{\o'^2\sin^2\theta}{c_1^2} + \frac{\o'^2\cos^2\theta}{c_1^2}\right)^{1/2}$$

$$= \o'/c_1 \tag{7.76}$$

For the shear waves the displacement is

$$u_x = -\psi_{y,z} \quad \text{and} \quad u_z = \psi_{y,x}$$

so the amplitude of a shear wave is

$$(u_x^2 + u_z^2)^{1/2} = \left(\frac{\psi'^2_y\sin^2\alpha}{c_2^2} + \frac{\psi'^2_y\cos^2\alpha}{c_2^2}\right)^{1/2}$$

$$= \psi'_y/c_1 \tag{7.77}$$

From equations (7.76) and (7.77) the relative amplitudes of the reflected waves may be found directly from the ratios of the potential functions.

For an incident shear wave the procedure is identical to that for the incident dilatational wave. Here the functions are

$$\psi_i = \varnothing_i\left[t - \left(\frac{x}{c_2}\sin\theta_i + \frac{z}{c_2}\cos\theta_i\right)\right] \tag{7.78}$$

$$\psi_r = \varnothing_r\left[t - \left(\frac{x}{c_2}\sin\theta_r + \frac{z}{c_2}\cos\theta_r\right)\right] \tag{7.79}$$

and

$$\varnothing_r = \psi_r\left[t - \left(\frac{x}{c_1}\sin\alpha_r - \frac{z}{c_1}\cos\alpha_r\right)\right] \tag{7.80}$$

Using these functions in conjunction with equations (7.71) and (7.72) the surface stresses can be equated to zero to produce

$$\begin{bmatrix} \sin 2\theta & -(c_2/c_1)^2(\frac{\lambda}{\mu} + 2\cos^2\alpha) \\ \cos 2\theta & (c_2/c_1)^2\sin 2\alpha \end{bmatrix} \begin{bmatrix} \psi_r'' \\ \varnothing_r'' \end{bmatrix} = \begin{bmatrix} \sin 2\theta \\ -\cos 2\theta \end{bmatrix} \tag{7.81}$$

Figure 7.9 shows a plot of the relative amplitudes. Here it is noticed that no waves are propagated if the angle of incidence exceeds a value of 30° to 35°, the exact value depending on Poisson's ratio. Although the reflected amplitudes appear to be large the energy in these waves can be shown to be equal to the incident energy.

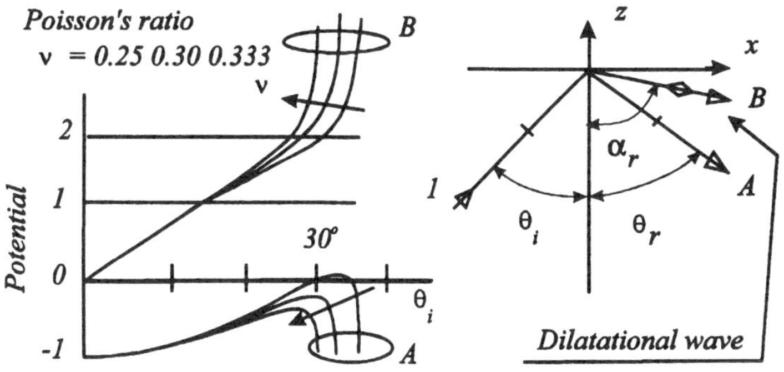

Fig. 7.9 Reflection of shear wave at free surface

7.9 Surface waves (Rayleigh waves)

Here we are seeking a solution to the wave equations of a form in which the amplitudes decay with distance from a free surface. In this case we shall postulate a running wave solution with potential functions

$$\varnothing = A(z)e^{j(\omega t - kx)} \tag{7.82}$$

and

$$\psi_y = B(z)e^{j(\omega t - kx)} \tag{7.83}$$

which is compatible with particle movement in the xz plane, as in the previous section. Also we have the same boundary condition of zero stress at the free surface.

The wave equation (7.51) can be written

$$c_1^2(\varnothing_{,xx} + \varnothing_{,zz}) = \ddot{\varnothing}$$

Substituting from equation (7.82) and dividing through by the common factor gives

$$c_1^2\left(A(-k^2) + \frac{\mathrm{d}^2A}{\mathrm{d}z^2}\right) = A(-\omega^2)$$

or

$$\frac{\mathrm{d}^2A}{\mathrm{d}z^2} - \left(k^2 - \frac{\omega^2}{c_1^2}\right)A = 0 \qquad (7.84)$$

Let

$$p_1 = \left(k^2 - \frac{\omega^2}{c_1^2}\right)^{1/2} \qquad (7.85)$$

so that

$$\frac{\mathrm{d}^2A}{\mathrm{d}z^2} - p_1^2A = 0 \qquad (7.86)$$

The general solution to this equation is

$$A = a\mathrm{e}^{-p_1z} + b\mathrm{e}^{p_1z} \qquad (7.87)$$

where a and b are arbitrary constants.

We require A to tend to zero as z tends to infinity and therefore $b = 0$ and at $z = 0$, $A = A_0$ and thus $a = A_0$. Hence

$$A(z) = A_0e^{p_1z} \qquad (7.88)$$

The wave equation (7.52) can be written as

$$c_2^2(\psi_{y,xx} + \psi_{y,zz}) = \ddot{\psi}_y$$

so substituting equation (7.83) yields

$$B(z) = B_0e^{-p_2z} \qquad (7.89)$$

where

$$p_2 = \left(k^2 - \frac{\omega^2}{c_2^2}\right)^{1/2} \qquad (7.90)$$

The potential functions are now

$$\varnothing = A_0e^{-p_1z}\,e^{\mathrm{j}(\omega t - kx)} \qquad (7.91)$$

and

$$\psi_y = B_0e^{-pz}\,e^{\mathrm{j}(\omega t - kx)} \qquad (7.92)$$

Equation (7.65) gives the stresses as

$$\sigma_{zz} = 2\mu(\varnothing_{,zz} + \psi_{y,xz}) + \lambda(\varnothing_{,xx} + \varnothing_{,zz}) = 0$$

and

$$\sigma_{xz} = 2\mu\left[\varnothing_{,xz} - \frac{1}{2}(\psi_{y,zz} - \psi_{y,xx})\right] = 0$$

Thus at $z = 0$

$$0 = 2\mu(p_1^2 A_0 - jkp_2 B_0) + \lambda(-k^2 ø + p_1^2 B_0)$$

$$0 = 2\mu\left[-jkp_1 A_0 - \frac{1}{2}(p_2^2 + k^2)B_0\right]$$

or

$$\begin{bmatrix} (\lambda + 2\mu)p_1^2 - \lambda k^2 & -jkp_2 2\mu \\ -jkp_1 2\mu & -(p_2^2 + k^2)2\mu \end{bmatrix}\begin{bmatrix} A_0 \\ B_0 \end{bmatrix} = \begin{bmatrix} 0 \\ 0 \end{bmatrix} \tag{7.93}$$

From equations (7.38) and (7.39)

$$\frac{\lambda}{\mu} = \frac{2\upsilon}{1 - 2\upsilon}$$

and

$$\frac{c_1^2}{c_2^2} = \frac{\lambda + 2\mu}{\mu} = \frac{\lambda}{\mu} + 2 = \frac{2 - 2\upsilon}{1 - \upsilon}$$

Let

$$(c_1/c_2) = R \tag{7.94}$$

Thus

$$\lambda/\mu = R - 2$$

Dividing all terms in equation (7.93) by μk^2 gives

$$\begin{bmatrix} \{R[1 - (c/c_1)^2] - (R - 2)\} & -j2\surd[1 - (c/c_2)^2] \\ -j2\surd[1 - (c/c_1)^2] & -[1 - (c/c_2)^2 + 1] \end{bmatrix}\begin{bmatrix} A_0 \\ B_0 \end{bmatrix} = \begin{bmatrix} 0 \\ 0 \end{bmatrix} \tag{7.95}$$

For a non-trivial solution the determinant of the square matrix must be zero. Thus

$$-[2 - (c/c_2)^2]^2 + 4\surd\{[1 - (c/c_2)^2/R][1 - (c/c_2)^2]\} = 0$$

A little more algebra leads to the following cubic in $(c/c_2)^2$

$$\left(\frac{c}{c_2}\right)^6 - 8\left(\frac{c}{c_2}\right)^4 + \left(24 - \frac{16}{R}\right)\left(\frac{c}{c_2}\right)^2 + \left(\frac{16}{R} - 16\right) = 0 \tag{7.96}$$

where

$$R = (c_1/c_2)^2 = \frac{1 - \upsilon}{1 - 2\upsilon} \tag{7.97}$$

It is well known that for a positive Poisson's ratio the value cannot exceed 0.5. Analysis of the cubic shows that there is always one real root with $c < c_2$. For $\upsilon < 0.263$ there are three real roots but the upper two are for wave speeds greater than c_1 which are not admissible. The speed of the Rayleigh wave does not depend on frequency but only on the elastic constants.

From the first equation of (7.95) the ratio of the amplitudes is

$$\frac{B_0}{A_0} = -j\left(\frac{2 - (c/c_2)^2}{2\surd[1 - (c/c_2)^2]}\right) \tag{7.98}$$

For ease of reference let $B_0/A_0 = -i\beta$. (7.99)

The computed values of wave speed and relative amplitudes of the potentials are given in the following table:

υ	0.250	0.300	0.333
c/c_2	0.919	0.927	0.933
B_0/A_0	−j1.47	−j1.52	−j1.56

From equations (7.61) and (7.62) we have

$$u_x = \text{ø}_{,x} - \psi_{y,z}$$

and

$$u_z = \text{ø}_{,z} + \psi_{y,x}$$

Substituting from equations 7.91 and 7.92

$$u_x = A_0 e^{-p_1 z} - jk\, e^{j(\omega t - kx)} - B_0 e^{-p_2 z} - p2\, e^{j(\omega t - kx)}$$

$$u_z = A_0 e^{-p_1 z} - p\, e^{j(\omega t - kx)} + B_0 e^{-p_2 z} - jk\, e^{j(\omega t - kx)}$$

At $x = 0$, $z = 0$ we get

$$u_x = (-jkA_0 + p_2 B_0)e^{j\omega t} \quad (7.100)$$

$$u_z = (-p_1 A_0 - jkB_0)e^{j\omega t} \quad (7.101)$$

Using equation (7.99) the ratio of the displacements may be written as

$$\frac{u_z}{u_x} = \frac{-jk\{1 + \beta\sqrt{[1 - (c/c_2)^2]}\}}{-k\{\sqrt{[1 - (c/c_2)^2(c_2/c_1)^2]} + \beta\}} \quad (7.102)$$

For $\upsilon = 0.3$ we have that $c/c_2 = 0.927$ and $\beta = 1.52$. Thus

$$\frac{u_z}{u_x} = j1.52 \quad (7.103)$$

This means that the amplitude of the displacement in the z direction is 1.52 times that in the x direction and is leading by 90°. That is, the motion is elliptical with the major axis vertical and the particle motion anti-clockwise.

The Rayleigh waves are similar to deep-water gravity waves except that the speed does not depend on wavelength. When dealing with high-frequency waves in a bar the Rayleigh wave is often the form of propagation, the motion being concentrated in the region near the surface. The exact solution for waves in an infinite cylindrical bar was developed by Pochammer and Chree and here the Rayleigh wave was the form for high-frequency axial and bending waves.

7.10 Conclusion

In these last two chapters we have attempted to bring out the most important physical aspects of wave propagation in elastic solids. Many of the ideas are new when compared with rigid body mechanics and to normal mode vibration theory. Wave methods are most useful for short-duration phenomena such as impact. The time scale is judged by comparing the impact time with the time taken for a disturbance to be reflected back from a boundary. Large-scale events like earthquakes require wave study and so do small-scale events where the wave speed is relatively low, as in problems involving springs.

The three-dimensional waves have been described using Cartesian co-ordinates but many interesting problems are best solved using cylindrical or spherical co-ordinates. The fundamental equations have been developed using the vector operator

$$\boldsymbol{\nabla} = \boldsymbol{i}\frac{\partial}{\partial x} + \boldsymbol{j}\frac{\partial}{\partial y} + \boldsymbol{k}\frac{\partial}{\partial z}$$

$$= (e)^{\mathrm{T}}(\nabla) = (\nabla)^{\mathrm{T}}(e)$$

The operations on a typical scalar ø and a typical vector ψ can be summarized.

The matrix form $(\nabla)ø$ has its vector counterpart

$$(e)^{\mathrm{T}}(\nabla)ø = \boldsymbol{\nabla}ø = \text{gradient } ø$$

Similarly $(\nabla)^{\mathrm{T}}(\psi) = (\nabla)^{\mathrm{T}}(e)\cdot(e)^{\mathrm{T}}(\psi)$ is the scalar

$$\boldsymbol{\nabla}\cdot\psi = \text{divergence } \psi$$

Also $[\nabla]^{\times}(\psi) = [\nabla]^{\times}(e)\cdot(e)^{\mathrm{T}}(\psi)$ has its vector form

$$\{(e)^{\mathrm{T}}[\nabla]^{\times}(e)\}\cdot\{(e)^{\mathrm{T}}(\psi)\} = \boldsymbol{\nabla}\times\psi = \text{curl(or rot) } \psi$$

Because all the equations derived in this chapter assume a common basis for the vectors (i.e. ***i***,***j*** and ***k***) the following identities can be made

$$(\nabla)ø \Rightarrow \text{grad}\,ø$$

$$(\nabla)^{\mathrm{T}}(\psi) \Rightarrow \text{div}\,\psi$$

$$[\nabla]^{\times}(\psi) \Rightarrow \text{curl}\,\psi$$

The expressions for div, grad and curl in cylindrical and spherical co-ordinates are given in Appendix 3. Also included are the relevant expressions for stress and strain.

8
Robot Arm Dynamics

8.1 Introduction

In this chapter we examine the way in which three-dimensional dynamics is applied to a system of rigid bodies connected by various types of joints. Initially we shall describe some typical arrangements of robot arms together with their end effectors. We shall only be concerned with the overall dynamics and not with the detail. This is a vast subject area of which dynamics is a substantial and vital part.

8.2 Typical arrangements

8.2.1 CARTESIAN CO-ORDINATES

Figure 8.1 shows the arrangement of a rectangular robot arm where the position of the end effector is located by specifying the x, y, z co-ordinates. Each joint responds to one co-ordinate, and all joints in this arrangement are sliding joints. An end effector is usually a gripper or hand-like mechanism; these will be briefly described later.

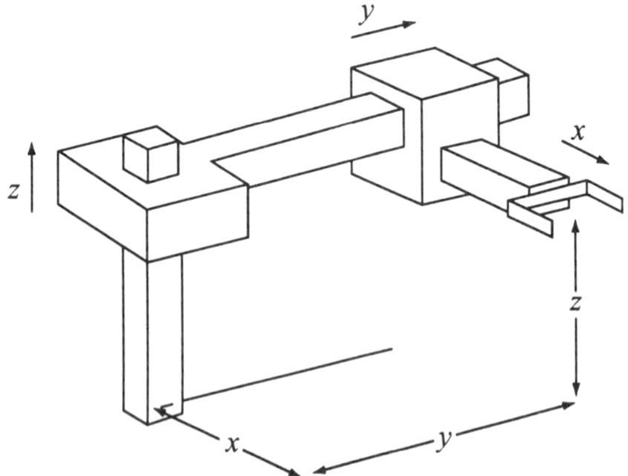

Fig. 8.1

8.2.2 CYLINDRICAL CO-ORDINATES

A typical cylindrical co-ordinate arm is shown in Fig. 8.2. In this case the joints respond to r, θ and z co-ordinates with the joints being sliding, revolute and sliding respectively.

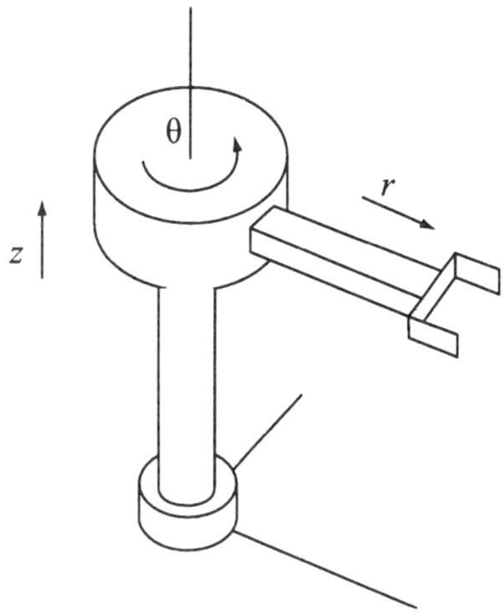

Fig. 8.2

8.2.3 SPHERICAL CO-ORDINATES

As can be seen from Fig. 8.3 this arm is controlled by specifying r, θ and ø with the joints being sliding and two revolute.

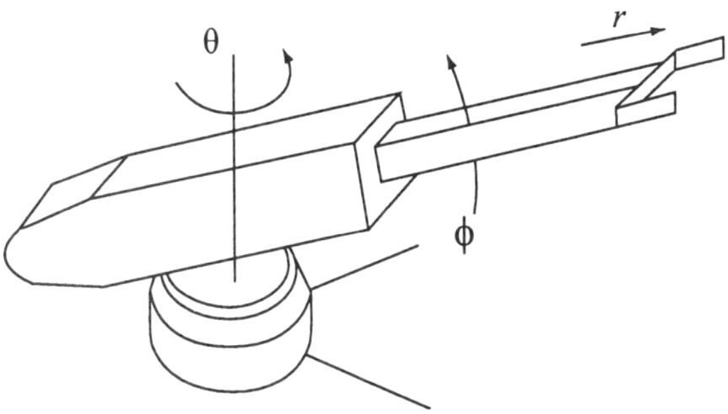

Fig. 8.3

8.2.4 REVOLUTE ARM

A very common layout is shown in Fig. 8.4(a) in which all joints are revolute; this is a versatile system and more akin to the human arm.

8.2.5 END EFFECTOR

A simple end effector in the form of a gripper is shown in Fig. 8.5. This example has three degrees of freedom plus a gripping action. The movements at the wrist are often referred to

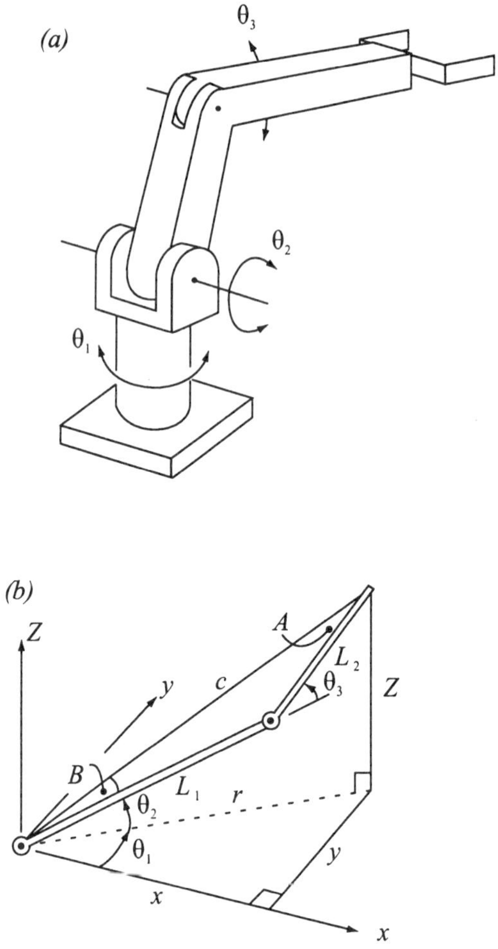

Fig. 8.4 (a) and **(b)**

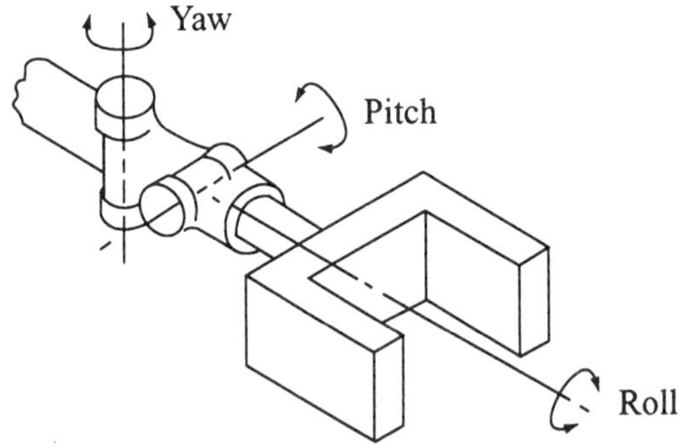

Fig. 8.5

as roll, pitch and yaw. It is quite common to find that for some end effectors only roll and pitch are provided.

8.3 Kinematics of robot arms

In this section we shall first revise and extend the study of the kinematics of a rigid body with particular reference to rotation about a point and change of reference axes. The concept of homogeneous transformation matrices will then be introduced so that a systematic description of arm position and displacement can be made.

The most common task to be performed is: given the path of the end effector, determine the magnitudes of the joint displacements as functions of time. This is referred to as the inverse kinematic problem and is usually more difficult than the forward problem of calculating the path of the end effector given the joint positions. The obvious exception is the case of the Cartesian system.

For one position of a cylindrical system

$$\begin{aligned} r &= \sqrt{(x^2 + y^2)} \\ z &= z \\ \theta &= \arctan (y/x) \end{aligned} \tag{8.1}$$

and for a spherical system

$$\begin{aligned} r &= \sqrt{(x^2 + y^2 + z^2)} \\ \theta &= \arctan [z/\sqrt{(x^2 + y^2)}\,] \\ \phi &= \arctan (y/x) \end{aligned} \tag{8.2}$$

For the revolute arm of Fig. 8.4(b)

$$\begin{aligned} \theta_1 &= \arctan (y/x) \\ r &= \sqrt{(x^2 + y^2)} \\ c &= \sqrt{(r^2 + z^2)} \\ A &= \arccos [\,(L_1^2 + c^2 - L_1^2) / (2L_2c)\,] \\ B &= \arcsin [\,(L_2/L_1)\ \sin A\,] \\ \theta_2 &= \arctan (z/r) - B \\ \theta_3 &= A + B \end{aligned} \tag{8.3}$$

8.3.1 VECTOR-MATRIX REPRESENTATION

A position vector $\boldsymbol{p}$ (shown in Fig. 8.6) has scalar components p_x, p_y and p_z when referred to the xyz frame. This is written

$$\boldsymbol{p} = \boldsymbol{i}p_x + \boldsymbol{j}p_y + \boldsymbol{k}p_z \tag{8.4}$$

which, in matrix form, becomes

$$\boldsymbol{p} = (\boldsymbol{i}\ \boldsymbol{j}\ \boldsymbol{k}) \begin{pmatrix} p_x \\ p_y \\ p_z \end{pmatrix}$$

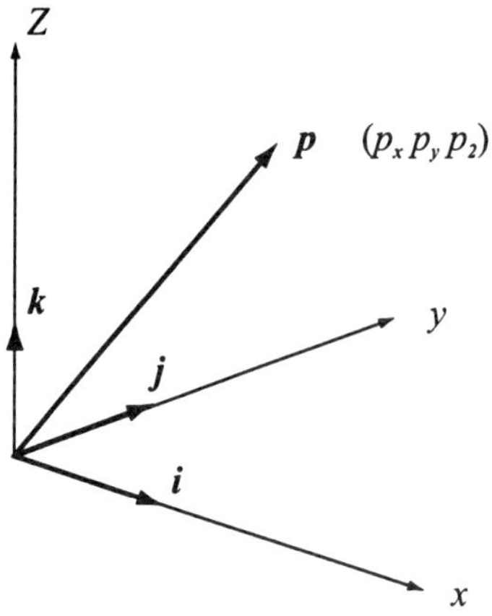

Fig. 8.6

If we let

$$(p) = (p_x\, p_y\, p_z)^{\mathrm{T}}$$

and

$$(e) = (\boldsymbol{i}\, \boldsymbol{j}\, \boldsymbol{k})^{\mathrm{T}}$$

then

$$\boldsymbol{p} = (e)^{\mathrm{T}}(p) \tag{8.5}$$

If the same vector is viewed from the set of primed axes as shown in Fig. 8.7

$$\boldsymbol{p} = (e')^{\mathrm{T}}(p') \tag{8.6}$$

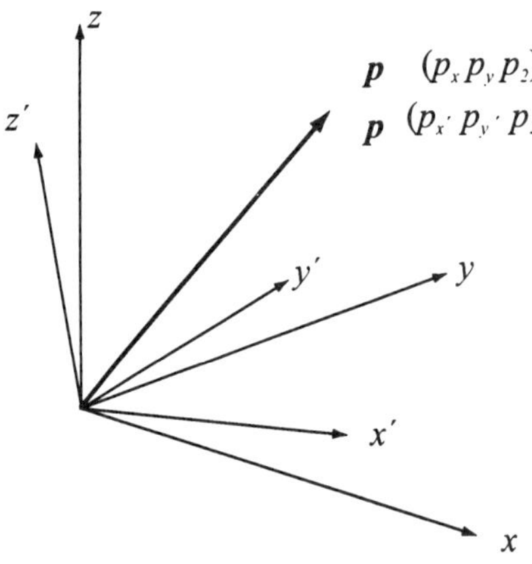

Fig. 8.7

8.3.2 CO-ORDINATE TRANSFORMATION

Since

$$\boldsymbol{p} = (e')^{\mathrm{T}}(p') = (e)^{\mathrm{T}}(p) \tag{8.7}$$

let us premultiply both sides by (e'), it being understood that the products of the unit vectors shall be the scalar products.

Thus

$$(e')(e')^{\mathrm{T}}(p') = (e')(e)^{\mathrm{T}}(p) \tag{8.8}$$

Now

$$(e')(e')^{\mathrm{T}} = \begin{pmatrix} \boldsymbol{i}' \\ \boldsymbol{j}' \\ \boldsymbol{k}' \end{pmatrix} (\,\boldsymbol{i}'\,\boldsymbol{j}'\,\boldsymbol{k}'\,) = \begin{bmatrix} \boldsymbol{i}'\cdot\boldsymbol{i}' & \boldsymbol{i}'\cdot\boldsymbol{j}' & \boldsymbol{i}'\cdot\boldsymbol{k}' \\ \boldsymbol{j}'\cdot\boldsymbol{i}' & \boldsymbol{j}'\cdot\boldsymbol{j}' & \boldsymbol{j}'\cdot\boldsymbol{k}' \\ \boldsymbol{k}'\cdot\boldsymbol{i}' & \boldsymbol{k}'\cdot\boldsymbol{j}' & \boldsymbol{k}'\cdot\boldsymbol{k}' \end{bmatrix}$$

$$= \begin{bmatrix} 1 & 0 & 0 \\ 0 & 1 & 0 \\ 0 & 0 & 1 \end{bmatrix} \tag{8.9}$$

so that equation (8.8) reads

$$(p') = (e')(e)^{\mathrm{T}}(p) \tag{8.8a}$$

and

$$(e')(e)^{\mathrm{T}} = \begin{pmatrix} \boldsymbol{i}' \\ \boldsymbol{j}' \\ \boldsymbol{k}' \end{pmatrix} (\,\boldsymbol{i}\,\boldsymbol{j}\,\boldsymbol{k}\,) = \begin{bmatrix} \boldsymbol{i}'\cdot\boldsymbol{i} & \boldsymbol{i}'\cdot\boldsymbol{j} & \boldsymbol{i}'\cdot\boldsymbol{k} \\ \boldsymbol{j}'\cdot\boldsymbol{i} & \boldsymbol{j}'\cdot\boldsymbol{j} & \boldsymbol{j}'\cdot\boldsymbol{k} \\ \boldsymbol{k}'\cdot\boldsymbol{i} & \boldsymbol{k}'\cdot\boldsymbol{j} & \boldsymbol{k}'\cdot\boldsymbol{k} \end{bmatrix}$$

$$= \begin{bmatrix} l_x & m_x & n_x \\ l_y & m_y & n_y \\ l_z & m_z & n_z \end{bmatrix} \tag{8.10}$$

where l_x, m_x, n_x are the direction cosines between the x' axis and the x, y, z axes, as shown in Fig. 8.8.

Now let $[\mathfrak{T}] = (e')(e)^{\mathrm{T}}$ so that equation (8.8a) becomes $(p') = [\mathfrak{T}](p)$; therefore $[\mathfrak{T}]$ is a transformation matrix. From the definition of the inverse of a matrix $(p) = [\mathfrak{T}]^{-1}(p')$ but by

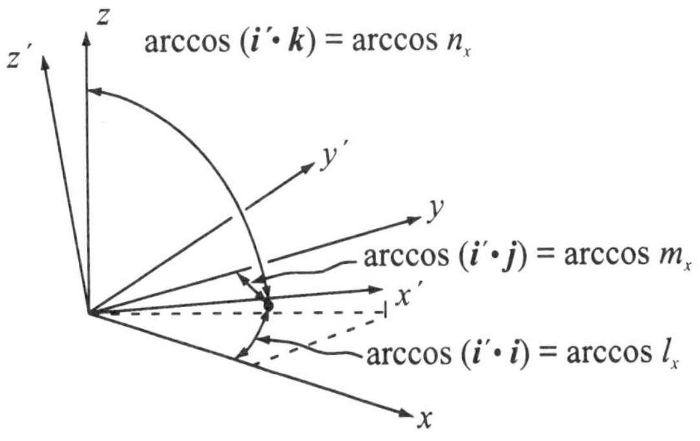

Fig. 8.8

premultiplying equation (8.7) by (e) and noting that $(e)(e)^{\mathrm{T}} = [I]$, the identity matrix, we obtain $(p) = (e)(e')^{\mathrm{T}}(p')$. By inspection it is seen that $(e)(e')^{\mathrm{T}}$ is the transpose of $(e')(e)^{\mathrm{T}}$. This is also seen from the rule for transposing the product of matrices, that is $[(e')(e)^{\mathrm{T}}]^{\mathrm{T}} = (e)(e')^{\mathrm{T}}$.

From this argument it is apparent that $[\mathfrak{T}]^{-1} = [\mathfrak{T}]^{\mathrm{T}}$, so by definition $[\mathfrak{T}]$ is an orthogonal matrix.

8.3.3 FINITE ROTATION

We shall now consider a closely related problem, that of rotating a vector.

Consider a vector $\boldsymbol{p}_1$ relative to fixed axes X, Y, Z. A further set of axes, U, V, W, moves with $\boldsymbol{p}_1$ and may be regarded as rigidly fixed to $\boldsymbol{p}_1$. If the UVW axes are rotated about the origin then relative to the fixed axes $\boldsymbol{p}_1$ moves to $\boldsymbol{p}_2$ as shown in Fig. 8.9.

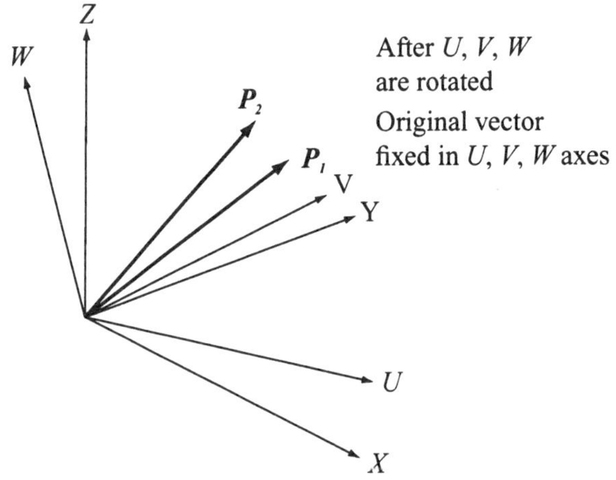

Fig. 8.9

Using the prime to indicate components seen from the UVW axes we have that initially $(p_1') = (p_1)$. We now look at $\boldsymbol{p}_2$ from the rotated axes UVW so that its components $(p_2') = [\mathfrak{T}](p_2)$, but because the vector is fixed relative to the UVW axes, $(p_2') = (p_1') = (p_1)$ and thus $(p_2) = [\mathfrak{T}]^{-1}(p_1)$.

If we define the rotation matrix $[R]$ by $(p_2) = [R](p_1)$ then

$$[R] = [\mathfrak{T}]^{-1} = [\mathfrak{T}]^{\mathrm{T}} = (e)(e')^{\mathrm{T}} \tag{8.11}$$

8.3.4 ROTATION ABOUT *X*, *Y* AND *Z* AXES

In general the rotation matrix is given by

$$[R] = \begin{pmatrix} \boldsymbol{i} \\ \boldsymbol{j} \\ \boldsymbol{k} \end{pmatrix} (\, \boldsymbol{i}' \, \boldsymbol{j}' \, \boldsymbol{k}' \,) = \begin{bmatrix} \boldsymbol{i}\cdot\boldsymbol{i}' & \boldsymbol{i}\cdot\boldsymbol{j}' & \boldsymbol{i}\cdot\boldsymbol{k}' \\ \boldsymbol{j}\cdot\boldsymbol{i}' & \boldsymbol{j}\cdot\boldsymbol{j}' & \boldsymbol{j}\cdot\boldsymbol{k}' \\ \boldsymbol{k}\cdot\boldsymbol{i}' & \boldsymbol{k}\cdot\boldsymbol{j}' & \boldsymbol{k}\cdot\boldsymbol{k}' \end{bmatrix} \tag{8.12}$$

So for rotation of the UVW axes by an angle α about the X axis, referring to Fig. 8.10, and noting that $\boldsymbol{i} = \boldsymbol{i}'$ and that $\boldsymbol{j}\cdot\boldsymbol{j}' = \cos(\text{angle between the } Y \text{ axis and the } V \text{ axis}) = \cos\alpha$, etc., the rotation matrix is

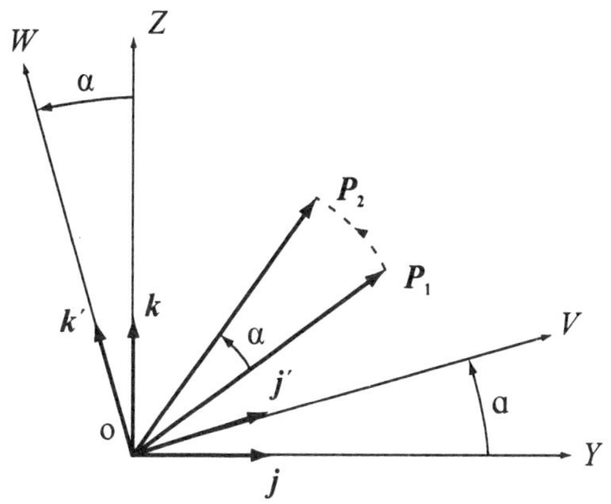

Fig. 8.10

$$[R]_{x,\alpha} = \begin{bmatrix} 1 & 0 & 0 \\ 0 & \cos\alpha & -\sin\alpha \\ 0 & \sin\alpha & \cos\alpha \end{bmatrix} \tag{8.12a}$$

This result should be verified by simple trigonometry.

Similarly for a rotation of β about the Y axis

$$[R]_{y,\beta} = \begin{bmatrix} \cos\beta & 0 & \sin\beta \\ 0 & 1 & 0 \\ -\sin\beta & 0 & \cos\beta \end{bmatrix} \tag{8.13}$$

and for a rotation of γ about the Z axis

$$[R]_{z,\gamma} = \begin{bmatrix} \cos\gamma & -\sin\gamma & 0 \\ \sin\gamma & \cos\gamma & 0 \\ 0 & 0 & 1 \end{bmatrix} \tag{8.14}$$

Note that by inspection

$$[R]_{x,\alpha}^{-1} = [R]_{x,\alpha}^{T} = [R]_{x,-\alpha} \tag{8.15}$$

That is, the transpose is the same as the inverse which is also the same as rotation by a negative angle.

8.3.5 SUCCESSIVE ROTATIONS ABOUT FIXED AXES

In this section we shall adopt a simpler notation for rotation matrices, replacing $[R]_{x,\alpha}$ by $[X,\alpha]$ to mean a rotation of α about the fixed X axis.

If a vector with components (p_1) as seen from the fixed axes is rotated about the X axis by an angle α, then the new components are

$$(p_2) = [X,\alpha](p_1) \tag{8.16}$$

If now this vector is rotated about the Y axis by an angle β then the components will be

$$(p_3) = [Y,\beta](p_2) = [Y,\beta][X,\alpha](p_1) \tag{8.17}$$

It follows that any further rotations result in successive premultiplications by the appropriate rotation matrix.

In the above case the new composite rotation matrix is

$$[R] = [Y,\beta][X,\alpha] = \begin{bmatrix} \cos\beta & 0 & \sin\beta \\ 0 & 1 & 0 \\ -\sin\beta & 0 & \cos\beta \end{bmatrix} \begin{bmatrix} 1 & 0 & 0 \\ 0 & \cos\alpha & -\sin\alpha \\ 0 & \sin\alpha & \cos\alpha \end{bmatrix}$$

$$= \begin{bmatrix} C\beta & S\beta S\alpha & S\beta C\alpha \\ 0 & C\alpha & -S\alpha \\ -S\beta & C\beta S\alpha & C\beta C\alpha \end{bmatrix}$$

where the usual abbreviations are made by writing C for cosine and S for sine.

It must be emphasized that reversing the order of the rotations produces a different result because $[X,\alpha][Y,\beta] \neq [Y,\beta][X,\alpha]$.

8.3.6 ROTATION ABOUT AN ARBITRARY AXIS

If we wish to form the rotation matrix for a rotation of ø about an axis defined by the unit vector $\boldsymbol{n}$ as shown in Fig. 8.11, one method is given in the following steps:

1. Rotate the axis of rotation so that it coincides with one of the fixed axes.
2. Rotate the body by ø about that axis.
3. Rotate the axis back to its original position.

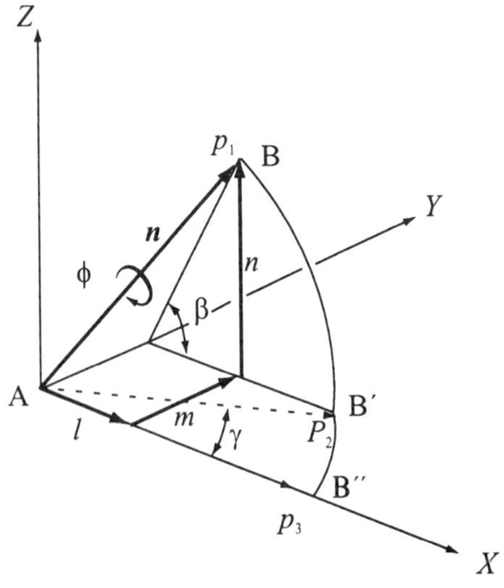

Fig. 8.11

Referring again to Fig. 8.11,

Step 1: Rotate the axis about the Y axis by β followed by a rotation of γ about the Z axis; $\tan\beta = n/l$ and $\sin\gamma = m$, where l, m and n are the components of the unit vector $\boldsymbol{n}$. Note that in this example γ would be numerically negative.
Step 2: Rotate by ø about the X axis.
Step 3: Rotate back.

In matrix form

$$[R]_{n,ø} = \underbrace{\{[Y,-\beta]\,[Z,\gamma]\}}_{\{\text{step 3}\}}\ \underbrace{\{[X,ø]\}}_{\{\text{step 2}\}}\ \underbrace{\{[Z,-\gamma][Y,\beta]\}}_{\{\text{step 1}\}} \qquad (8.18)$$

(Remember that $[Y,\beta]^{-1} = [Y,-\beta]$.)

Alternative method

A vectorial relationship can be achieved as is shown in Fig. 8.12. Here $\boldsymbol{n}$ is the unit vector in the direction of the rotation and ø is the finite angle of rotation. Owing to the rotation the vector $\boldsymbol{r}$ becomes $\boldsymbol{r}'$. The vector $\boldsymbol{r}$ generates the surface of a right circular cone; the head of the vector moves on a circular arc PQ. N is the centre of the circular arc so

$$\boldsymbol{n}\cdot\boldsymbol{r} = |\boldsymbol{r}|\cos\alpha = \text{ON}$$

and

$$|\boldsymbol{n}\times\boldsymbol{r}| = |\boldsymbol{r}|\sin\alpha = \text{NP} = \text{NQ}$$

Note also that the direction of $\boldsymbol{n}\times\boldsymbol{r}$ is that of $\overrightarrow{\text{VQ}}$.

Now

$$\boldsymbol{r}' = \overrightarrow{\text{ON}} + \overrightarrow{\text{NV}} + \overrightarrow{\text{VQ}}$$

$$= \boldsymbol{n}(\boldsymbol{n}\cdot\boldsymbol{r}) + [\boldsymbol{r} - \boldsymbol{n}(\boldsymbol{n}\cdot\boldsymbol{r})]\cos ø + (\boldsymbol{n}\times\boldsymbol{r})\sin ø$$

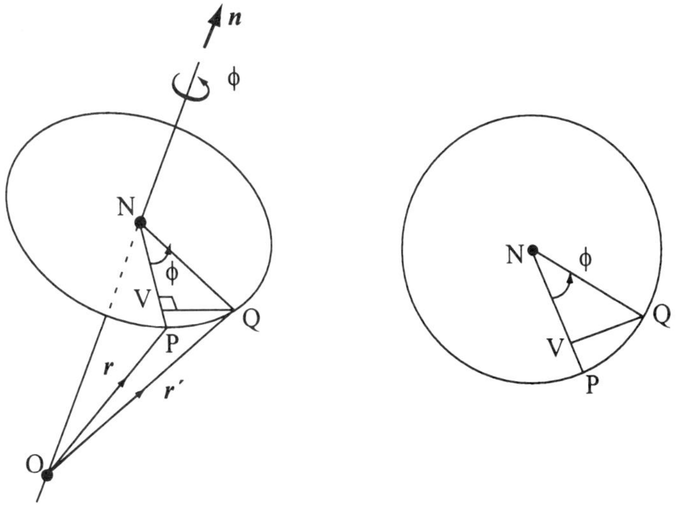

Fig. 8.12

$$= \boldsymbol{r}\cos\phi + \boldsymbol{n}(\boldsymbol{n}\cdot\boldsymbol{r})(1 - \cos\phi) + (\boldsymbol{n}\times\boldsymbol{r})\sin\phi \tag{8.19}$$

If we use the same basis for all vectors then the above vector equation may be written in matrix form (*see* Appendix 1 on vector–matrix algebra) as

$$(r)' = (r)\cos\phi + (n)(n)^{T}(r)(1 - \cos\phi) + [n]^{\times}(r)\sin\phi \tag{8.20}$$

where

$$(n) = (lmn)^{T} \quad (r) = (r_x r_y r_z)^{T}$$

$$[n]^{\times} = \begin{bmatrix} 0 & -n & m \\ n & 0 & -l \\ -m & l & 0 \end{bmatrix}$$

(l, m and n are the components of $\boldsymbol{n}$ referred to the chosen set of axes).

8.3.7 ROTATION ABOUT BODY AXES

It is very common for rotation to take place about axes which are fixed to the body and not to axes which are fixed in space. For example, with the end effector, or hand, the axes of pitch, roll and yaw are fixed with respect to the hand.

Let us first consider a simple case of just two successive rotations. In Fig. 8.13 a body with body axes *UVW* is initially lined up with the fixed *XYZ* axes. The body is first rotated by α about the *X* axis and then by γ about the *Z* axis. Exactly the same result can be obtained by a rotation of γ about the *W* axis followed by a rotation of α about the *U* axis. This can best be demonstrated by using a marked box as shown in Fig. 8.13(a).

The rotation matrix for the first case is

$$[R] = [Z,\gamma][X,\alpha]$$

The form of a transformation matrix for rotation about the *X* axis is identical to that for rotation by the same angle about the *U* axis, similarly for the *Y* and *V* axes and also the *Z* and *W* axes. So $[Z,\gamma][X,\alpha]$ must be equivalent to $[W,\gamma][U,\alpha]$. Note that the matrix for the second rotation now postmultiplies the matrix for the first rotation rather than premultiplying as it

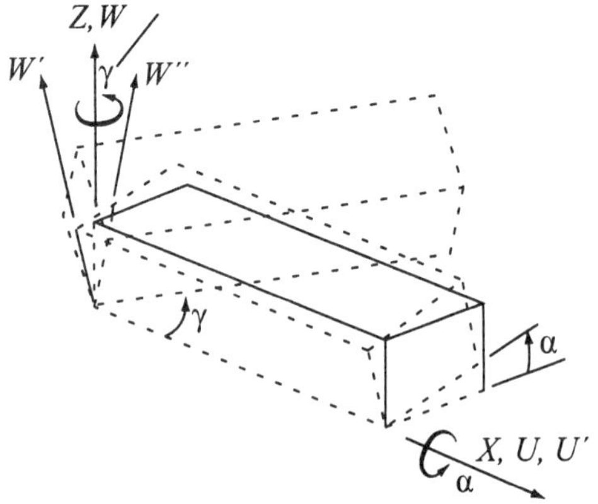

Fig. 8.13 (a)

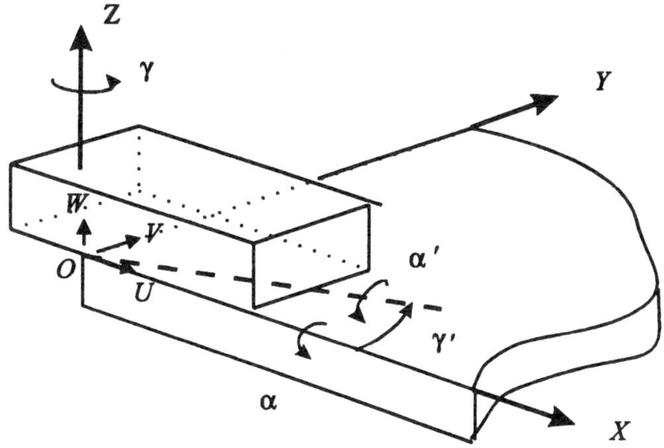

Fig. 8.13 (b)

did in the case of rotation about fixed axes. Because the first two rotations were completely abitrary it follows that the rule is general. However, further justification will now be given.

After the two rotations just made a further rotation β is now made about the V axis. This could be treated as a rotation about an arbitrary axis by rotating the body back to the initial position, rotating about the Y (or V) axis and then rotating the body back again. That is,

$$\begin{aligned} & \qquad\quad \{\text{rotate back}\}\ \{[Y,\beta]\}\ \{\text{return to base}\}\ \{\text{first two rotations}\} \\ [R] &= [Z,\gamma][X,\alpha] \quad [Y,\beta]\ [X,-\alpha][Z,-\gamma] \quad [Z,\gamma][X,\alpha] \\ &= [W,\gamma][U,\alpha][V\beta] \end{aligned} \tag{8.21}$$

Note that $[Z,-\gamma][Z,\gamma] = [X,-\alpha][X,\alpha] = [I]$, the identity matrix. This process can clearly be repeated for any further rotations about body axes.

In summary, for rotation about a fixed axis the new rotation matrix premultiplies the existing rotation matrix and for rotation about a body axis the new rotation matrix postmultiplies the existing rotation matrix.

8.3.8 HOMOGENEOUS CO-ORDINATES

The objective of this section is to find a way of producing transformation matrices which will allow for translation of a body as well as rotation.

For a pure translation $\boldsymbol{u}$ of a body, a point defined by a vector $\boldsymbol{p}_1$ from some origin will be transformed to a vector $\boldsymbol{p}_2$ where $\boldsymbol{p}_2 = \boldsymbol{p}_1 + \boldsymbol{u}$, or in terms of their components

$$p_{2x} = p_{1x} + u_x$$

$$p_{2y} = p_{1y} + u_y$$

$$p_{2z} = p_{1z} + u_z$$

or

$$(p_2) = (p_1) + (u) \tag{8.22}$$

For a combined rotation followed by a translation

$$(p_2) = [R](p_1) + (u) \tag{8.23}$$

If we now introduce an equation

$$1 = (0)^{\mathrm{T}}(p_1) + 1 \tag{8.24}$$

(where $(0) = (0\ 0\ 0)^{\mathrm{T}}$, a null vector), we may now combine equations (8.23) and (8.24) to give

$$\begin{bmatrix} (p_2) \\ 1 \end{bmatrix} = \begin{bmatrix} [R] & (u) \\ (0) & 1 \end{bmatrix} \begin{bmatrix} (p_1) \\ 1 \end{bmatrix} \tag{8.25}$$

or, in abbreviated form,

$$(\hat{p}_2) = [T](\hat{p}_1) \tag{8.26}$$

Here $(\hat{p})$ is the 4×1 homogeneous vector and $[T]$ is the 4×4 homogeneous transformation matrix. In projective geometry the null vector and unity are replaced by variables so that the transformation can also accomplish scaling and perspective, but these features are not required in this application.

For pure rotation $(u) = (0)$ and for a pure translation $[R] = [I]$ (the identity matrix). Therefore if we carry out the translation first (which is simply the vector addition of $\boldsymbol{p}_1$ and $\boldsymbol{u}$) and then perform the rotation the combined transformation matrix will be

$$\begin{bmatrix} [R] & (0) \\ (0)^{\mathrm{T}} & 1 \end{bmatrix} \begin{bmatrix} [I] & (u) \\ (0)^{\mathrm{T}} & 1 \end{bmatrix} = \begin{bmatrix} [R] & [R]\ (u) \\ (0)^{\mathrm{T}} & 1 \end{bmatrix}$$

so the transformed vector is

$$(p_2) = [R](p_1) + [R](u) = [R]((p_1) + (u))$$

as would be expected. Note that rotation followed by translation produces a different result. This is because the rotation is about the origin and not a point fixed on the body.

8.3.9 TRANSFORMATION MATRIX FOR SIMPLE ROBOT ARM

Figure 8.14 shows a Cartesian co-ordinate robot arm. It is required to express co-ordinates in *UVW* axes in terms of the *XYZ* axes. This can be achieved by starting with the *UVW* axes coincident with the *XYZ* axes and then moving the axes by a displacement *L* parallel to the *Z* axis, *M* parallel to the *X* axis and then by *N* parallel to the *Y* axis. (The order of events in this case is not important.) Writing this out in full we obtain the overall transformation matrix

$$\begin{bmatrix} 1 & 0 & 0 & 0 \\ 0 & 1 & 0 & N \\ 0 & 0 & 1 & 0 \\ 0 & 0 & 0 & 1 \end{bmatrix} \begin{bmatrix} 1 & 0 & 0 & M \\ 0 & 1 & 0 & 0 \\ 0 & 0 & 1 & 0 \\ 0 & 0 & 0 & 1 \end{bmatrix} \begin{bmatrix} 1 & 0 & 0 & 0 \\ 0 & 1 & 0 & 0 \\ 0 & 0 & 1 & L \\ 0 & 0 & 0 & 1 \end{bmatrix} = \begin{bmatrix} 1 & 0 & 0 & M \\ 0 & 1 & 0 & N \\ 0 & 0 & 1 & L \\ 0 & 0 & 0 & 1 \end{bmatrix}$$

This result is equivalent to a single displacement of $(M\ N\ L)^{\mathrm{T}}$.

We now consider a spherical co-ordinate arm as shown in Fig. 8.15 again starting with the two sets of axes in coincidence. First we could translate by d_x along the *X* axis, then rotate

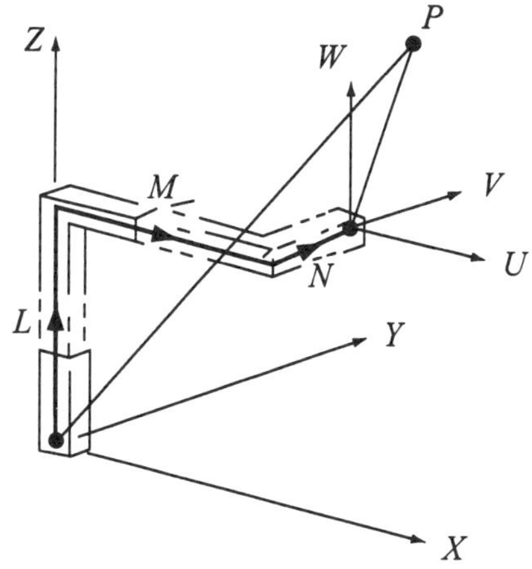

Fig. 8.14

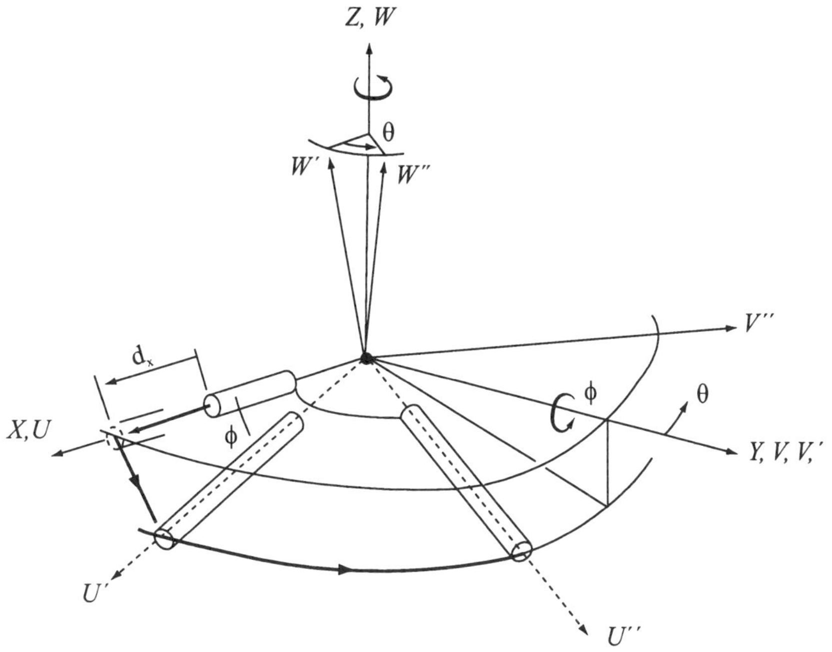

Fig. 8.15

by ø about the *Y* axis followed by θ about the *Z* axis. The overall transformation matrix is [*Z*,θ][*Y*,ø][d_x]

$$\begin{bmatrix} C\theta & -S\theta & 0 & 0 \\ S\theta & C\theta & 0 & 0 \\ 0 & 0 & 1 & 0 \\ 0 & 0 & 0 & 1 \end{bmatrix}\begin{bmatrix} C\phi & 0 & S\phi & 0 \\ 0 & 1 & 0 & 0 \\ -S\phi & 0 & C\phi & 0 \\ 0 & 0 & 0 & 1 \end{bmatrix}\begin{bmatrix} 1 & 0 & 0 & d_x \\ 0 & 1 & 0 & 0 \\ 0 & 0 & 1 & 0 \\ 0 & 0 & 0 & 1 \end{bmatrix}$$

$$= \begin{bmatrix} C\theta C\phi & -S\theta & C\theta S\phi & d_x C\theta C\phi \\ S\theta C\phi & C\theta & S\theta S\phi & d_x S\theta C\phi \\ -S\phi & 0 & C\phi & -d_x S\phi \\ 0 & 0 & 0 & 1 \end{bmatrix}$$

The above sequence could be interpreted as a rotation of θ about the *W* axis, a rotation of ø about the *V* axis and finally a translation along the *U* axis, as shown in Fig. 8.16.

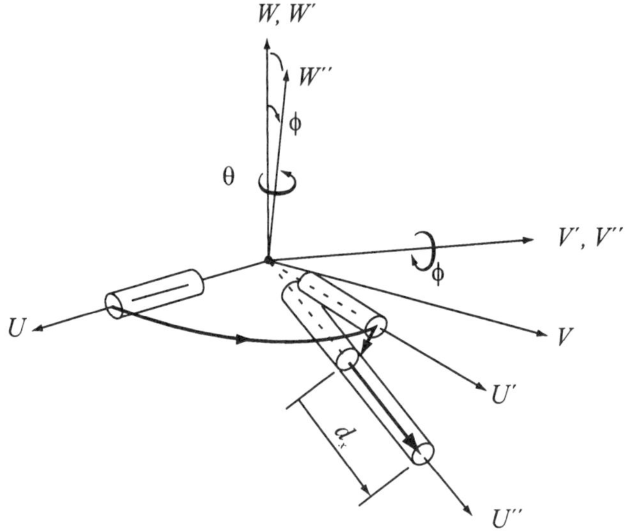

Fig. 8.16

8.3.10 THE DENAVIT–HARTENBERG REPRESENTATION

For more complicated arrangements it is preferable to use a standardized notation describing the geometry of a robot arm. Such a scheme was devised in 1955 by Denavit and Hartenberg and is now almost universally adopted.

Figure 8.17 shows an arbitrary rigid link with a joint at each end. The joint axis is designated the *z* axis and the joint may either slide parallel to the axis or rotate about the axis. To make the scheme general the joint axes at each end are taken to be two skew lines. Now it is a fact of geometry that a pair of skew lines lie in a unique pair of parallel planes; a clear visualization of this fact is very helpful in following the definitions of the notation. The *i*th link is defined to have joints which are labelled $(i - 1)$ at one end and (i) at the other. It is another geometric fact that there is a unique line which is the shortest distance between the two *z* axes, shown as N_{i-1} to O_i on Fig. 8.17, and is normal to both axes (and both planes).

If the joint axes are parallel then there is not a unique pair of planes, so choose the pair which are normal to N_{i-1} O_i. The origin of the $(i - 1)$ set of axes by definition lies on the z_{i-1} axis but the location along this axis and the orientation of the x_{i-1} axis have been determined by the previous links.

The *i*th set of axes have their origin at N_i and the x_i axis is the continuation of the line N_{i-1} to O_i. If the joint axes are in the same plane it follows that x_i is normal to that plane. This can be seen if the two planes are almost coincident.

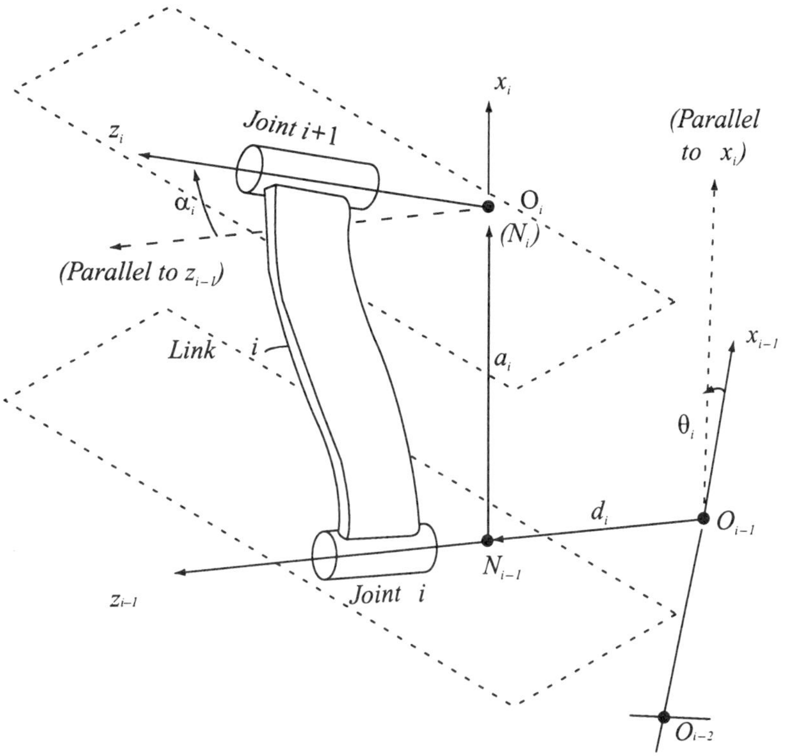

Fig. 8.17

The parameters which are used to define the link geometry and motion are:

θ_i is the joint angle from the x_{i-1} axis to the x_i axis about the z_{i-1} axis. A change in θ_i indicates a rotation of the ith link about the z_{i-1} axis.

d_i is the distance from the origin of the $(i - 1)$th co-ordinate origin to the intersection of the z_{i-1} axis with the x_i axis along the z_{i-1} axis (i.e. O_{i-1} to N_{i-1}). A change in d_i indicates a translation of the ith link along the z_{i-1} axis.

a_i is the offset distance from the intersection of the z_{i-1} axis with the x_i axis to the origin of the ith frame along the x_i axis (i.e. N_{i-1} O_i, the shortest distance between the two joint axes).

α_i is the offset angle from the z_{i-1} axis to the z_i axis about the x_i axis.

As in the simple cases considered previously we wish to perform a transformation of co-ordinates expressed in the ith co-ordinate system to those expressed in the $(i - 1)$th system. We achieve this by first lining up the ith to the $(i - 1)$th frame and consider the operations required to return the ith frame to its proper position. This is achieved by the following actions:

1. Rotate by θ_i about the z_{i-1} axis.
2. Translate by d_i along the z_{i-1} axis.
3. Translate by a_i along the x_i axis.
4. Rotate by α_i about the x_i axis.

The first two transformations are relative to the current base frame $(xyz)_{i-1}$ while the last two are relative to the body axes $(xyz)_i$. Thus the overall transformation matrix is

$$_{i-1}[A]_i = [T,z_{i-1},d_i][T,z_{i-1},\theta_i][T,x_i,a_i][T,x_i,\alpha_i]$$

Note the order of multiplication. This overall transformation matrix is often called the *A* matrix. In full we have

$$_{i-1}[A]_i = \begin{bmatrix} 1 & 0 & 0 & 0 \\ 0 & 1 & 0 & 0 \\ 0 & 0 & 1 & d_i \\ 0 & 0 & 0 & 1 \end{bmatrix}\begin{bmatrix} C\theta_i & -S\theta_i & 0 & 0 \\ S\theta_i & C\theta_i & 0 & 0 \\ 0 & 0 & 1 & 0 \\ 0 & 0 & 0 & 1 \end{bmatrix}\begin{bmatrix} 1 & 0 & 0 & a_i \\ 0 & 1 & 0 & 0 \\ 0 & 0 & 1 & 0 \\ 0 & 0 & 0 & 1 \end{bmatrix}\begin{bmatrix} 1 & 0 & 0 & 0 \\ 0 & C\alpha_i & -S\alpha_i & 0 \\ 0 & S\alpha_i & C\alpha_i & 0 \\ 0 & 0 & 0 & 1 \end{bmatrix}$$

$$= \begin{bmatrix} C\theta_i & -S\theta_i C\alpha_i & S\theta_i S\alpha_i & a_i C\theta_i \\ S\theta_i & C\theta_i C\alpha_i & -C\theta_i S\alpha_i & a_i S\theta_i \\ 0 & S\alpha_i & C\alpha_i & d_i \\ 0 & 0 & 0 & 1 \end{bmatrix} \tag{8.27}$$

So we may now write

$$(\hat{p}_{i-1}) = {}_{i-1}[A]_i\,(\hat{p}_i) \tag{8.28}$$

For a complete robot arm the *A* matrices for each link can be computed. The complete transformation is

$$(\hat{p}_0) = {}_0[A]_1\ {}_1[A]_2\ {}_2[A]_3 \cdots {}_{n-1}[A]_n\,(\hat{p})_n \tag{8.29}$$

8.3.11 APPLICATION TO A SIMPLE MANIPULATOR

Figure 8.18 shows a simple manipulator consisting of three links, all with revolute joints, and a gripper. We shall consider the problem in two stages, initially being concerned with the positioning of the end effector and later with its orientation and use.

The first task is to assign the axes; this has to be done carefully as it must obey the rules given in the previous section. The first link, a vertical pillar, rotates about the z_0 axis and the x_0 axis is chosen, arbitrarily, to be normal to the pin axis at the bottom of the pillar. The other two links are pin jointed as shown. The z_1 axis is the pin axis at the top of the pillar and the x_1 axis is normal to the plane containing the z_0 and z_1 axes. The z_2 and z_3 axes are both parallel to the z_1 axis. The x_2 axis lies along link 2 and the x_3 axis is along link 3.

We can now produce a table showing the parameters for each of the three links:

Link	θ	d	a	α
1	θ_1	d_1	0	+90°
2	θ_2	0	a_2	0
3	θ_3	0	a_3	0

The three variables are θ_1, θ_2 and θ_3; all other parameters are constant.

Notice that *a* is the shortest distance between the *z* axes and *d* is the shortest distance between the *x* axes.

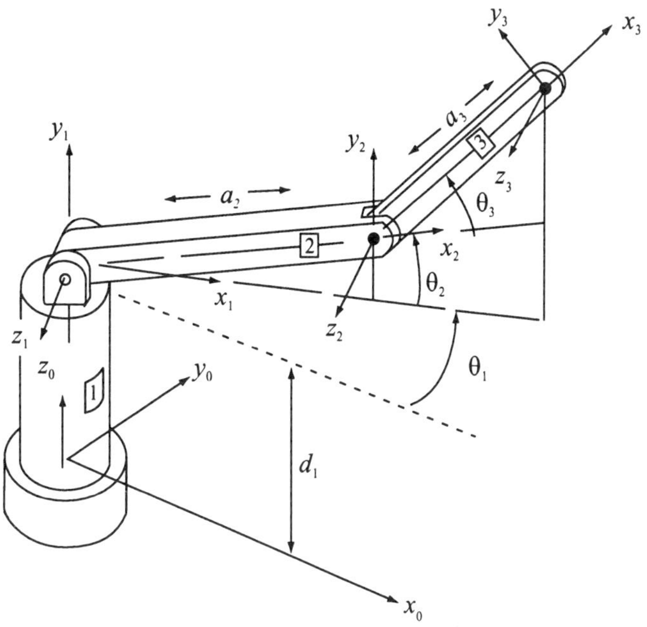

Fig. 8.18

In this arrangement d_1, a_2 and a_3 are constant so that the relevant A matrices are

$$ {}_0[A]_1 = \begin{bmatrix} C\theta_1 & 0 & S\theta_1 & 0 \\ S\theta_1 & 0 & -C\theta_1 & 0 \\ 0 & 1 & 0 & d_1 \\ 0 & 0 & 0 & 1 \end{bmatrix} $$

$$ {}_1[A]_2 = \begin{bmatrix} C\theta_2 & -S\theta_2 & 0 & a_2C\theta_2 \\ S\theta_2 & C\theta_2 & 0 & a_2S\theta_2 \\ 0 & 0 & 1 & 0 \\ 0 & 0 & 0 & 1 \end{bmatrix} $$

$$ {}_2[A]_3 = \begin{bmatrix} C\theta_3 & -S\theta_3 & 0 & a_3C\theta_3 \\ S\theta_3 & C\theta_3 & 0 & a_3S\theta_3 \\ 0 & 0 & 1 & 0 \\ 0 & 0 & 0 & 1 \end{bmatrix} $$

The overall transformation matrix is

$$ {}_0[A]_3 = {}_0[A]_1 \; {}_1[A]_2 \; {}_2[A]_3 $$

The elements of this matrix are

$$ A_{11} = \cos\theta_1 \cos(\theta_2 + \theta_3) $$

$$A_{12} = -\cos\theta_1 \sin(\theta_2 + \theta_3)$$

$$A_{13} = \sin\theta_1$$

$$A_{14} = \cos\theta_1 \,[a_3 \cos(\theta_2 + \theta_3) + a_2 \cos\theta_2]$$

$$A_{21} = \sin\theta_1 \cos(\theta_2 + \theta_3)$$

$$A_{22} = -\sin\theta_1 \sin(\theta_2 + \theta_3)$$

$$A_{23} = -\cos\theta_1$$

$$A_{24} = \sin\theta_1 \,[a_3 \cos(\theta_2 + \theta_3) + a_2 \cos\theta_2]$$

$$A_{31} = \sin(\theta_2 + \theta_3)$$

$$\mathrm{A}_{32} = \cos(\theta_2 + \theta_3)$$

$$\mathrm{A}_{33} = 0$$

$$\mathrm{A}_{34} = a_3 \sin(\theta_2 + \theta_3) + a_2 \sin\theta_2 + d_1$$

$$\mathrm{A}_{41} = 0$$

$$\mathrm{A}_{42} = 0$$

$$\mathrm{A}_{43} = 0$$

$$\mathrm{A}_{44} = 1$$

The origin of the $(xyz)_3$ axes is found from

$$(\hat{p}_0) = {}_0[A]_3(\hat{p}_3)$$

with $(\hat{p}_3) = (0\ 0\ 0\ 1)^{\mathrm{T}}$, *see* equation (8.25). Thus

$$x_0 = A_{14}$$

$$y_0 = A_{24}$$

$$z_0 = A_{34}$$

In this case these may easily be checked by trigonometry.

8.3.12 THE END EFFECTOR

To demonstrate the kinematic aspects of an end effector we shall consider a simple gripper, as shown in Fig. 8.19. The arrangement here is one of many possibilities; in this one the z_3 axis could be termed the pitch axis, z_4 the yaw axis and z_5 the roll axis. This is one example of the use of Eulerian angles which were discussed fully in Chapter 4.

The parameters for links 4, 5 and 6 are given in the following table:

Link	θ	d	a	α
4	θ_4	0	a_4	+90°
5	θ_5	0	0	+90°
6	θ_6	d_6	0	0

from which the homogeneous matrices are

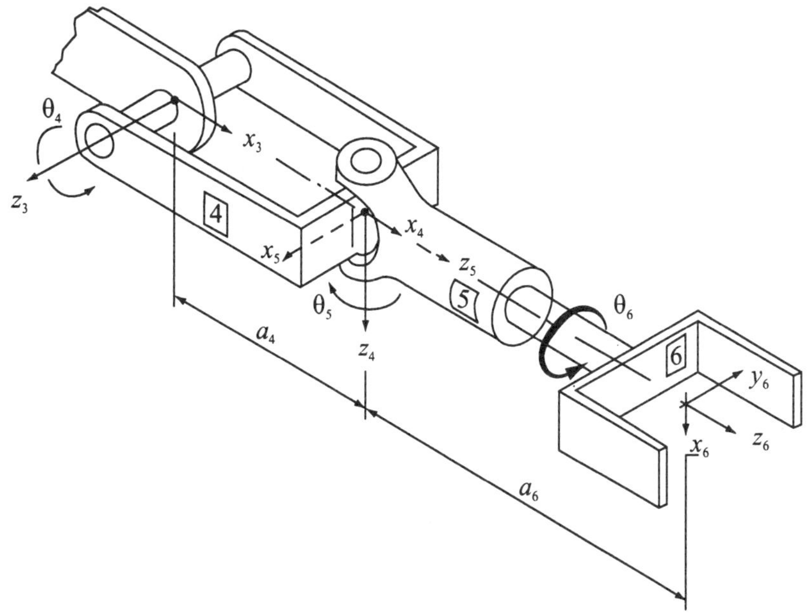

Fig. 8.19

$$
{}_3[A]_4 = \begin{bmatrix} C\theta_4 & 0 & S\theta_4 & a_4C\theta_4 \\ S\theta_4 & 0 & -C\theta_4 & a_4S\theta_4 \\ 0 & 1 & 0 & 0 \\ 0 & 0 & 0 & 1 \end{bmatrix}
$$

$$
{}_4[A]_5 = \begin{bmatrix} C\theta_5 & 0 & S\theta_5 & 0 \\ S\theta_5 & 0 & -C\theta_5 & 0 \\ 0 & 1 & 0 & 0 \\ 0 & 0 & 0 & 1 \end{bmatrix}
$$

$$
{}_5[A]_6 = \begin{bmatrix} C\theta_6 & -S\theta_6 & 0 & 0 \\ S\theta_6 & C\theta_6 & 0 & 0 \\ 0 & 0 & 1 & d_6 \\ 0 & 0 & 0 & 1 \end{bmatrix}
$$

The overall transformation matrix for the end effector is

$$
{}_3[A]_6 = {}_3[A]_4 \; {}_4[A]_5 \; {}_5[A]_6
$$

The components of ${}_3[A]_6$ are

$$
A_{11} = \cos\theta_4 \cos\theta_5 \cos\theta_6 + \sin\theta_4 \sin\theta_6
$$

$$
A_{12} = -\cos\theta_4 \cos\theta_5 \sin\theta_6 + \sin\theta_4 \cos\theta_6
$$

$$
A_{13} = \cos\theta_4 \sin\theta_6
$$

$$A_{14} = a_4 \cos\theta_4 + d_6 \cos\theta_4 \sin\theta_5$$

$$A_{21} = \sin\theta_4 \cos\theta_5 \cos\theta_6 - \cos\theta_4 \sin\theta_6$$

$$A_{22} = -\sin\theta_4 \cos\theta_5 \sin\theta_6 - \cos\theta_4 \cos\theta_6$$

$$A_{23} = \sin\theta_4 \sin\theta_5$$

$$A_{24} = a_4 \sin\theta_4 + d_6 \sin\theta_4 \sin\theta_5$$

$$A_{31} = \sin\theta_5 \cos\theta_6$$

$$A_{32} = -\sin\theta_5 \sin\theta_6$$

$$A_{33} = -\cos\theta_5$$

$$A_{34} = -d_6 \cos\theta_5$$

$$A_{41} = 0$$

$$A_{42} = 0$$

$$A_{43} = 0$$

$$A_{44} = 1$$

In this example the end effector is shown in a position for which $\theta_4 = 90°$, $\theta_5 = 90°$ and $\theta_6 = 0$. The complete transformation from the $(xyz)_6$ axes to the $(xyz)_0$ set of axes is

$$(\hat{p})_0 = {}_0[A]_6(\hat{p})_6 = {}_0[A]_3 \; {}_3[A]_6(\hat{p})_6$$

or, in general terms,

$$\begin{bmatrix} x_0 \\ y_0 \\ z_0 \\ 1 \end{bmatrix} = \begin{bmatrix} n_x & s_x & a_x & r_x \\ n_y & s_y & a_y & r_y \\ n_z & s_z & a_z & r_z \\ 0 & 0 & 0 & 1 \end{bmatrix} \begin{bmatrix} x_6 \\ y_6 \\ z_6 \\ 1 \end{bmatrix} \tag{8.30}$$

It is seen by putting $x_6 = 0$, $y_6 = 0$ and $z_6 = 0$ that $(r_x r_y r_z)$ is the location of the origin of the $(xyz)_6$ axes. If we put $x_6 = 1$ with $y_6 = 0$ and $z_6 = 0$ we have

$$x_0 - r_x = n_x$$

$$y_0 - r_y = n_y$$

$$z_0 - r_z = n_z$$

Therefore $(n_x n_y n_z)$ are the direction cosines of the x_6 axis. Similarly the components of (s) are the direction cosines of the y_6 axis and the components of (a) are the direction cosines of the z_6 axis. These directions are referred to as normal, sliding and approach respectively, the sliding axis being the gripping direction.

8.3.13 THE INVERSE KINEMATIC PROBLEM

For the simpler cases the inverse case can be solved by geometric means, *see* equations (8.1) to (8.3); that is, the joint variables may be expressed directly in terms of the co-ordinates and orientation of the end effector. For more complicated cases approximate techniques may be used. An iterative method which is found to converge satisfactorily is first to locate the end effector by a trial and error approach to the first three joint variables followed by a similar method on the last three variables for the orientation of the end effector. Further adjustments of the position of the arm will be necessary because moving the end effector will alter the reference point. This adjustment will then have a small effect upon the orientation.

Small variations of the joint variables can be expressed in terms of small variations of the co-ordinates. For example, if (p) is a function of (θ) then

$$\begin{bmatrix} \Delta x_0 \\ \Delta y_0 \\ \Delta z_0 \end{bmatrix} \begin{bmatrix} \frac{\partial x_0}{\partial \theta_1} & \frac{\partial x_0}{\partial \theta_2} & \frac{\partial x_0}{\partial \theta_3} \\ \frac{\partial y_0}{\partial \theta_1} & \frac{\partial y_0}{\partial \theta_2} & \frac{\partial y_0}{\partial \theta_3} \\ \frac{\partial z_0}{\partial \theta_1} & \frac{\partial z_0}{\partial \theta_2} & \frac{\partial z_0}{\partial \theta_3} \end{bmatrix} \begin{bmatrix} \Delta \theta_1 \\ \Delta \theta_2 \\ \Delta \theta_3 \end{bmatrix}$$

or

$$(\Delta p)_0 = [D]\,(\Delta\theta) \tag{8.31}$$

where $[D]$ is the matrix of partial differential coefficients which are dependent on position. It is referred to by some authors as the Jacobian. The partial differential coefficients can be obtained by differentiation of the respective A matrices. The matrix, if not singular, can be inverted to give

$$(\Delta\theta) = [D]^{-1}\,(\Delta p)_0 \tag{8.32}$$

since

$$(\Delta\theta) = (\theta)_{n+1} - (\theta)_n$$

$$(\theta)_{n+1} = (\theta)_n + [D]_n^{-1}\,(\Delta p)_0 \tag{8.33}$$

Repeated use enables the joint positions to be evaluated.

In general since

$$(\hat{p})_0 = {}_0[A]_1 \ldots {}_{n-1}[A]_n\,(\hat{p})_n$$

then

$$(\Delta\hat{p})_0 = \sum_{i=1}^{i-n} \left[{}_0[A]_1 \ldots \frac{\partial}{\partial q_i}\left({}_{i-1}[A]_i\right)\Delta q_i \ldots {}_{n-1}[A]_n \right] (\hat{p})_n$$

where q_i is one of the variables, that is θ or d.

It should be noted that in this context $(\hat{p})_n$ is constant and that ${}_{i-1}[A]_i$ is a function of θ_i for a revolute joint or of d_i for a prismatic/sliding joint.

For the general case

$${}_{i-1}[A]_i = \begin{bmatrix} C\theta_i & -S\theta_i C\alpha_i & S\theta_i S\alpha_i & a_i C\theta_i \\ S\theta_i & C\theta_i C\alpha_i & -C\theta_i S\alpha_i & a_i S\theta_i \\ 0 & S\alpha_i & C\alpha_i & d_i \\ 0 & 0 & 0 & 1 \end{bmatrix}$$

so differentiating with respect to θ_i

$$\frac{\partial}{\partial\theta_i}\,{}_{i-1}[A]_i = \begin{bmatrix} -S\theta_i & -C\theta_i C\alpha_i & C\theta_i S\alpha_i & -a_i S\theta_i \\ C\theta_i & -S\theta_i C\alpha_i & S\theta_i S\alpha_i & a_i C\theta_i \\ 0 & 0 & 0 & 0 \\ 0 & 0 & 0 & 0 \end{bmatrix} \tag{8.34}$$

Note that the right hand side of equation (8.34) may be written

$$\begin{bmatrix} 0 & -1 & 0 & 0 \\ 1 & 0 & 0 & 0 \\ 0 & 0 & 0 & 0 \\ 0 & 0 & 0 & 0 \end{bmatrix} \begin{bmatrix} C\theta_i & -S\theta_i C\alpha_i & S\theta_i S\alpha_i & a_i C\theta_i \\ S\theta_i & C\theta_i C\alpha_i & -C\theta_i S\alpha_i & a_i S\theta_i \\ 0 & S\alpha_i & C\alpha_i & d_i \\ 0 & 0 & 0 & 1 \end{bmatrix}$$

or in symbol form

$$\frac{\partial}{\partial\theta_i}\,{}_{i-1}[A]_i = [Q]\,{}_{i-1}[A]_i \tag{8.35}$$

where

$$[Q] = \begin{bmatrix} 0 & -1 & 0 & 0 \\ 1 & 0 & 0 & 0 \\ 0 & 0 & 0 & 0 \\ 0 & 0 & 0 & 0 \end{bmatrix}$$

In a similar manner

$$\frac{\partial}{\partial d_i}\,{}_{i-1}[A]_i = [P]\,{}_{i-1}[A]_i \tag{8.36}$$

where

$$[P] = \begin{bmatrix} 0 & 0 & 0 & 0 \\ 0 & 0 & 0 & 0 \\ 0 & 0 & 0 & 1 \\ 0 & 0 & 0 & 0 \end{bmatrix}$$

8.3.14 LINEAR AND ANGULAR VELOCITY OF A LINK

The basis for determining the joint velocities given the motion of a particular point has already been established in section 8.3.13 where the matrix $[D]$ was discussed. If we consider the variations to take place in time Δt and then make $\Delta t \to 0$ then, by definition of velocity,

$$(\dot{p})_0 = [D](\dot{\theta}) \tag{8.37}$$

so the joint velocities can found by inversion

$$(\dot{\theta}) = [D]^{-1}(\dot{p})_0 \tag{8.38}$$

Referring to Fig. 8.20 we can also write

$$(\hat{p})_0 = {}_0[A]_n(\hat{p})_n$$

Thus

$$\frac{\mathrm{d}}{\mathrm{d}t}(\hat{p})_0 = {}_0[\dot{A}]_n(\hat{p})_n \tag{8.39}$$

where $(\hat{p})_n$ is constant.

Consider the product of two A matrices

$$\begin{bmatrix} [R]_1 & (u)_1 \\ (0) & 1 \end{bmatrix}\begin{bmatrix} [R]_2 & (u)_2 \\ (0) & 1 \end{bmatrix} = \left[\begin{array}{c|c} [R]_1[R]_2 & [R]1(u)_2 + (u)_1 \\ \hline (0) & 1 \end{array}\right]$$

It is readily seen that for any number of multiplications the top left submatrix will be the product of all the rotation matrices and the top right submatrix is a function of $[R]$ and (u) submatrices. So in general the product of A matrices is of the form

$$[A] = \begin{bmatrix} [R] & (r) \\ (0) & 1 \end{bmatrix}$$

We have already shown that the column matrix (r) is the position of the origin of the final set of axes and the three columns of $[R]$ are the direction cosines of these axes.

So $(p)_0 = [R](p)_n + (r)$ and the position of a point relative to the base axes is

$$(p)_0 - (r) = (\Delta p) = [R](p)_n$$

Differentiating with respect to time gives the relative velocity

$$(\Delta\dot{p}) = [\dot{R}](p)_n \tag{8.40}$$

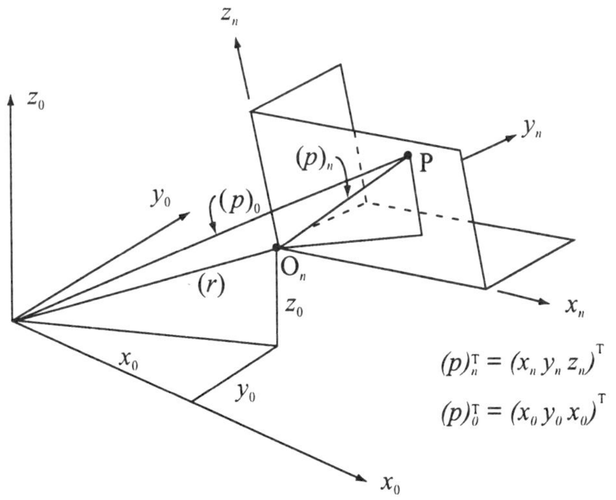

Fig. 8.20

so it is seen that $[\dot{R}]$ is related to the angular velocity of the nth link. Now $(\Delta\dot{p})$ is the velocity of P relative to O_n referred to the fixed base axes. We can find the components referred to the $(xyz)_n$ axes by premultiplying $(\Delta\dot{p})$ by $[R]^{-1} = [R]^T$ thus

$$(\Delta\dot{p})_n = [R]^T(\Delta\dot{p}) = [R]^T[\dot{R}](p)_n \tag{8.41}$$

We know that $[R]^T[R] = [I]$ so by differentiating with respect to time we have $[R]^T[\dot{R}] + [\dot{R}]^T[R] = [0]$ and as the second term is the transpose of the first it follows that $[R]^T[\dot{R}]$ is a skew symmetric matrix. This matrix will have the form

$$\begin{bmatrix} 0 & -\omega_z & \omega_y \\ \omega_z & 0 & -\omega_x \\ -\omega_y & \omega_x & 0 \end{bmatrix} = [\omega]^x = \begin{bmatrix} \omega_x \\ \omega_y \\ \omega_z \end{bmatrix}^x$$

where $(\omega_x \omega_y \omega_z)^T$ is the angular velocity vector of the nth link.

Now

$$[R]_n = \prod_{i=1}^{i=n} [R]_i$$

so

$$[\omega]_n^x = [R]_n^T[\dot{R}]_n = [R]_n^T \sum_{i=1}^{i=n} [R]_1[R]_2 \ldots [Q][R]_i \ldots [R]_n\dot{q}_i \tag{8.42}$$

where

$$[Q] = \begin{bmatrix} 0 & -1 & 0 \\ 1 & 0 & 0 \\ 0 & 0 & 0 \end{bmatrix}$$

for the 3×3 rotation matrices. Each term in the above series is equivalent to the change in $[\omega]^x$ as we progress from link to link.

8.3.15 LINEAR AND ANGULAR ACCELERATION

The second differentiation can be found by simply reapplying the rules developed for the first differentiation with respect to time.

We see that since

$$(\hat{p})_0 = {}_0[A]_n(\hat{p})_n$$

$$\frac{d^2}{dt^2}(\hat{p})_0 = \frac{d^2}{dt^2}\{{}_0[A]_n\}(\hat{p})_n \tag{8.43}$$

In order to see the operation let us look at a two-dimensional case and with $n = 2$, as shown in Fig. 8.21, for which the A matrices are functions of θ_1 and θ_2 so that

$$(\hat{p})_0 = {}_0[A]_1\,{}_1[A]_2(\hat{p})_2$$

where

$$(\hat{p})_0 = \begin{bmatrix} x_0 \\ y_0 \\ 1 \end{bmatrix} \quad \text{and} \quad (\hat{p})_2 = \begin{bmatrix} x_2 \\ y_2 \\ 1 \end{bmatrix}$$

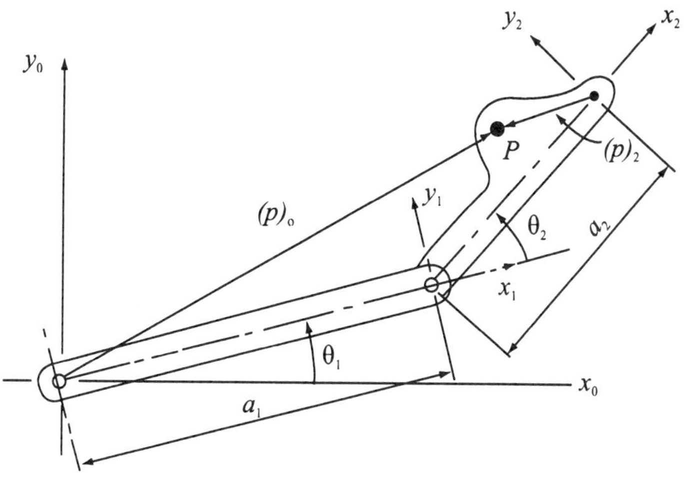

Fig. 8.21

and the A matrix for the two-dimensional case (for which $\alpha = 0$) may be written

$$[A] = \begin{bmatrix} \cos\theta & -\sin\theta & a\cos\theta \\ \sin\theta & \cos\theta & a\sin\theta \\ 0 & 0 & 1 \end{bmatrix}$$

Therefore

$$(\dot{p})_0 = [Q]_0[A]_1\ {}_1[A]_2\ \dot{\theta}_1(\hat{p})_2 + {}_0[A]_1[Q]_1[A]_2\ \dot{\theta}_2(\hat{p})_2$$

and

$$(\ddot{p})_0 = [\mathrm{Q}]_0^2[A]_1\ {}_1[A]_2\ \dot{\theta}_1^2(\hat{p})_2 + [Q]_0[A]_1[Q]_1[A]_2\ \dot{\theta}_1\dot{\theta}_2(\hat{p})_2$$
$$+\ [Q]_0[A]_1\ {}_1[A]_2\ \ddot{\theta}_1(\hat{p})_2$$
$$+\ [Q]_0[A]_1[Q]_1[A]_2\ \dot{\theta}_1\dot{\theta}_2(\hat{p})_2 + {}_0[A]_1[Q]_1^2[A]_2\ \dot{\theta}_2^2(\hat{p})_2 + {}_0[A]_1[Q]_1[A]_2\ \ddot{\theta}_1(\hat{p})_2$$

If we require the origin of the $(x\ y)_2$ axes to follow a specified path then for each point on the path, $(x\ y)_0$, the corresponding values of θ_1 and θ_2 can be found. Also if the velocities and accelerations of the point are prescribed the derivatives of the angles can be calculated using the above equations.

Once the values of θ_1, θ_2 and their derivatives are known any linear or angular velocity and linear or angular acceleration can be found.

The above scheme can in principle readily be extended to the three-dimensional case and any number of links can be considered.

The general form of the equations is

$$(\hat{p})_0 = \prod_{i=1}^{i=n}\ {}_{i-1}[A]_i\,(\hat{p})_n = [U]_n\,(\hat{p})_n \tag{8.44}$$

where $[U]_n$ is defined by the above equation (8.44).

The velocity is

$$(\dot{\hat{p}})_0 = \sum_{i=1}^{i=n} [U]_{n,i} (\hat{p})_n \dot{\theta}_i \tag{8.45}$$

where $[U]_{n,i}$ is a function of the A matrices and hence of the joint variables, that is

$$[U]_{n,i} = {}_0[A]_1 \ldots [Q]_{i-1}[\mathrm{A}]_i \ldots {}_{n-1}[\mathrm{A}]_n \tag{8.46}$$

For the acceleration

$$(\ddot{\hat{p}})_0 = \sum_{i=1}^{i=n} [U]_{n,i}(\hat{p})_n\ddot{\theta}_i + \sum_{j=1}^{j=n}\sum_{i=1}^{i=n} \frac{\partial}{\partial\theta_j} [U]_{n,i}(\hat{p})_n\dot{\theta}_i\dot{\theta}_j$$

$$= \sum_{i=1}^{i=n} [U]_{n,i}(\hat{p})_n\ddot{\theta}_i + \sum_{j=1}^{j=n}\sum_{i=1}^{i=n} [U]_{n,ij}(\hat{p})_n\dot{\theta}_i\dot{\theta}_j \tag{8.47}$$

where

$$[U]_{n,ij} = {}_0[A]_1\ {}_1[A]_2 \ldots [Q]_{i-1}[A]_i \ldots [Q]_{j-1}[A]_j \ldots {}_{n-1}[A]_n \tag{8.48}$$

We have shown previously that in general

$$[A] = \begin{bmatrix} [R] & (r) \\ 0 & 1 \end{bmatrix} \quad \text{and} \quad [\dot{A}] = \begin{bmatrix} [\dot{R}] & (\dot{r}) \\ 0 & 0 \end{bmatrix}$$

so

$$[\ddot{A}] = \begin{bmatrix} [\ddot{R}] & (\ddot{r}) \\ 0 & 0 \end{bmatrix}$$

Since, for constant $(\hat{p})_n$, we have

$$(\ddot{\hat{p}}) = [\ddot{A}]\,(\hat{p})_n$$

$$(\ddot{\hat{p}}) = \begin{bmatrix} [\ddot{R}] & (\ddot{r}) \\ 0 & 0 \end{bmatrix} (\hat{p})_n$$

then we have

$$(\ddot{p}) = [\ddot{R}](p)_n + (\ddot{r})$$

and thus the acceleration of P relative to the origin of the $(xyz)_n$ axes, $(\Delta\ddot{p})_n = (\ddot{p}) - (\ddot{r}) = [\ddot{R}](p)_n$.

If we now refer the components of this vector to the $(xyz)_n$ axes we have

$$[R]^{\mathrm{T}}(\Delta\ddot{p})_n = [R]^{\mathrm{T}}[\ddot{R}](p)_n.$$

Now

$$[\omega]^{\times} = [R]^{\mathrm{T}}[\dot{R}]$$

so

$$[\dot{\omega}]^{\times} = [\dot{R}]^{\mathrm{T}}[\dot{R}] + [R]^{\mathrm{T}}[\ddot{R}]$$

and therefore

$$[R]^{\mathrm{T}}[\ddot{R}] = [\dot{\omega}]^{\times} - [\dot{R}]^{\mathrm{T}}[\dot{R}]$$

Also

$$
\begin{aligned}
[\dot{R}]^{\mathrm{T}}[\dot{R}] &= [\dot{R}]^{\mathrm{T}}([R][R]^{\mathrm{T}})\,[\dot{R}] \\
&= ([R]^{\mathrm{T}}[\dot{R}])^{\mathrm{T}}([R]^{\mathrm{T}}[\dot{R}]) \\
&= [\omega]^{\mathrm{xT}}[\omega]^{\mathrm{x}} \\
&= -[\omega]^{\mathrm{x}}[\omega]^{\mathrm{x}}
\end{aligned}
$$

Finally

$$
\begin{aligned}
[R]^{\mathrm{T}}(\Delta\ddot{p})_n &= [R]^{\mathrm{T}}[\ddot{R}](p)_n \\
&= \{[\dot{\omega}]^{\mathrm{x}} + [\omega]^{\mathrm{x}}[\omega]^{\mathrm{x}}\}(p)_n \\
&= [\dot{\omega}]^{\mathrm{x}}(p)_n + [\omega]^{\mathrm{x}}[\omega]^{\mathrm{x}}(p)_n \qquad (8.49)
\end{aligned}
$$

where $(p)_n$ is of constant magnitude. This result is seen to agree with that obtained from direct vector analysis as shown in the next section.

8.3.16 DIRECT VECTOR ANALYSIS

It is possible to derive expressions for the velocity and acceleration of each link by vector analysis. The computation in this case uses only 3 × 1 and 3 × 3 matrices rather than the full 4 × 4 *A* matrices used in the last section.

Referring to Fig. 8.22 we see that the position vector of the origin of the $(xyz)_i$ axes is

$$\boldsymbol{r}_i = \boldsymbol{r}_{i-1} + \boldsymbol{d}_i + \boldsymbol{a}_i \qquad (8.50)$$

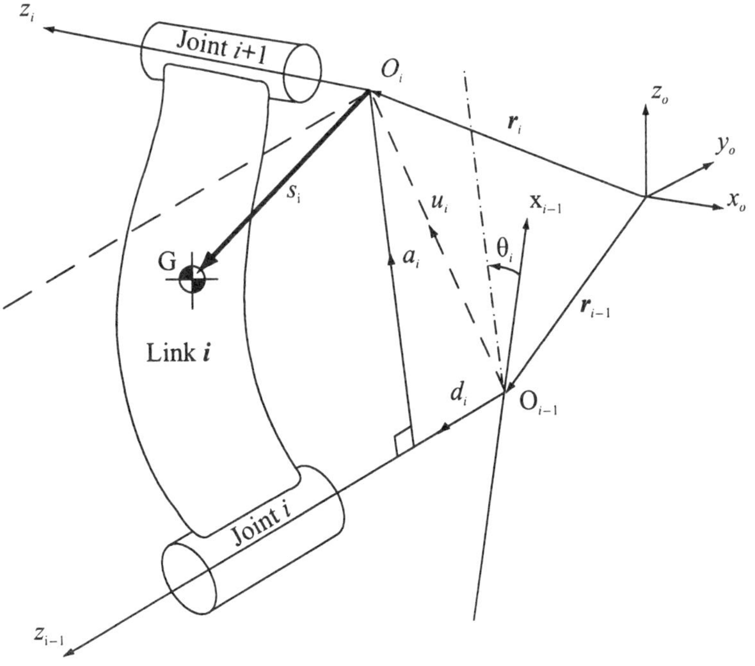

Fig. 8.22

and

$$\boldsymbol{\Omega}_i = \boldsymbol{\Omega}_{i-1} + \dot{\theta}_i \boldsymbol{k}_{i-1} \tag{8.51}$$

The velocity of O is

$$\boldsymbol{v}_i = \frac{\mathrm{d}\boldsymbol{r}_i}{\mathrm{d}t} = \frac{\mathrm{d}\boldsymbol{r}_{i-1}}{\mathrm{d}t} + \frac{\partial}{\partial t}(\boldsymbol{d}_i + \boldsymbol{a}_i) + \boldsymbol{\omega}_i \times (\boldsymbol{d}_i + \boldsymbol{a}_i) \tag{8.52}$$

We require our reference axes to be fixed to the ith link in order that when the moment of inertia of the ith link is introduced it shall be constant. For the second term on the right the partial differentiation means that only changes in magnitude as seen from the $(xyz)_i$ axes are to be considered. For a revolute joint where both $\boldsymbol{d}$ and $\boldsymbol{a}$ are constant in magnitude this term will be zero. For a prismatic joint the term will be $\dot{d}_i \boldsymbol{k}_{i-1}$.

To simplify the appearance of the subsequent equations we again use

$$\boldsymbol{u}_i = \boldsymbol{d}_i + \boldsymbol{a}_i \tag{8.53}$$

and write $\boldsymbol{v}_{i-1}$ for $\mathrm{d}(\boldsymbol{r}_{i-1})/(\mathrm{d}t)$ so equation (8.52) becomes

$$\boldsymbol{v}_i = \boldsymbol{v}_{i-1} + \frac{\partial \boldsymbol{u}_i}{\partial t} + \boldsymbol{\omega}_i \times \boldsymbol{u}_i \tag{8.54}$$

For a revolute joint the second term on the right hand side is zero.

The acceleration of O_i is

$$\boldsymbol{a}_i = \boldsymbol{a}_{i-1} + \frac{\partial^2 \boldsymbol{u}_i}{\partial t^2} + \frac{\partial \boldsymbol{\omega}_i}{\partial t} \times \boldsymbol{u}_i + \boldsymbol{\omega}_i \times \frac{\partial \boldsymbol{u}_i}{\partial t} + \boldsymbol{\omega}_i \times (\boldsymbol{\omega}_i \times \boldsymbol{u}_i) \tag{8.55}$$

For a revolute joint the second and fourth terms on the right hand side are zero whilst for a prismatic joint the third term is zero.

Both of these equations may be expressed in matrix form assuming that the base vectors for all terms are the unit vectors of the $(xyz)_i$ axes. Thus we may write

$$(v)_i = (v)_{i-1} + \frac{\partial}{\partial t}(u)_i + [\omega]_i^{\mathrm{x}}(u)_i \tag{8.56}$$

and

$$(a)_i = (a)_{i-1} + \frac{\partial^2}{\partial t^2}(u)_i + \left(\frac{\partial}{\partial t}[\omega]_i^{\mathrm{x}}\right)(u)_i + [\omega]_i^{\mathrm{x}}\frac{\partial}{\partial t}(u)_i + [\omega]_i^{\mathrm{x}}[\omega]_i^{\mathrm{x}}(u)_i \tag{8.57}$$

where

$$[\omega]_i^{\mathrm{x}} = [R]_i^{\mathrm{T}}[\dot{R}]_i \quad \text{and} \quad \frac{\partial}{\partial t}[\omega]_i^{\mathrm{x}} = [\dot{R}]_i^{\mathrm{T}}[\dot{R}]_i + [R]_i^{\mathrm{T}}[\ddot{R}]_i$$

also

$$[\dot{R}]_n = \sum_{i=1}^{i=n} [R]_1 \ldots [Q][R]_i \ldots [R]_n \dot{\theta}_i$$

and

$$[\ddot{R}]_n = \sum_{i=1}^{i=n} [R]_1 \ldots [Q][R]_i \ldots [R]_n \ddot{\theta}_i$$

$$+ \sum_{i=1}^{i=n}\sum_{j=1}^{j=n} [R]_1 \ldots [Q][R]_i \ldots [Q][R]_j \ldots [R]_n \dot{\theta}_i \dot{\theta}_j$$

8.3.17 TRAJECTORY PLANNING AND CONTROL

In a practical pick and place type of operation an object is to be moved from point A to point B and, for example, has to avoid an obstacle C, as indicated in Fig. 8.23. The problem is often tackled by planning for the end effector to move from A to B so as to arrive with low speed at B and then to align the gripper. The exact path is not important apart from the three specified points, so there are an infinite number of possible paths that the arm can follow. The many factors which affect the choice of path are outside the scope of this book as we are concerned only with the pure dynamics of the problem.

One technique used is to consider the path to be constructed from short segments passing through a number of prescribed points. Usually a polynomial of third or fourth order is chosen to represent the path between the specified points.

Another powerful method is to use position sensors to give feedback to an automatic control system. These control systems are frequently digital, which makes adaptive control easier.

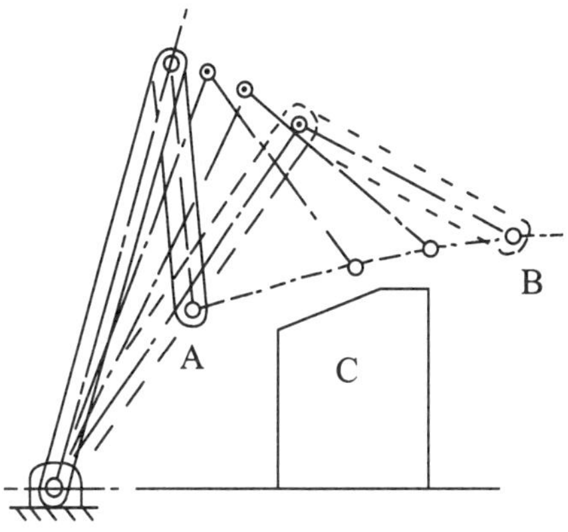

Fig. 8.23

8.4 Kinetics of a robot arm

Our next task is to determine the forces and couples associated with the prescribed motion. In the practical case it is not always possible to generate the required forces so the motion which ensues from given forces may need to be calculated.

A dynamical model may be used for the prediction of performance or for forming part of a real-time control algorithm.

We shall use Lagrange's equation in conjunction with homogeneous transformation matrices and the Newton–Euler approach using a vector algebra method. It should be noted that both Lagrange and Newton–Euler could be associated with either the homogeneous matrix formulation or vector algebra.

8.4.1 LAGRANGE'S EQUATIONS

Here we only generate one equation for each degree of freedom of the system and the basic formulation only needs expressions for velocities and not for acceleration. However, differentiation of the velocity- and position-dependent functions is required.

For link n we have

$$(\dot{\hat{p}}) = \sum_{i=1}^{i=n} [U]_{n,i} (\hat{p})_n \dot{\theta}_i$$

where

$$[U]_{n,i} = {}_0[A]_1 \; \ldots \; [Q]_{i-1}[A]_i \; \ldots \; {}_{n-1}[A]_n$$

so

$$(\dot{\hat{p}})_0^{\mathrm{T}}(\dot{\hat{p}})_0 = (\dot{x}_0 \;\; \dot{y}_0 \;\; \dot{z}_0 \;\; 0)\begin{bmatrix} \dot{x}_0 \\ \dot{y}_0 \\ \dot{z}_0 \\ 0 \end{bmatrix} = \dot{x}_0^2 + \dot{y}_0^2 + \dot{z}_0^2 + 0$$

$$= v_0^2 = \left(\sum_{i=1}^{i=n} (\hat{p})_n^{\mathrm{T}} [U]_{n,i}^{\mathrm{T}} \dot{\theta}_i\right)\left(\sum_{i=1}^{i=n} [U]_{n,i} (\hat{p})_n \dot{\theta}_i\right)$$

Thus the kinetic energy of link n will be

$$T_n = \sum_{r=1}^{r=R}\left[\frac{m_r}{2} (\hat{p}_r)_n^{\mathrm{T}} \left(\sum_{i=1}^{i=n} [U]_{n,i}^{\mathrm{T}} \dot{\theta}_i\right)\left(\sum_{i=1}^{i=n} [U]_{n,i} \dot{\theta}_i\right)(\hat{p}_r)_n\right] \tag{8.58}$$

where R is the number of point masses used to represent the rigid body. For an exact representation $R \geq 3$.

The total kinetic energy will be just the sum of the energies for each link in the chain.

It would be convenient to be able to reverse the order of summation since $(\hat{p}_r)_n$ and m_r are properties of the link and do not depend on its location. This can be achieved by noting that

$$\text{Trace } (\dot{\hat{p}})_n(\dot{\hat{p}})_n^{\mathrm{T}} = \text{Trace}\begin{bmatrix} \dot{x}_0\dot{x}_0 & \dot{x}_0\dot{y}_0 & \dot{x}_0\dot{z}_0 & 0 \\ \dot{y}_0\dot{x}_0 & \dot{y}_0\dot{y}_0 & \dot{y}_0\dot{z}_0 & 0 \\ \dot{z}_0\dot{x}_0 & \dot{z}_0\dot{y}_0 & \dot{z}_0\dot{z}_0 & 0 \\ 0 & 0 & 0 & 0 \end{bmatrix}$$

$$= \dot{x}_0^2 + \dot{y}_0^2 + \dot{z}_0^2$$

$$= v_0^2 \tag{8.59}$$

So we may now write

$$T_n = \text{Trace}\left(\sum_{i=1}^{i=n} [U]_{n,i} \dot{\theta}_i\right)\left(\sum_{r=1}^{r=R} \frac{m_r}{2} (\hat{p}_r)_n (\hat{p}_r)_n^{\mathrm{T}}\right)\left(\sum_{i=1}^{i=n} [U]_{n,i}^{\mathrm{T}} \dot{\theta}_i\right) \tag{8.60}$$

The link variables, θ or d, satisfy the requirements for generalized co-ordinates and so will be designated by q, as is usual in Lagrange's equations. Also the centre term depicts the inertia properties of the nth link and will be abbreviated to $\frac{1}{2}[J]_n$. Note that $[J]_n$ is a symmetric matrix. The top left 3 x 3 submatrix is related to the moment of inertia matrix of the nth link relative to the nth joint, but is not identical to it. Thus

$$[J] = \left[\begin{array}{ccc|c} \Sigma mx^2 & \Sigma mxy & \Sigma mxz & \Sigma mx \\ \Sigma myx & \Sigma my^2 & \Sigma myz & \Sigma my \\ \Sigma mzx & \Sigma mzy & \Sigma mz^2 & \Sigma mz \\ \hline \Sigma mx & \Sigma my & \Sigma mz & \Sigma m \end{array}\right]$$

In terms of $[J]$ the kinetic energy of the nth link is

$$T_n = \text{Trace}\left(\sum_{i=1}^{i=n}[U]_{n,i}\dot{q}_i\right)\frac{1}{2}[J]_n\left(\sum_{i=1}^{i=n}[U]^{\mathrm{T}}_{n,i}\dot{q}_i\right)$$

We now need to carry out the differentiations as prescribed by Lagrange's equations,

$$\frac{\partial T_n}{\partial \dot{q}_k} = \frac{1}{2}\text{Trace}\left[[U]_{n,k}[J]_n\left(\sum_{i=1}^{i=n}([U]^{\mathrm{T}}_{n,i}\dot{q}_i)\right)\left(\sum_{i=1}^{i=n}([U]_{n,i}\dot{q}_i)\right)[J]_n[U]^{\mathrm{T}}_{n,k}\right]$$

but since the second term is the transpose of the first

$$\frac{\partial T_n}{\partial \dot{q}_k} = 2\times\frac{1}{2}\text{Trace}\left([U]_{n,k}[J]_n\sum_{i=1}^{i=n}([U]^{\mathrm{T}}_{n,i}\dot{q}_i)\right)$$

Therefore

$$\frac{\mathrm{d}}{\mathrm{d}t}\left(\frac{\partial T_n}{\partial \dot{q}_k}\right) = \text{Trace}\Bigg[[U]_{n,kk}\dot{q}_k[J]_n\sum_{i=1}^{i=n}([U]^{\mathrm{T}}_{n,i}\dot{q}_i)$$

$$+ [U]_{n,k}[J]_n\sum_{i=1}^{i=n}([U]^{\mathrm{T}}_{n,i}\ddot{q}_i)$$

$$+ [U]_{n,k}[J]_n\sum_{i=1}^{i=n}\left(\sum_{j=1}^{j=n}([U]^{\mathrm{T}}_{n,ij}\dot{q}_j\right)\dot{q}_i\Bigg] \tag{8.61}$$

In a similar manner we can obtain

$$\left(\frac{\partial T_n}{\partial q_k}\right) = \text{Trace}\left([U]_{n,kk}\dot{q}_k[J]_n\sum_{i=1}^{i=n}([U]^{\mathrm{T}}_{n,i}\dot{q}_i)\right) \tag{8.62}$$

So for the nth link

$$\frac{\mathrm{d}}{\mathrm{d}t}\left(\frac{\partial T_n}{\partial \dot{q}_k}\right) - \left(\frac{\partial T_n}{\partial q_k}\right) = \text{Trace}\Bigg[[U]_{n,k}[J]_n\sum_{i=1}^{i=n}([U]^{\mathrm{T}}_{n,i}\ddot{q}_i)$$

$$+ [U]_{n,k}[J]_n\sum_{i=1}^{i=n}\left(\sum_{j=1}^{j=n}([U]^{\mathrm{T}}_{n,ij}\dot{q}_j\right)\dot{q}_i\Bigg] \tag{8.63}$$

We are now in a position to sum over the whole arm of N links. For clarity the constant terms are now taken to the right of the summation sign to give

$$\frac{\mathrm{d}}{\mathrm{d}t}\left(\frac{\partial T}{\partial \dot{q}_k}\right) - \left(\frac{\partial T}{\partial q_k}\right) = \text{Trace}\Bigg(\sum_{n=1}^{n=N}\sum_{i=1}^{i=n}[U]_{n,k}[J]_n[U]^{\mathrm{T}}_{n,i}\ddot{q}_i$$

$$+ \sum_{n=1}^{n=N}\sum_{i=1}^{i=n}\sum_{j=1}^{j=n}[U]_{n,k}[J]_n[U]^{\mathrm{T}}_{n,ij}\dot{q}_i\dot{q}_j\Bigg)$$

$$= Q_k \tag{8.64}$$

the generalised force, where $T = \Sigma T_n$.

For revolute joints Q will be the torque at the pivot and for a prismatic joint Q will be the sliding force.

Equation (8.64) may be written in a more compact form by reversing the order of summation. This requires adjusting the limits. The form given below can be justified by expansion. The term max(i,j,k) means the highest value of i,j, or k

$$Q_k = \sum_{i=1}^{i=N}M_{ki}\ddot{q}_i + \sum_{i=1}^{i=N}\sum_{j=1}^{j=N}G_{kij}\dot{q}_i\dot{q}_j \quad 1 \le k \le N \tag{8.65}$$

where

$$M_{ki} = \sum_{n=\max(k,i)}^{n=N} (\text{Trace}[U]_{n,k}[J]_n [U]^{\mathrm{T}}_{n,i}) \tag{8.66}$$

and

$$G_{kij} = \sum_{n=\max(k,i,j)}^{n=N} (\text{Trace}[U]_{n,k}[J]_n [U]^{\mathrm{T}}_{n,ij}) \tag{8.67}$$

For a two-link arm the above equations become

$$Q_1 = M_{11}\ddot{q}_1 + M_{12}\ddot{q}_2 + G_{111}\dot{q}_1^2 + 2G_{112}\dot{q}_1\dot{q}_2 + G_{112}\dot{q}_2^2$$

$$Q_2 = M_{21}\ddot{q}_1 + M_{22}\ddot{q}_2 + G_{211}\dot{q}_1^2 + 2G_{212}\dot{q}_1\dot{q}_2 + G_{222}\dot{q}_2^2$$

(note that $G_{k12} = G_{k21}$). It should be remembered that the coefficients M and G are functions of q_i because the $[U]$ are functions of q_i, but $[J]$ is constant. Then

$$M_{11} = \text{Trace}\{[U]_{1,1}[J]_1[U]^{\mathrm{T}}_{1,1}\} + \text{Trace}\{[U]_{2,1}[J]_2[U]^{\mathrm{T}}_{2,1}\}$$

$$M_{12} = \text{Trace}\{[U]_{2,1}[J]_1[U]^{\mathrm{T}}_{1,2}\}$$

$$G_{111} = \text{Trace}\{[U]_{1,1}[J]_1[U]^{\mathrm{T}}_{1,11}\} + \text{Trace}\{[U]_{2,1}[J]_2[U]^{\mathrm{T}}_{2,11}\}$$

$$G_{122} = \text{Trace}\{[U]_{2,1}[J]_2[U]_{2,22}\}$$

$$G_{112} = \text{Trace}\{[U]_{2,1}[J]_2[U]_{2,12}\} = G_{121}$$

$$M_{21} = \text{Trace}\{[U]_{1,2}[J]_1[U]^{\mathrm{T}}_{1,1}\} + \text{Trace}\{[U]_{2,1}[J]_2[U]^{\mathrm{T}}_{2,1}\}$$

$$M_{22} = \text{Trace}\{[U]_{2,2}[J]_2[U]^{\mathrm{T}}_{2,2}\}$$

$$G_{211} = \text{Trace}\{[U]_{1,2}[J]_1[U]^{\mathrm{T}}_{1,11}\} + \text{Trace}\{[U]_{2,2}[J]_2[U]^{\mathrm{T}}_{2,11}\}$$

$$G_{222} = \text{Trace}\{[U]_{2,2}[J]_2[U]_{2,22}\}$$

$$G_{212} = \text{Trace}\{[U]_{2,2}[J]_2[U]_{2,12}\} = G_{221}$$

8.4.2 NEWTON–EULER METHOD

This method will involve all internal forces between each link as indicated on a free-body diagram, as shown in Fig. 8.24. Expressions for the accelerations of the individual centres of mass and the angular velocities and accelerations can be found as discussed in the previous sections.

So for each link, treated as a rigid body, the six equations of motion can be formulated. With reference to Figs 8.22 and 8.24 the equations of motion for the ith link are

$$\boldsymbol{F}_i + \boldsymbol{F}_{i-1} = m_i \boldsymbol{a}_{Gi} \tag{8.68}$$

and

$$\boldsymbol{C}_i + \boldsymbol{C}_{i-1} + \boldsymbol{r}_{Gi} \times \boldsymbol{F}_i - (\boldsymbol{p}_i - \boldsymbol{r}_{Gi}) \times \boldsymbol{F}_{i-1} = \frac{\mathrm{d}}{\mathrm{d}t}(\mathsf{J}_{Gi} \cdot \boldsymbol{\omega}_i) \tag{8.69}$$

where J_{Gi} is the moment of inertia dyadic referred to the centre of mass of the ith link.

The acceleration of the centre of mass, $\boldsymbol{a}_{Gi}$, may be found from

$$\boldsymbol{a}_{Gi} = \boldsymbol{a}_i + \frac{\partial}{\partial t}(\boldsymbol{\omega}_i) \times \boldsymbol{r}_{Gi} + \boldsymbol{\omega}_i \times (\boldsymbol{\omega}_i \times \boldsymbol{r}_{Gi}) \tag{8.70}$$

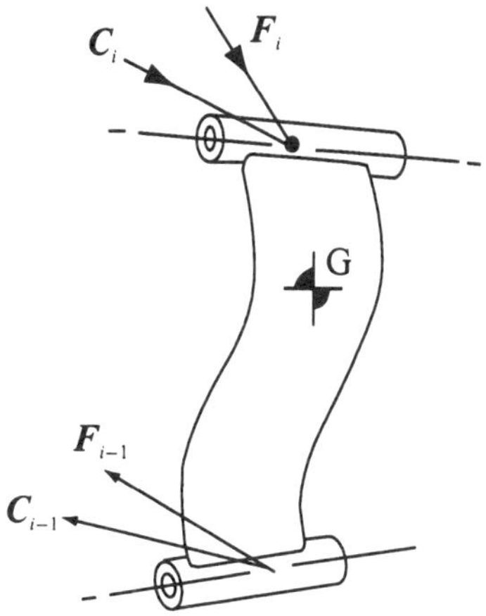

Fig. 8.24

The above three equations may also be written in matrix form using $(xyz)_i$ axes as the basis

$$(F)i + (F)_{i-1} = m_i(a)_{Gi} \tag{8.71}$$

$$(C)_i + (C)_{i-1} + (r)^{\times}_{Gi}(F)_i - [(p)_i - (r)_{Gi}]^{\times}(F)_{i-1}$$
$$= [J]_{Gi}\frac{\partial}{\partial t}(\omega)_i + [J]_{Gi}[\omega]^{\times}_i(\omega)_i \tag{8.72}$$

$$(a)_{Gi} = (a)_i + \frac{\partial}{\partial t}[\omega]^{\times}(r)_{Gi} + [\omega]^{\times}_i[\omega]^{\times}_i(r)_{Gi} \tag{8.73}$$

Discussion example

A spherical robot is shown in Fig. 8.25(a). During operation the position co-ordinates are: $\theta_1 = 90°$, $\theta_2 = 60°$ and $d_3 = 0.8$ m. Also $d\theta_2/dt = 2$ rad/s constant and $dd_3/dt = 3$ m/s constant.

The inertial data are

for link 2: mass = 20 kg moment of inertia about $G_{2y} = 8\,\text{kgm}^2$
for link 3: mass = 10 kg moment of inertia about $G_{3y} = 4\,\text{kgm}^2$
$OG_2 = 0.2$ m and $G_3E = 0.3$ m (see Fig. 8.26)

a) Using the A matrices calculate the position and velocity of the end of the arm E.
b) Using Lagrange's equation obtain the equations of motion for co-ordinates θ_2 and d_3.

(a) The table of link parameters is first constructed. Referring to Fig. 8.25(a) the origin of the $(xyz)_2$ axes has been chosen to coincide with the origin of the $(xyz)_1$ axes. By using the data sheet given at the end of the chapter we see that as a is the distance between successive z axes all the a dimensions are zero.

The distance between successive x axes is d and hence $d_2 = 0$.

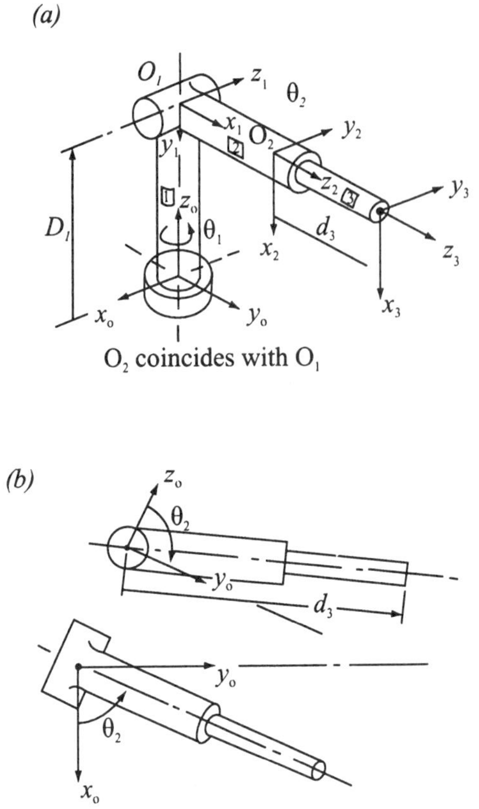

Fig. 8.25 (a) and **(b)**

Now α is the rotation of one *z* axis relative to the previous *z* axis so $\alpha_1 = -90^\circ$ and $\alpha_2 = 90^\circ$. Special care is needed to ensure that the signs are correct.

The table is as follows:

Link	θ	α	*a*	*d*
1	90°	−90°	0	$D_1 = 0$
2	60°	90°	0	0
3	0	0	0	0.8

With reference to the data sheet the three *A* matrices are

$$
{}_0[A]_1 = \begin{bmatrix} C\theta_1 & 0 & S\theta_1 & 0 \\ S\theta_1 & 0 & C\theta_1 & 0 \\ 0 & -1 & 0 & 0 \\ 0 & 0 & 0 & 1 \end{bmatrix}
$$

$$
{}_1[A]_2 = \begin{bmatrix} C\theta_2 & 0 & S\theta_2 & 0 \\ S\theta_2 & 0 & -C\theta_2 & 0 \\ 0 & 1 & 0 & 0 \\ 0 & 0 & 0 & 1 \end{bmatrix}
$$

$$
{}_2[A]_3 = \begin{bmatrix} 1 & 0 & 0 & 0 \\ 0 & 1 & 0 & 0 \\ 0 & 0 & 1 & d_3 \\ 0 & 0 & 0 & 1 \end{bmatrix}
$$

The overall transformation matrix is

$$
{}_0[T]_3 = {}_0[A]_3 = {}_0[A]_1\, {}_1[A]_2\, {}_2[A]_3
$$

$$
= \begin{bmatrix} C\theta_1 C\theta_2 & S\theta_1 & C\theta_1 S\theta_2 & d_3 C\theta_1 S\theta_2 \\ S\theta_1 C\theta_2 & -C\theta_1 & S\theta_1 S\theta_2 & d_3 S\theta_1 S\theta_2 \\ -S\theta_2 & 0 & C\theta_2 & d_3 C\theta_2 \\ 0 & 0 & 0 & 1 \end{bmatrix}
$$

$$
= \begin{bmatrix} (n) & (s) & (a) & (p) \\ 0 & 0 & 0 & 1 \end{bmatrix}
$$

Here the last column gives the co-ordinates of the $(xyz)_3$ axes

$$
\begin{aligned}
x_{\mathrm{EO}} &= d_3 \cos(\theta_1) \sin(\theta_2) = 0 \\
y_{\mathrm{EO}} &= d_3 \sin(\theta_1) \sin(\theta_2) = 0.8 \times 1 \times 0.866\,\mathrm{m} \\
z_{\mathrm{EO}} &= d_3 \cos(\theta_2) = 0.8 \times 0.5\,\mathrm{m}
\end{aligned}
$$

These results can easily be verified by simple trigonometry.

We shall now use the A matrices to evaluate the velocity of the point E. Now

$$
(\hat{\dot{p}}E)_0 = \frac{\mathrm{d}}{\mathrm{d}t}\, {}_0[A]_3 (\hat{p}_E)_3 = {}_0[\dot{A}]_3 (\hat{p}_E)_3
$$

There is only one term on the right hand side of the equation because the position vector of point E as seen from the $(xyz)_2$ axes is $(0\ 0\ 0\ 1)^{\mathrm{T}}$ for all time.

For brevity let us write ${}_0[A]_3 = A_1 A_2 A_3$ so, with reference to equations (8.35) and (8.36) we have

$$
{}_0[\dot{A}]_3 = Q A_1 A_2 A_3 \dot{\theta}_1 + A_1 Q A_2 A_3 \dot{\theta}_2 + A_1 A_2 P A_3 \dot{d}_3
$$

As $\dot{\theta}_1 = 0$ the first term is zero and by direct multiplication the other two terms are

$$
A_1 Q A_2 A_3 \dot{\theta}_2 = \begin{bmatrix} 0 & 0 & 0 & 0 \\ -0.866 & 0 & 0.5 & 0.4 \\ -0.500 & 0 & -0.866 & -0.779 \\ 0 & 0 & 0 & 0 \end{bmatrix} \times 2
$$

$$A_1A_2PA_3\dot{d}_3 = \begin{bmatrix} 0 & 0 & 0 & 0 \\ 0 & 0 & 0 & 0.866 \\ 0 & 0 & 0 & 0.500 \\ 0 & 0 & 0 & 0 \end{bmatrix} \times 3$$

The velocity of E is given by summing the last columns of the preceding two matrices

$$\dot{x}_{EO} = 0$$

$$\dot{y}_{EO} = 0.4\text{x}2 + 0.866\text{x}3 = 3.398\,\text{m/s}$$

$$\dot{z}_{EO} = -0.693\text{x}2 + 0.500\text{x}3 = -0.114\,\text{m/s}$$

Again these results are readily confirmed by direct means.

The matrix multiplications involved in the above calculations can be carried out using any convenient mathematical computer package.

(b) Lagrange's equations are

$$\frac{\mathrm{d}}{\mathrm{d}t}\left(\frac{\partial T}{\partial \dot{q}_i}\right) - \frac{\partial T}{\partial q_i} + \frac{\partial V}{\partial q_i} = Q_i$$

The virtual work done by the active forces and couples is

$$\delta W = Q_1\delta q_1 + Q_2\delta q_2$$

$$\delta W = C_2\delta\theta_2 + F_3\delta \mathrm{d}_3$$

where C_2 is the torque about the z_0 axis acting on link 2 and F_3 is the force acting on link 3 along the z_3 axis.

The kinetic energy (*see* Fig. 8.26) is

$$T = \frac{1}{2}\left\{m_2(a\dot{\theta}_2)^2 + I_2\dot{\theta}_2^2 + m_3[(d_3 - b)^2\dot{\theta}_2^2 + \dot{d}_3^2] + I_3\dot{\theta}_2^2\right\}$$

For $q_i = \theta_2$

$$\frac{\partial T}{\partial \dot{\theta}_2} = [m_2a^2 + I_2 + I_3 + m_2[(d_3 - b)^2]\dot{\theta}_2$$

$$\frac{\mathrm{d}}{\mathrm{d}t}\left(\frac{\partial T}{\partial \dot{\theta}_2}\right) = [m_2 + I_2 + I_3 + m_3(d_3 - b)^2]\ddot{\theta}_2 + 2m_3(d_3 - b)\dot{d}_3\dot{\theta}_2$$

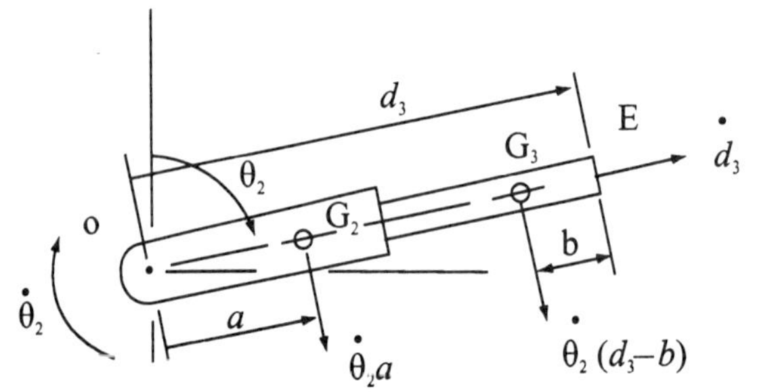

Fig. 8.26

$$\frac{\partial T}{\partial \theta_2} = 0, \quad \frac{\partial V}{\partial \theta_2} = 0 \quad \text{and} \quad Q_1 = C_2$$

so the equation of motion for θ_2 is

$$[m_2 + I_2 + I_3 + m_3(d_3 - b)^2]\ddot{\theta}_2 + 2m_3(d_3 - b)\dot{d}_3\dot{\theta}_2 = C_2$$

Inserting the numerical values gives

$$(15.3)\ddot{\theta}_2 + 10\dot{d}_3\dot{\theta}_2 = C_2$$

and as $\ddot{\theta}_2 = 0, \dot{d}_3 = 3\,\text{m/s}$ and $\dot{\theta}_2 = 2\,\text{rad/s}$ we have

$$C_2 = 60\,\text{Nm}$$

Similarly for $q_i = d_3$

$$\frac{\mathrm{d}}{\mathrm{d}t}\left(\frac{\partial T}{\partial \dot{d}_3}\right) = m_3\ddot{d}_3 \quad \text{and} \quad \frac{\partial T}{\partial d_3} = m_3(d_3 - b)\dot{\theta}_2^2$$

so the equation of motion for d_3 is

$$m_3\ddot{d}_3 - m_3(d_3 - b)\dot{\theta}_2^2 = F_3$$

The same set of equations can be formed using D'Alembert's principle. We refer to Fig. 8.27 where the accelerations have been determined by direct means. Also shown with heavy arrows are the virtual displacements.

D'Alembert's principle states that the virtual work done by the active forces less that done by the 'inertia forces' equals zero.

For virtual displacement $\delta r_j = \delta\theta_2$, $(\delta d_3 = 0)$

$$C_2\delta\theta_2 - I_2\ddot{\theta}_2\delta\theta_2 - m_2a\ddot{\theta}_2a\delta\theta_2 - m_3[(d_3 - b)\ddot{\theta}_2 + 2\dot{d}_3\dot{\theta}_2](d_3 - b)\delta\theta_2 - I_3\ddot{\theta}_2\delta\theta_2 = 0$$

and now with $\delta r_j = \delta d_3$, $(\delta\theta_2 = 0)$

$$F_3\delta d_3 + m_3[(d_3 - b)\dot{\theta}_2^2 - \ddot{d}_3]\delta d_3 = 0$$

Thus

$$C_2 - [I_2 + I_3 + m_2a^2 + m_3(d_3 - b)^2]\ddot{\theta}_2 - 2m_3\dot{d}_3\dot{\theta}_2(d_3 - b) = 0$$

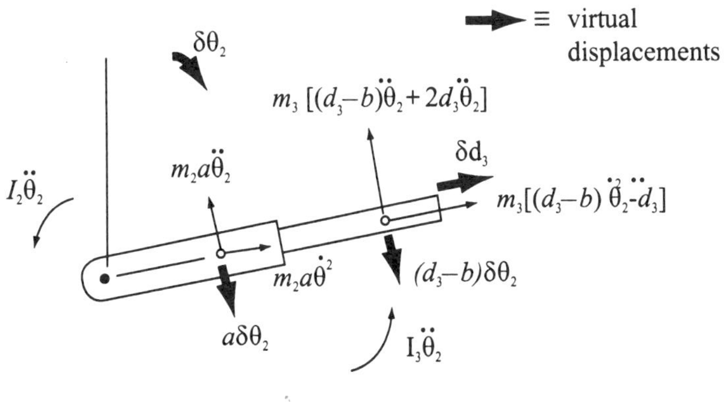

Fig. 8.27

and

$$F_3 + m_3[(d_3 - b)\ddot{\theta}_2 - \ddot{d}_3] = 0$$

Let us now return to the Lagrange method but this time make use of the 4x4 homogeneous matrix methods. Using equation (8.65) we have $N = 3$ but since link 1 is stationary it is not involved in the kinetics, although it still affects the geometry.

The inertia data is not in the form needed to generate the [J] matrices. If we define $I_p = \Sigma mx^2 + \Sigma my^2 + \Sigma mz^2$ then it is easy to show that $2I_p = I_{xx} + I_{yy} + I_{zz}$. Therefore

$$\Sigma mx^2 = I_p - I_{xx}$$

$$\Sigma y^2 = I_p - I_{yy}$$

$$\Sigma z^2 = I_p - I_{zz}$$

For link 2 we use the parallel axes theorem to to evaluate $I_{0y} = I_{Gy} + m_2a^2 = 8.0 + 20 \times 0.2^2 = 8.8\,\text{kgm}^2$. We will assume that $I_{0x} = I_{0y}$ and that I_{0z} is negligible. Therefore $I_p = 8.8\,\text{kgm}^2$. It now follows that $\Sigma mx^2 = 0$, $\Sigma my^2 = 0$ and $\Sigma mz^2 = 8.8\,\text{kgm}^2$. The term $\Sigma mz = 20 \times 0.2 = 4$.

The inertia matrix for link 2 is

$$[J]_2 = \begin{bmatrix} 0 & 0 & 0 & 0 \\ 0 & 0 & 0 & 0 \\ 0 & 0 & 8.8 & 4 \\ 0 & 0 & 4 & 20 \end{bmatrix}$$

similarly for link 3

$$[J]_3 = \begin{bmatrix} 0 & 0 & 0 & 0 \\ 0 & 0 & 0 & 0 \\ 0 & 0 & 4.9 & -3 \\ 0 & 0 & -3 & 10 \end{bmatrix}$$

From equation (8.66)

$$M_{22} = \sum_{n=2}^{n=3} \text{Trace}\{[U]_{n,2}[J]_n[U]_{n,2}^{T}\}$$

$$M_{32} = \sum_{n=3}^{n=3} \text{Trace}\{[U]_{n,3}[J]_n[U]_{n,3}^{T}\}$$

or

$$M_{22} = \text{Trace}\{[U]_{2,2}[J]_2[U]_{2,2}^{T}\} + \text{Trace}\{[U]_{3,2}[J]_3[U]_{3,2}^{T}\}$$

$$M_{23} = \text{Trace}\{[U]_{3,2}[J]_3[U]_{3,3}\}$$

$$G_{222} = \text{Trace}\{[U]_{2,2}[J]_2[U]_{2,22}\} + \text{Trace}\{[U]_{3,2}[J]_3[U]_{3,22}\}$$

$$G_{233} = \text{Trace}\{[U]_{3,2}[J]_3[U]_{3,33}^{r}\}$$

$$G_{223} = \text{Trace}\{[U]_{3,2}[J]_3[U]_{3,23}\}$$

Inserting the numerical values and using a matrix multiplication program it is found that

$$M_{22} = 8.8 + 6.5 = 15.3\,\text{kgm}^2$$

and

$$G_{223} = 5.0\,\text{kgm}$$

All other terms are zero. Thus

$$15.3\ddot{\theta}_2 + 5.0\dot{\theta}_2\dot{d}_3 = C_2$$

In this problem the geometry is particularly simple so that the 4x4 matrix methods can readily be compared with other methods.

It is left as an exercise for the reader to obtain the equations of motion using the Newtonian approach. The free-body diagrams and kinematics are shown in Fig. 8.28.

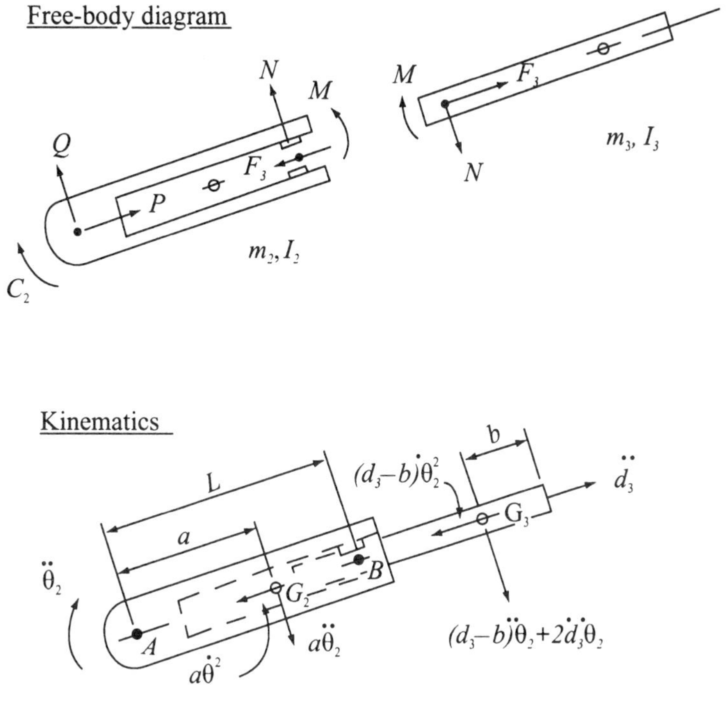

Fig. 8.28

Robotics data sheet

LINK CO-ORDINATE SYSTEM

The $z(i)$ axis is the axis of rotation or sliding of joint $(i+1)$. The $x(i)$ axis lies along the common normal to the $z(i)$ and $z(i-1)$ axes. This locates the origin of the $(xyz)_i$ axes except when the z axes are parallel; in this case choose the normal which passes through the origin of the $(xyz)_{i-1}$ axes.

$\theta(i)$ is the joint angle from the $x(i-1)$ axis to the $x(i)$ axis about the $z(i-1)$ axis.
$d(i)$ is the distance from the origin of the $(i-1)$ co-ordinate frame to the intersection of the $z(i-1)$ axis with the $x(i)$ axis along the $z(i-1)$ axis.
$a(i)$ is the offset distance from the intersection of the $z(i-1)$ axis with the $x(i)$ axis and the $z(i)$ axis (i.e. the shortest distance between the $z(i-1)$ axis and the $z(i)$ axis).
$\alpha(i)$ is the offset angle from the $z(i-1)$ axis to the $z(i)$ axis.

HOMOGENEOUS TRANSFORMATION MATRIX FOR A SINGLE LINK

This matrix is

$$[A] = \begin{bmatrix} C\theta & -C\alpha S\theta & S\alpha S\theta & aC\theta \\ S\theta & C\alpha C\theta & -S\alpha C\theta & aS\theta \\ 0 & S\alpha & C\alpha & d \\ 0 & 0 & 0 & 1 \end{bmatrix}$$

The overall transformation matrix is

$$[T] = \prod_{i=1}^{i=n} {}_{i-1}[A]_i$$

and the position vector

$$(\hat{p}) = (p_x \; p_y \; p_z \; 1)$$

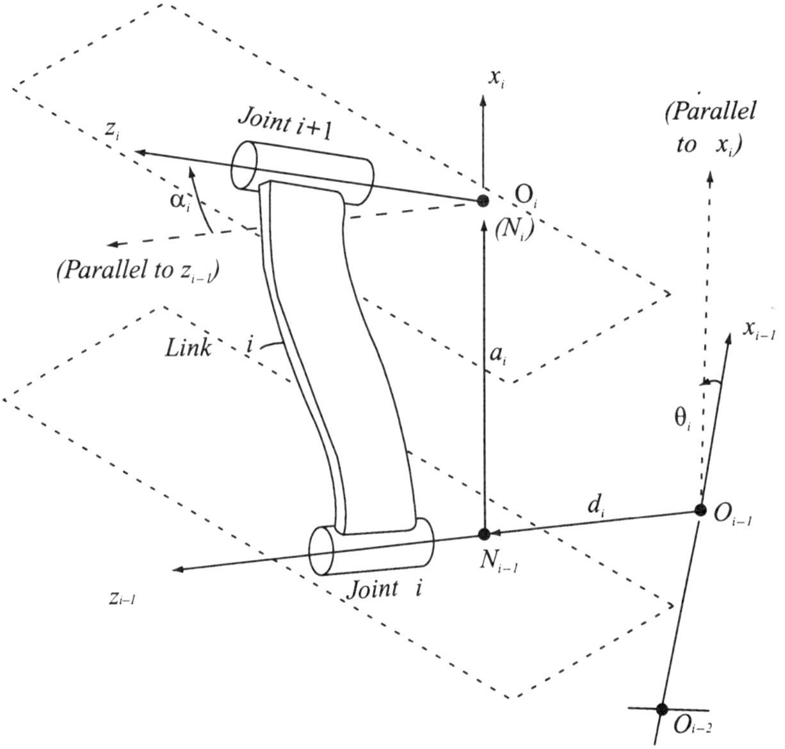

Fig. 8.29

9
Relativity

9.1 Introduction

In this chapter we shall reappraise the foundations of mechanics taking into account Einstein's special theory of relativity. Although it does not measurably affect the vast majority of problems encountered in engineering, it does define the boundaries of Newtonian dynamics. Confidence in the classical form will be enhanced as we shall be able to quantify the small errors introduced by using Newtonian theory in common engineering situations.

The laser velocity transducer employs the Doppler effect which, for light, requires an understanding of special relativity. The form of the equations derived for cases where the velocities of the transmitter and/or the receiver are small compared with that of the signal is the same for both sound and light. This will be discussed later.

We shall also consider the definition of force. It is of note that relativistic definitions are such that they encompass the Newtonian. The general theory of relativity raises some interesting questions regarding the nature of force, but these do not materially affect the equations of motion already derived.

9.2 The foundations of the special theory of relativity

It is not our intention to retrace the steps leading to the theory other than to mention the most significant milestones. In the same way that Isaac Newton crystallized the laws of mechanics which have formed the basis for the previous chapters in this book, Albert Einstein provided the genius that solved the riddle of the constancy of the speed of light.

James Clerk Maxwell's equations for electrodynamics predicted that all electromagnetic waves travelled at a constant speed in a vacuum. If the value of the speed of light, c, is evaluated for what we shall assume to be an inertial frame of reference then, according to Maxwell, the same speed is predicted for all other inertial frames. This means that a ray of light emitted from a source and received by an observer moving at a constant speed relative to the source would still record the same speed for the ray of light.

Light was supposed to be transmitted through some medium called the ether. In order to accommodate the constancy of the speed of light various schemes of dragging of the ether were put forward and also the notion of contraction in the direction of motion of moving bodies. Lorentz proposed a transformation of co-ordinates which went some way to solving the problem. The real breakthrough came when Einstein, instead of trying to justify the constancy of the speed of light, raised it to the status of a law. He also made it clear that the concept of simultaneity had to be abandoned.

The two basic tenets of special relativity are

the laws of physics are identical for all inertial frames

and

the speed of light is the same for all inertial observers

Figure 9.1 shows two frames of reference, the primed system moving at a constant speed v relative to the the first frame which, for ease of reference, will be regarded as the fixed frame. The x axis is chosen to be in the same direction as the relative velocity. An event E is defined by four co-ordinates: three spatial and one of time. In the original frame the event can be represented by a vector having four components, so in matrix form

$$(E) = (ct \ x \ y \ z)^{\mathrm{T}} \tag{9.1}$$

and in the primed system

$$(E') = (ct' \ x' \ y' \ z')^{\mathrm{T}} \tag{9.2}$$

The factor c could be any arbitrary speed simply to make all terms dimensionally equivalent, but as it is postulated that the speed of light is constant then this is chosen as the parameter.

If a pulse of light is generated when O was coincident with O$'$ and $t = t' = 0$ then, at a later time, the square of the radius of the spherical wavefront is

$$(ct)^2 = x^2 + y^2 + z^2 \tag{9.3}$$

and in the moving frame

$$(ct')^2 = x'^2 + y'^2 + z'^2 \tag{9.4}$$

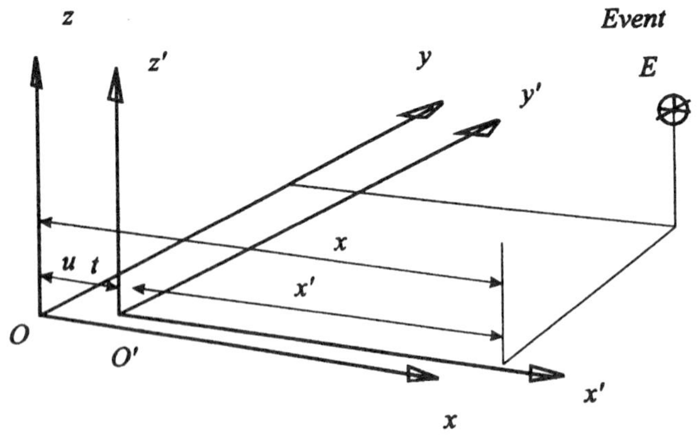

Fig. 9.1

We define the conjugate of (E) as

$$(\tilde{E}) = (ct \; -x \; -y \; -z)^{\mathrm{T}} \tag{9.5}$$

so that

$$(E)^{\mathrm{T}}(\tilde{E}) = (ct)^2 - x^2 - y^2 - z^2 \tag{9.6}$$

From equation (9.3)

$$(E)^{\mathrm{T}}(\tilde{E}) = 0 \tag{9.7}$$

Similarly

$$(E')^{\mathrm{T}}(\tilde{E}') = 0 \tag{9.8}$$

Also we define

$$[\eta] = \begin{bmatrix} 1 & 0 & 0 & 0 \\ 0 & -1 & 0 & 0 \\ 0 & 0 & -1 & 0 \\ 0 & 0 & 0 & -1 \end{bmatrix} \tag{9.9}$$

so we can write

$$(\tilde{E}) = [\eta](E) \tag{9.10}$$

Equation (9.7) can be written as

$$(E)^{\mathrm{T}}(\tilde{E}) = (E)^{\mathrm{T}}[\eta](E') = 0 \tag{9.11}$$

and equation (9.8) can be written as

$$(E')^{\mathrm{T}}(\tilde{E}') = (E')^{\mathrm{T}}[\eta](E') = 0 \tag{9.12}$$

We now assume that a linear transformation, $[T]$, exists between the two co-ordinate systems, that is

$$(E') = [T](E) \tag{9.13}$$

with the proviso that as $v \to 0$ the transformation tends to the Galilean.

Thus we can write

$$(E')^{\mathrm{T}}[\eta](E') = (E)^{\mathrm{T}}[T]^{\mathrm{T}}[\eta][T](E) \tag{9.14}$$

and because (E') is arbitrary it follows that

$$[T]^{\mathrm{T}}[\eta][T] = [\eta] \tag{9.15}$$

Now by inspection $[\eta][\eta] = [I]$, the unit matrix, so premultiplying both sides of equation (9.15) by $[\eta]$ gives

$$[\eta][T]^{\mathrm{T}}[\eta][T] = [I] \tag{9.16}$$

From symmetry

$$y' = y \tag{9.17}$$

and

$$z' = z \tag{9.18}$$

Consider the transformation of two co-ordinates only, namely ct and x. Let

$$\begin{bmatrix} ct' \\ x' \end{bmatrix} = \begin{bmatrix} A & B \\ C & D \end{bmatrix} \begin{bmatrix} ct \\ x \end{bmatrix} = [T] \begin{bmatrix} ct \\ x \end{bmatrix} \tag{9.19}$$

and in this case

$$[\eta] = \begin{bmatrix} 1 & 0 \\ 0 & -1 \end{bmatrix} \tag{9.20}$$

Substituting into equation (9.16) gives

$$\begin{bmatrix} 1 & 0 \\ 0 & -1 \end{bmatrix} \begin{bmatrix} A & C \\ B & D \end{bmatrix} \begin{bmatrix} 1 & 0 \\ 0 & -1 \end{bmatrix} \begin{bmatrix} A & B \\ C & D \end{bmatrix} = \begin{bmatrix} 1 & 0 \\ 0 & 1 \end{bmatrix}$$

or

$$\begin{bmatrix} A^2 - C^2 & AB - CD \\ CD - AB & D^2 - B^2 \end{bmatrix} = \begin{bmatrix} 1 & 0 \\ 0 & 1 \end{bmatrix}$$

Thus

$$A^2 - C^2 = 1 \tag{9.21}$$

$$D^2 - B^2 = 1 \tag{9.22}$$

$$AB = CD \tag{9.23}$$

Substituting equations (9.21) and (9.22) into equation (9.23) squared gives

$$A^2(1 - D^2) = (1 - A^2)D^2$$

so

$$\frac{A^2}{(1 - A^2)} = \frac{D^2}{(1 - D^2)}$$

This equation is satisfied by putting $A = \pm D$, we choose the positive value to ensure that as $v \rightarrow 0$ the transformation is Galilean. Let

$$A = D = \gamma \text{ (say)} \tag{9.24}$$

Hence it follows from equation (9.23) that if

$$A = D \quad \text{then} \quad B = C \tag{9.25}$$

We can now write equation (9.19) as

$$ct' = \gamma ct + Bx \tag{9.26}$$

$$x' = Bct + \gamma x \tag{9.27}$$

Now for $x' = 0$, $x = vt$. Therefore equation (9.27) reads

$$0 = Bct + \gamma vt$$

or

$$B = -\gamma v/c$$

Letting

$$\beta = v/c \tag{9.28}$$

gives

$$B = -\beta\gamma$$

From equation (9.22)

$$D^2 = 1 + B^2$$

and therefore

$$\gamma^2 = 1 + \gamma^2\beta^2$$

Thus

$$\gamma = \frac{1}{\sqrt{(1 - \beta^2)}} \tag{9.29}$$

which is known as the *Lorentz factor.*

The transformation for x' and ct' is

$$ct' = \gamma ct - \gamma\beta x \tag{9.30}$$

$$x' = -\beta\gamma ct + \gamma x \tag{9.31}$$

Inverting,

$$ct = \gamma ct' + \gamma\beta x' \tag{9.32}$$

$$x = \beta\gamma ct' + \gamma x' \tag{9.33}$$

The sign change is expected since the velocity of the original frame relative to the primed frame is $-v$.

The complete transformation equation is

$$(E') = [T](E)$$

$$\begin{bmatrix} ct' \\ x' \\ y' \\ z' \end{bmatrix} = \begin{bmatrix} \gamma & -\beta\gamma & 0 & 0 \\ -\beta\gamma & \gamma & 0 & 0 \\ 0 & 0 & 1 & 0 \\ 0 & 0 & 0 & 1 \end{bmatrix} \begin{bmatrix} ct \\ x \\ y \\ z \end{bmatrix} \tag{9.34}$$

This is known as the *Lorentz* (or *Fitzgerald–Lorentz*) *transformation*. For small β (i.e. $v \to 0$), $\gamma \to 1$ and equations (9.34) become

$$t' = t$$

$$x' = -vt + x$$

$$y' = y$$

$$z' = z$$

which is the Galilean transformation, as required.

For an arbitrary event E we can write

$$(E)^{\mathrm{T}}(\tilde{E}) = (E)^{\mathrm{T}}[\eta](E) = R^2 \tag{9.35}$$

a scalar, and

$$(E')^{\mathrm{T}}(\tilde{E}') = (E')^{\mathrm{T}}[\eta](E') = R'^2 \tag{9.36}$$

Note that in equations (9.7) and (9.8) R^2 and R'^2 are both zero because the event is a ray of light which originated at the origin.

Now $(E') = [T](E)$ so equation (9.36) becomes

$$(E)^{\mathrm{T}}[T][\eta][T](E) = R'^2$$

but from equation (9.15) we see that

$$R'^2 = (E')^T[\eta](E') = (E)^T[T][\eta][T](E) = (E)^T[\eta](E) = R^2 \tag{9.37}$$

Thus we have the important result that

$$(E)^T(\tilde{E}) = (E')^T(\tilde{E}') = R^2$$

an invariant. In full

$$(ct)^2 - x^2 - y^2 - z^2 = (ct')^2 - x'^2 - y'^2 - z'^2 = R^2 \tag{9.38}$$

Let us now write $E = E_2 - E_1$ and substitute into equation (9.37) giving

$$(E_2^T - E_1^T)[\eta](E_2 - E_1) = (E_2'^T - E_1'^T)[\eta](E_2' - E_1')$$

which expands to

$$(E_2^T)[\eta](E_2) + (E_1^T)[\eta](E_1) - (E_2^T)[\eta](E_1) - (E_1^T)[\eta](E_2^T)$$

$$= (E_2'^T)[\eta](E_2') + (E_1'^T)[\eta](E_1') - (E_2'^T)[\eta](E_1') - (E_1'^T)[\eta](E_2')$$

The first term on the left of the equation is equal to the first term on the right, because of equation (9.37), and similarly the second term on the left is equal to the second term on the right. Because $[\eta]$ is symmetrical the fourth terms are the transposes of the respective third terms and since these are scalars they must be equal. From this argument we have that

$$(E_1)[\eta](E_2) = (E_1')[\eta](E_2') \tag{9.37a}$$

and is another invariant.

9.3 Time dilation and proper time

It follows from equations (9.37) and (9.37a) that if $(\Delta E) = (E_2 - E_1)$ then the product

$$(\Delta E)^T(\Delta\tilde{E}) = (\Delta E')^T(\Delta\tilde{E}') = (\Delta R)^2$$

is an invariant. In full

$$(\Delta ct)^2 - (\Delta x)^2 - (\Delta y)^2 - (\Delta z)^2 = (\Delta ct')^2 - (\Delta x')^2 - (\Delta y')^2 - (\Delta z')^2$$
$$= (\Delta R)^2 \tag{9.39}$$

Because the relative motion is wholly in the x direction

$$(\Delta y) = (\Delta y') \quad \text{and} \quad (\Delta z) = (\Delta z')$$

so equation (9.39) can be written as

$$(\Delta ct)^2 - (\Delta x)^2 = (\Delta ct')^2 - (\Delta x')^2 = (\Delta R)^2 + (\Delta y)^2 + (\Delta z)^2 \tag{9.40}$$

which is invariant.

If $\Delta ct'$ is the difference in time between two events which occur at the same location in the moving frame, that is $\Delta x' = 0$, equation (9.40) tells us that

$$(\Delta ct') = \sqrt{[(\Delta ct)^2 - (\Delta x)^2]}$$

But $x = vt = \beta ct$ and therefore

$$(\Delta ct') = \sqrt{[(\Delta ct)^2 - \beta^2(\Delta ct)^2]}$$
$$= (\Delta ct)\sqrt{(1 - \beta^2)}$$

and by the definition of γ, equation (9.29), we have that

$$(\Delta ct') = \frac{1}{\gamma}(\Delta ct) \tag{9.41}$$

The two events could well be the ticks of a standard clock which is at rest relative to the moving frame.

Because $\gamma > 1$, $(\Delta ct) > (\Delta ct')$; that is, the time between the ticks of the moving clock as seen from the fixed frame is greater than reported by the moving observer. This time dilation is independent of the direction of motion so it is seen that an identical result is obtained if a stationary clock is viewed from the moving frame. It is paramount to realize that the dilation is only apparent; there is no reason why a clock should run slow just because it is being observed.

For example, if the speed of the moving frame is 86.5% of the speed of light (i.e. $\beta = 0.865$) then $\gamma = 2$. If the standard clock attached to the moving frame ticks once every second (i.e. $\Delta ct' = 1$) then the time interval as seen from the fixed frame will be $\Delta ct = \gamma(\Delta ct') = 2$ seconds, and the moving clock appears to run slow. Looking at it the other way, when the 'fixed' clock indicates 1 second the moving clock indicates only half a second. The moving observer will still consider his or her clock to indicate 1 second intervals.

In order that the speed of light shall be constant it is necessary that the length of measuring rods in the moving frame must appear to contract in the x direction in the same proportion as the time dilates. Thus

$$(\Delta L') = \frac{1}{\gamma}(\Delta L) \tag{9.42}$$

Returning again to equation (9.40)

$$(\Delta ct)^2 - (\Delta x)^2 = (\Delta ct')^2 - (\Delta x')^2$$

if two events occur at the same location in primed frame, that is $\Delta x' = 0$, then

$$\Delta ct' = \sqrt{[(\Delta ct)^2 - (\Delta x)^2]} \tag{9.43}$$

from which it is seen that the time interval as seen from the frame which moves such that the two events occur at the same location, in that moving frame, is a minimum time. All other observers will see the events as occurring at different locations but by use of equation (9.43) they will be able to compute t'. This time is designated the *proper time* and given the symbol τ. In equation (9.43) Δx will be $v\Delta t$ and thus

$$\Delta c\tau = \Delta ct' = \sqrt{[(\Delta ct)^2 - (\Delta ct/c)^2]}$$

so

$$\Delta\tau = \Delta t/\gamma \tag{9.44}$$

9.4 Simultaneity

So far we have assumed that $(\Delta R)^2$ is positive, that is $(\Delta ct)^2 > (\Delta x)^2$, but it is quite possible that $(\Delta R)^2$ will be negative. This means that $|\Delta x| > |\Delta ct|$, and therefore no signal could pass between the two events, for it is postulated that no information can travel faster than light in a vacuum. In this case one event cannot have any causal effect on the other.

Figure 9.2 is a graph of ct against x on which a ray of light passing through the origin at $t = 0$ will be plotted as a line at 45° to the axes. The trace of the origin of the primed axes is shown as the line $x' = 0$ at an angle arctan(β) to the $x = 0$ line. The $ct' = 0$ line will be at an angle arctan(β) to the $ct = 0$ line so that the light ray is the same as for the fixed axes.

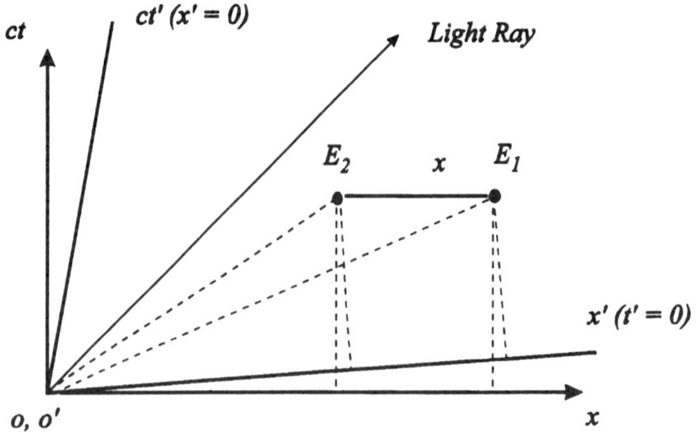

Fig. 9.2

Consider two events E_2 and E_1 which in the fixed axes are simultaneous and separated by a distance Δx. However, from the point of view of the moving axes event E_1 occurs first. If the moving frame reverses its direction of motion then the order of the two events will be reversed.

Equation (9.40) gives

$$0 - (x)^2 = (ct')^2 - (x')^2$$

and equation (9.30) shows that

$$ct' = -\beta\gamma x \tag{9.45}$$

Hence simultaneous events in one frame are not simultaneous in a second frame which is in relative motion with respect to the first.

From the above argument it follows that if there is a causal relationship between two events ($R^2 > 0$) then all observers will agree on the order of events. This is verified by writing

$$(\Delta ct)^2 - (\Delta x)^2 > 0 \tag{9.46}$$

and

$$(\Delta ct') = \gamma[(\Delta ct) - \beta(\Delta x)] \tag{9.47}$$

Because $|\Delta x| < |\Delta ct|$ and (by definition) $\beta < 1$ it follows that if (Δct) is positive then so is $(\Delta ct')$, and hence the order of events is unaltered.

9.5 The Doppler effect

The Doppler effect in acoustics is well known so we shall review this topic first. Here we shall look at the implications of Galilean relativity. Figure 9.3 shows two inertial frames of reference; set 2 is moving at constant speed v_2 and $v_1 = 0$. The Galilean equations are

$$t_2 = t_1 \tag{9.48}$$

$$x_2 = x_1 - v_2 t \tag{9.49}$$

$$y_2 = y_1 \tag{9.50}$$

$$z_2 = z_1 \tag{9.51}$$

Differentiating equation (9.49) we have

$$\dot{x}_2 = \dot{x}_1 - v_2 \tag{9.52}$$

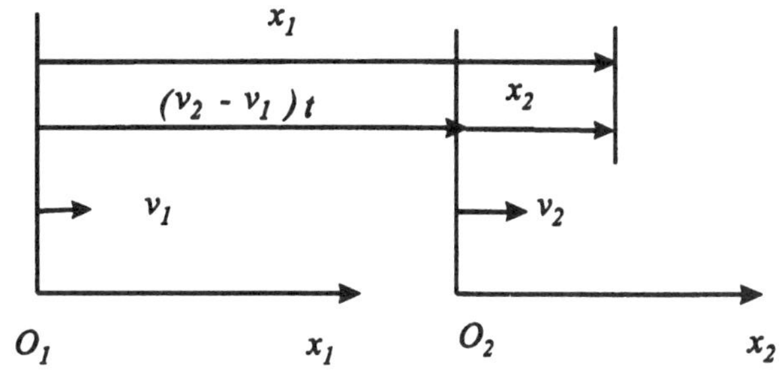

Fig. 9.3

If the velocity of sound relative to the fixed frame is c_1 then

$$c_2 = c_1 - v_2 \tag{9.53}$$

Now because both observers agree on the value of time and hence agree on simultaneity they will both agree on the wavelength. (Both frames could be equipped with pressure transducers and at a given instant measure the pressure variation along the respective x axes.) The wavelength λ is related to the wave speed and the periodic time T by

$$\lambda = c_1/T_1 = c_2/T_2 \tag{9.54}$$

Hence, using equations (9.53) and (9.54)

$$\frac{T_1}{T_2} = \frac{c_1}{c_2} = \frac{c_1}{c_1 - v_2} \tag{9.55}$$

and since frequency $\upsilon = 1/T$

$$\frac{\upsilon_2}{\upsilon_1} = 1 - v_2/c_1 \tag{9.56}$$

Thus if a sound wave is generated by a source at O_1 then the frequency measured in the moving frame, when $v_2 > v_1$, will be less.

Now let us suppose that both frames are moving in the positive x direction. The first frame has a velocity v_1 relative to a fixed frame in which the air is stationary and the second frame has a velocity v_2 also relative to the fixed frame. We now have that

$$c_1 = c - v_1 \tag{9.57}$$

and

$$c_2 = c - v_2 \tag{9.58}$$

Thus

$$\frac{T_1}{T_2} = \frac{c_2}{c_1} = \frac{c - v_2}{c - v_1}$$

or

$$\frac{\upsilon_2}{\upsilon_1} = \frac{c - v_2}{c - v_1} \tag{9.59}$$

Here we have the Doppler equation for both source and receiver moving.

If frame 2 reflects the sound wave then equation (9.59) can be used for a wave moving in the opposite direction by simply replacing c by $-c$. The frequency of the sound received back in frame 1, υ_{1r}, is found from

$$\frac{\upsilon_{1r}}{\upsilon_2} = \frac{c + v_1}{c + v_2}$$

so that

$$\frac{\upsilon_{1r}}{\upsilon_1} = \frac{\upsilon_{1r}}{\upsilon_2}\frac{\upsilon_2}{\upsilon_1} = \frac{(c + v_1)(c - v_2)}{(c + v_2)(c - v_1)} \tag{9.60}$$

Now

$$\frac{\Delta\upsilon}{\upsilon} = (\upsilon_{1r} - \upsilon_1)/\upsilon_1$$
$$= \frac{(c + v_1)(c - v_2) - (c + v_2)(c - v_1)}{(c + v_2)(c - v_1)}$$
$$= \frac{2c(v_1 - v_2)}{(c + v_2)(c - v_1)}$$

which, for v small compared with c, reduces to

$$\frac{\Delta\upsilon}{\upsilon} \cong \frac{2(v_1 - v_2)}{c} \tag{9.61}$$

In dealing with light we start with the premise that the velocity of light c is constant and therefore the above analysis is not valid. However, we can start from the Lorentz transformation. Figure 9.4 depicts two frames of reference in relative motion. A wave of monochromatic light is travelling in the positive x direction and is represented by a wave function, W, in the 'fixed' frame where $\omega = 2\pi\upsilon$ is the circular frequency and $k = 2\pi/\lambda$ is the wavenumber. So for an arbitrary function f

$$W = f(\frac{\omega}{c}ct - kx) \tag{9.62}$$

and in the moving frame

$$W = f(\frac{\omega'}{c}ct' - k'x') \tag{9.63}$$

Note that again, without loss of generality, we have taken the axes to be coincident at $t = t' = 0$. Recalling the Lorentz equations

$$ct' = \gamma ct - \gamma\beta x$$

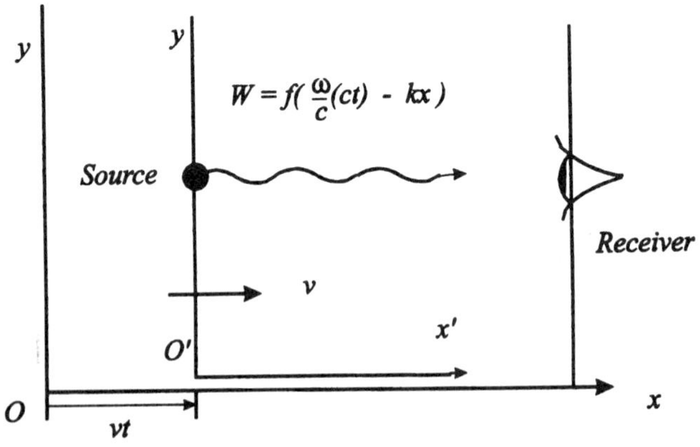

Fig. 9.4

$$x' = \gamma x - \gamma\beta ct$$

and substituting into equation (9.63) gives

$$W = f\left(\frac{\omega'}{c}(\gamma ct - \gamma\beta x) - k'(\gamma x - \gamma\beta ct)\right)$$

$$= f\left[\left(\frac{\omega'}{c} + \beta k'\right)\gamma ct - \left(k' + \frac{\omega'}{c}\beta\right)\gamma x\right]$$

but

$$c = \frac{\omega}{k} = \frac{\omega'}{k'}$$

so

$$W = f\left(\gamma(1 - \beta)\frac{\omega'}{c}ct - \gamma(1 - \beta)k'x\right) \tag{9.64}$$

Therefore comparing the arguments of equations (9.62) and (9.64) we see that

$$\frac{\omega}{c} = \frac{\omega'}{c}\gamma(1 + \beta) \tag{9.65}$$

and

$$k = k'\gamma(1 + \beta) \tag{9.66}$$

Now by definition

$$\gamma = \frac{1}{\sqrt{(1 - \beta^2)}} \tag{9.65}$$

so substituting this into equation (9.65) gives

$$\omega = \omega'\left(\frac{(1 + \beta^2)}{(1 - \beta^2)}\right)^{1/2}$$

$$= \omega'\left(\frac{1 + \beta}{1 - \beta}\right)^{1/2} \tag{9.67}$$

This wave is travelling to the right so it will only be received by an observer in the 'fixed' frame for which $x > vt$. Such an observer will see the source approaching and the frequency of the light will be higher than that recorded in the moving frame.

If the wave is travelling to the left then the wave functions will be of the form (note sign change)

$$W = f\left(\frac{\omega}{c}ct + kx\right) \tag{9.68}$$

Reworking the above analysis gives

$$\omega = \omega'\left(\frac{1 - \beta}{1 + \beta}\right)^{1/2} \tag{9.67a}$$

In this case an observer at the origin will record a frequency which will be lower than that recorded in the moving frame. That is, visible light generated in the moving frame will be shifted towards the red end of the spectrum.

The same result would have been obtained had the moving frame been moving to the left, in which case β would have had the opposite sign. The observer would then have to be at $x > vt$.

We see that if source and receiver are in relative motion the received light is red shifted if they are receding. If the receiver reflects the light back to the source then the frequency, ω'', of the received reflected light will be further red shifted, so

$$\frac{\omega''}{\omega'} = \left(\frac{1 - \beta}{1 + \beta}\right)^{1/2}$$

Therefore

$$\frac{\omega''}{\omega} = \frac{\omega''}{\omega'}\frac{\omega'}{\omega} = \left(\frac{1 - \beta}{1 + \beta}\right)$$

$$\frac{\Delta\omega}{\omega} = \frac{\omega''}{\omega} - 1 = \frac{-2\beta}{1 + \beta} \tag{9.69}$$

which for small β gives

$$\frac{\Delta\omega}{\omega} \cong -2\beta \tag{9.70}$$

or

$$\frac{\Delta\upsilon}{\upsilon} \cong -2\frac{v}{c}$$

Thus

$$\Delta\upsilon \cong -2\frac{v}{\lambda} \tag{9.70a}$$

(This principle is used in the Bruel & Kjaer Laser Velocity Set type 3544.)

9.6 Velocity

We have stated that the speed of light is not affected by the relative speed of the transmitter and receiver. It is now necessary to consider the effect of relative motion of the reference frames on the observed velocities. Starting with the Lorentz equations, equation (9.34), written in differential form

$$\Delta(ct)' = \gamma\Delta ct - \gamma\beta\Delta x \tag{9.71}$$

$$\Delta x' = \gamma\Delta x - \gamma\beta\Delta(ct) \tag{9.72}$$

$$\Delta y' = \Delta y \tag{9.73}$$

$$\Delta z' = \Delta z \tag{9.74}$$

and dividing equation (9.72) by (9.71) gives

$$\frac{\Delta x'}{c\Delta t'} = \frac{\Delta x - \beta c\Delta t}{c\Delta t - \beta\Delta x}$$

Dividing the numerator and denominator of the right hand side by $c\Delta t$ and going to the limit $\Delta t \to 0$ gives

$$\frac{u_x'}{c} = \frac{u_x/c - \beta}{1 - \beta u_x/c}$$

or

$$u_x' = \frac{u_x - v}{1 - vu_x/c^2} \tag{9.75}$$

The other two velocities are similarly found to be

$$u'_y = \frac{u_y}{\gamma(1 - vu_x/c^2)} \tag{9.76}$$

and

$$u'_z = \frac{u_z}{\gamma(1 - vu_x/c^2)} \tag{9.77}$$

For a Galilean transformation if u_x is close to c and $-v$ is close to c then u'_x could exceed c. Using equation (9.75) it can be shown that u'_x can never exceed c provided that the magnitudes of both u_x and v are less than c, as required by the special theory of relativity. This is proved by rewriting equation (9.75) as

$$\frac{u'_x}{c} = \frac{u_x/c - v/c}{1 - (v/c)(u_x/c)}$$

which has the form

$$z = \frac{x - y}{1 - yx}$$

so we require to show that if $|x| < 1$ and $|y| < 1$ then $|z| < 1$.
This is equivalent to proving

$$|1 - yx| > |x - y|$$

or

$$(1 - yx)^2 > (x - y)^2$$

$$1 + y^2x^2 - 2yx > x^2 + y^2 - 2yx$$

$$(1 - x^2) > y^2(1 - x^2)$$

Finally

$$1 > y^2$$

which satisfies our conditions. Because the above expressions are symmetrical in x and y it follows that $1 > x^2$ is also satisfied. Thus our statement that $|u'_x/c| < 1$ has been proved.

We now seek a four-vector form of velocity that transforms in the same way as the event four-vector. The difference between two events may be written as

$$(\Delta E) = (\Delta ct \ \Delta x \ \Delta y \ \Delta z)^{\mathrm{T}} \tag{9.78}$$

and we need to divide by a suitable time interval. The change in proper time $\Delta\tau$ is independent of the motion of the observer; hence dividing by this quantity will ensure that the velocity so defined will behave under a Lorentz transform identically to (ΔE).

The speed of the particle relative to the fixed frame will be

$$u = \sqrt{(u_x^2 + u_y^2 + u_z^2)} \tag{9.79}$$

so that

$$\gamma = (1 - u^2/c^2)^{-1/2} \tag{9.80}$$

for a frame of reference moving with the particle.

The proper time, as given in equation (9.45), is

$$\tau = t/\gamma = t'/\gamma'$$

where

$$\gamma' = (1 - u'^2/c^2)^{-1/2}$$

We can now write the proper velocity as

$$(U) = \left(\frac{\Delta ct}{\Delta\tau}\ \frac{\Delta x}{\Delta\tau}\ \frac{\Delta y}{\Delta\tau}\ \frac{\Delta z}{\Delta\tau}\right)^{\mathrm{T}}_{\Delta\tau \to 0}$$

or, since $\Delta\tau = \Delta\tau/\gamma$,

$$(U) = \gamma(c\ u_x\ u_y\ u_z)^{\mathrm{T}} \tag{9.81}$$

Similarly

$$(U') = \gamma'(c\ u_x'\ u_y'\ u_z')^{\mathrm{T}} \tag{9.82}$$

As a check we shall now transform equation (9.81) using the transformation equation (9.34) to give

$$\gamma' c = \gamma_0\gamma c - \gamma_0\beta_0\gamma u_x \tag{9.83}$$

$$\gamma' u_x' = -\gamma_0\beta_0\gamma c + \gamma_0\gamma u_x \tag{9.84}$$

$$\gamma' u_y' = \gamma u_y \tag{9.85}$$

$$\gamma' u_z' = \gamma u_z \tag{9.86}$$

where

$$\gamma_0 = (1 - v^2/c^2)^{-1/2}$$

and v is the relative velocity between the two frames.

From equation (9.83) we get

$$\gamma'/\gamma = \gamma_0(1 - \beta_0 u_x/c)$$

Thus from equation (9.84)

$$u_x' = \frac{u_x - v}{(1 - vu_x/c^2)}$$

and from equations (9.85) and (9.86)

$$u_y' = \frac{u_y}{(1 - vu_x/c^2)}$$

$$u_z' = \frac{u_z}{(1 - vu_x/c^2)}$$

which are the same as equations (9.75) to (9.77) derived previously. This result gives confidence in the method for obtaining the appropriate four-velocity.

Now let us evaluate the product of (U) and its conjugate

$$\begin{aligned}(U)^{\mathrm{T}}(\tilde{U}) &= \gamma^2(c^2 - u_x^2 - u_y^2 - u_z^2)\\ &= \gamma^2(c^2 - u^2)\\ &= \frac{c^2 - u^2}{1 - u^2/c^2} = c^2\end{aligned} \tag{9.87}$$

which is, of course, invariant.

9.7 The twin paradox

One of the most well known of the paradoxes which seem to defy logic is that known as the twin paradox. The story is that one twin stays on Earth whilst the other goes on a long journey into space and on returning to Earth it is found that the travelling twin has aged less than the one who remained. The apparent paradox is that as motion is relative then from the point of view of the travelling twin it could equally well be said that the one remaining on Earth should have aged less because, owing to time dilation, the Earth-bound clock would appear to run slow.

The essential difference between the two twins is that the traveller will have to change to another frame of reference in order to leave the Earth and yet another to return. The other twin remains in just one frame. Figure 9.5 depicts the series of events. Event A is when both axes are coincident and the primed set is moving with a velocity v in the x direction. Event B is when the traveller reaches the destination and changes to a double-primed frame which has a velocity $-v$ in the x direction relative to the 'fixed' set of axes.

In the usual notation, for the outward journey

$$ct' = \gamma ct - \gamma\beta x \tag{9.88}$$

$$x' = -\gamma\beta ct + \gamma x \tag{9.89}$$

and for the homeward journey

$$ct'' = \gamma ct + \gamma\beta x + ct_0'' \tag{9.90}$$

$$x'' = \gamma\beta ct + \gamma x + x_0'' \tag{9.91}$$

t_0'' and x_0'' are constants to be determined

If the length of the outward journey is L then when $x = L$, $x' = 0$, so from equation (9.89)

$$0 = -\gamma\beta ct + \gamma L$$

or

$$ct_B = L/\beta$$

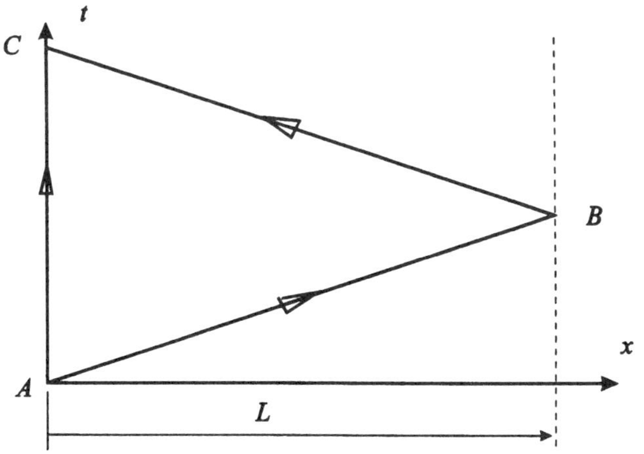

Fig. 9.5

From equation (9.88)

$$ct_B' = \gamma(L/\beta) - \gamma\beta L = L\gamma(1/\beta - \beta)$$

$$= L\gamma\frac{1 - \beta^2}{\beta} = \frac{L}{\gamma\beta}$$

For the return journey equation (9.90) gives

$$ct_B'' = \gamma(L/\beta) + \gamma\beta L + ct_0'' = ct_B'$$

Here we have arranged that the clocks in the double-primed set are synchronized with the primed set at event B. Thus

$$ct_0'' = L\gamma(\frac{1}{\beta} - \beta) - \gamma L(\frac{1}{\beta} + \beta)$$

$$= -2\gamma\beta L$$

so equation (9.90) becomes

$$ct'' = \gamma ct + \gamma\beta x - 2\gamma\beta L$$

At the end of the return journey, event C, when $x = 0$

$$ct_C'' = \gamma ct_C - 2\gamma\beta L$$

According to the fixed observer $ct_C = c2L/v = 2L/\beta$, but (9.92)

$$ct_C'' = \gamma 2L/\beta - 2\gamma L\beta = 2\gamma L(\frac{1}{\beta} - \beta)$$

$$= \frac{2L}{\gamma\beta} \tag{9.93}$$

Finally, comparing equations (9.92) and (9.93), we obtain

$$\frac{t_C''}{t_C} = \frac{1}{\gamma} \tag{9.94}$$

Because γ is never less than unity the total time read by the clocks which have travelled out and back will always be less than the clock which remained on Earth. Equation (9.94) is similar to equation (9.41) which defined time dilation, but if the process were simply time dilation then each twin would experience the same effects so that no difference in their ages would be observed.

There are many 'explanations' of the paradox but we shall resist the temptation to make the results of the above argument appear commonplace. They are not a matter of everyday experience and therefore comprehension is difficult. The difference in ageing arises from a logical mathematical development directly from the two postulates of special relativity with no additional assumptions. The ageing effect has been tested on many occasions and the results confirm the predictions.

9.8 Conservation of momentum

Having defined a four-velocity which transforms under the Lorentz transformation in the same way as the four-vector and the four-displacement we now seek an expression for the four-momentum. We shall retain the classical definition of mass and assume that it is measured in the frame where the body is at rest, or approximately so; therefore it is a quantity upon which all observers will agree. This could be called the proper mass in the same way that we defined proper time but the convention is to call it the rest mass m. We have used the symbol m since this is our classical definition, but we shall refer to it as the rest mass in order to distinguish it from the relativistic mass used in some texts.

Let us define the four-momentum as the product of the rest mass and the four-velocity. Clearly this will transform as the velocity thus

$$(P) = m(U) \tag{9.95}$$

$$(P) = m\gamma(c\ \ u_x\ \ u_y\ \ u_z)^{\mathrm{T}} \tag{9.96}$$

where

$$\gamma^2 = 1/(1 - u^2/c^2) \tag{9.97}$$

and

$$u^2 = (u)^{\mathrm{T}}(u) = (u_x^2 + u_y^2 + u_z^2) \tag{9.98}$$

Equation (9.95) expands to

$$P_0 = m\gamma c \tag{9.99}$$

$$P_x = m\gamma u_x \tag{9.100}$$

$$P_y = m\gamma u_y \tag{9.101}$$

$$P_z = m\gamma u_z \tag{9.102}$$

If we now consider a group of particles then the four-momentum will be

$$(P) = \Sigma m_i\gamma_i(c\ \ u_x\ \ u_y\ \ u_z)_i^{\mathrm{T}} \tag{9.103}$$

It is convenient to write this four-vector matrix in a partitioned form such as

$$(P) = \Sigma m_i\gamma_i(c\ (u)_i^{\mathrm{T}})^{\mathrm{T}} \tag{9.103a}$$

where $(u)_i$ is the usual velocity three-vector. Thus there is a scalar part

$$P_0 = \Sigma m_i\gamma_i c \tag{9.104}$$

and a vector part

$$(p) = \Sigma m_i\gamma_i(u)_i \tag{9.105}$$

If $|u_i| \ll c$ then $\gamma_i \cong 1$ and by assuming P_0 to be constant equation (9.104) is simply the conservation of mass. By assuming that (p) is constant equation (9.105) becomes the conservation of classical momentum.

It is easy to show that $(P)^{\mathrm{T}}(\tilde{P}) = (P)^{\mathrm{T}}[\eta](P) = M^2c^2$ where

$$M = \sqrt{(\Sigma m_i^2)} \tag{9.106}$$

and M is called the invariant mass of the system.

We now postulate that relativistic four-momentum is also conserved for an isolated system of particles. Thus

$$\Sigma m_i\gamma_i c = \text{constant} \tag{9.107}$$

and

$$\Sigma m_i\gamma_i(u)_i = \text{constant} \tag{9.108}$$

Now

$$\gamma_i = \frac{1}{\surd(1 - u_i^2/c^2)}$$

Therefore equation (9.107) becomes

$$\Sigma m_i \gamma_i c = \Sigma \frac{m_i c}{\surd(1 - u_i^2/c^2)}$$

$$= \Sigma m_i c + \Sigma m_i c \frac{1}{2} \frac{u_i^2}{c^2} + \Sigma m_i c \frac{1}{2} \times \frac{3}{2} \frac{u_i^4}{c^4} + \ldots$$

$$= cm + \frac{1}{c} \Sigma \frac{1}{2} m_i u_i^2 + \ldots = \text{constant} \tag{9.109}$$

where $\Sigma m_i = m$.

The implication here is that not only is mass conserved but, at least for moderate speeds, so is conventional kinetic energy. This suggests that the energy of a single body should be defined as

$$E = cP_0 = \gamma mc^2 \tag{9.110}$$

and the relativistic kinetic energy as

$$T = E - mc^2 \tag{9.111}$$

Frequently one sees $\gamma m = m'$, the relativistic mass, in which case equation (9.110) reads

$$E = m'c^2$$

and equation (9.111) reads

$$(p) = m'(u)$$

However, we feel that it is better to use the rest mass rather than the relativistic mass because this results in less confusion in the earlier stages of coming to terms with relativity. There is no reason why the mass of an object should change just because it is being observed by an observer in rapid motion. There seems to be no advantage in associating the Lorentz factor γ with the mass or with the velocity; it is clearer to leave γ exposed as a reminder of the origin of the equations. We shall return to equation (9.110) later.

We summarize by writing equation (9.104) as

$$(P) = \left(E/c, \Sigma\gamma_i m(u_i)^{\mathrm{T}}\right)^{\mathrm{T}} \tag{9.104a}$$

where

$$E = \Sigma\gamma_i m_i c^2$$

9.9 Relativistic force

Having established a plausible definition of four-momentum it is reasonable to attempt to define force in terms of some rate of change of four-momentum. Our requirements are, as before, that the four-vector shall transform as the position four-vector and that the definition shall agree with the Newtonian when the speeds are much less than that of light.

We shall now investigate the implications of defining the force four-vector as the rate of change of the four-momentum with respect to the proper time. Thus for a single particle the relativistic force could be

$$(F) = \frac{\mathrm{d}}{\mathrm{d}\tau}(P)$$

$$(F) = \gamma\frac{\mathrm{d}}{\mathrm{dt}}\gamma\,(mc \;\; mu_x \;\; mu_y \;\; mu_z)^{\mathrm{T}}$$

or

$$(F_0\,\gamma(f)^{\mathrm{T}}) = \gamma\frac{\mathrm{d}}{\mathrm{dt}}\,(\gamma mc \;\; \gamma m(u)^{\mathrm{T}})^{\mathrm{T}} \tag{9.112}$$

The form of (f) will be discussed in the following argument.

Now

$$\gamma = \frac{1}{\sqrt{(1 - u^2/c^2)}}$$

with

$$u^2 = (u)^{\mathrm{T}}(u)$$

so that

$$\frac{\mathrm{d}\gamma}{\mathrm{d}t} = \frac{(a)^{\mathrm{T}}(u)/c^2}{(1 - u^2/c^2)^{3/2}} = \gamma^3(a)^{\mathrm{T}}(u)/c^2 \tag{9.113}$$

where

$$(a) = \frac{\mathrm{d}}{\mathrm{d}t}(u)$$

the conventional acceleration.

Substituting equation (9.113) into equation (9.112) leads to

$$(F) = \gamma\left(\frac{\gamma^3(a)^{\mathrm{T}}(u)}{c^2}\,(mc \;\; m(u)^{\mathrm{T}})^{\mathrm{T}} + \gamma(0 \;\; m(a)^{\mathrm{T}}\right)^{\mathrm{T}}$$

$$= \left(\frac{\gamma^4 m(a)^{\mathrm{T}}(u)}{c}, \frac{\gamma^4 m(a)^{\mathrm{T}}(u)(u)}{c^2} + \gamma^2 m(a)^{\mathrm{T}}\right)^{\mathrm{T}} \tag{9.114}$$

In its component form

$$F_0 = \gamma^4 m(a)^{\mathrm{T}}(u)/c \tag{9.115}$$

$$F_x = \gamma^2 m[\gamma^2(a)^{\mathrm{T}}(u)u_x/c^2 + a_x] \tag{9.116}$$

$$F_y = \gamma^2 m[\gamma^2(a)^{\mathrm{T}}(u)u_y/c^2 + a_y] \tag{9.117}$$

$$F_z = \gamma^2 m[\gamma^2(a)^{\mathrm{T}}(u)u_z/c^2 + a_z] \tag{9.118}$$

It is apparent that if $|u| \ll c$, $F_x \to ma_x$, the Newtonian form, but this does not mean that F_x is the best choice for the definition of relativistic force.

Let us consider the scalar product of force and velocity with the force

$$(\tilde{f}) = (F_x \; F_y \; F_z) \tag{9.119}$$

Thus

$$\begin{aligned}(\tilde{f})^{\mathrm{T}}(u) &= \gamma^2 m[\gamma^2 (a)^{\mathrm{T}}(u)\,\frac{u^2}{c^2} + (a)^{\mathrm{T}}(u)] \\ &= \gamma^4 m\left[\frac{u^2}{c^2} + \frac{1}{\gamma^2}\right](a)^{\mathrm{T}}(u) \\ &= \gamma^3 m(a)^{\mathrm{T}}(u)\end{aligned} \tag{9.120}$$

Using equation (9.113)

$$(\tilde{f})^{\mathrm{T}}(u) = \gamma m\,\frac{\mathrm{d}\gamma}{\mathrm{d}t}\,c^2 = \gamma\,\frac{\mathrm{d}}{\mathrm{d}t}\,(\gamma mc^2) \tag{9.120a}$$

Here, as in all the preceding arguments, we have taken the rest mass m to be constant.

From equation (9.110) we see that equation (9.120a) becomes

$$(\tilde{f})^{\mathrm{T}}(u) = \gamma\,\frac{\mathrm{d}E}{\mathrm{d}t} \tag{9.121}$$

It would seem more appropriate, since we are dealing with three-vectors, that the right hand side of equation (9.121) should be simply the time rate of change of energy, in which case we need to redefine the relativistic three-force as

$$(f) = (F_x/\gamma \;\; F_y/\gamma \;\; F_z/\gamma)^{\mathrm{T}} \tag{9.122}$$

the components of which are

$$f_x = \gamma m[\gamma^2 (a)^{\mathrm{T}}(u) u_x/c^2 + a_x] \tag{9.123}$$

$$f_y = \gamma m[\gamma^2 (a)^{\mathrm{T}}(u) u_y/c^2 + a_y] \tag{9.124}$$

$$f_z = \gamma m[\gamma^2 (a)^{\mathrm{T}}(u) u_z/c^2 + a_z] \tag{9.125}$$

Equation (9.121) now reads

$$(f)^{\mathrm{T}}(u) = \frac{\mathrm{d}E}{\mathrm{d}t} \tag{9.126}$$

Hence the rate at which the relativistic force is doing work is equal to the time rate of change of energy, which is the familiar form. We conclude that equations (9.122) to (9.125) give the most convenient definition of relativistic force. (We have regarded force as a defined quantity.)

It is important to note that only contact forces are considered here. Long-range forces lead us into serious difficulties because of the relativity of simultaneity. We can no longer expect the forces on distant objects to be equal and opposite.

9.10 Impact of two particles

In Fig. 9.6 we show two particles moving along the x axis and colliding. This process is seen from the laboratory frame of reference but using the Lorentz transformation it is

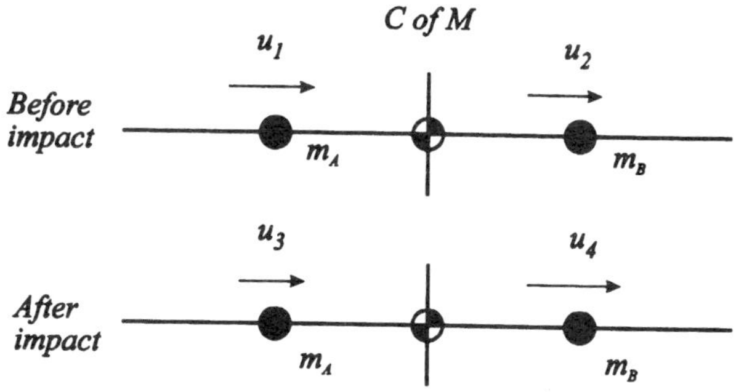

Fig. 9.6

possible to use a frame in which the momentum is zero. This is similar to the the use of co-ordinates referred to the centre of mass but this time we use the centre of momentum, also abbreviated to COM. In this frame the components of total momentum are

$$P_0 = \Sigma\gamma_i m_i = E/c \tag{9.127}$$

$$P_x = \Sigma\gamma_i m_i u_{xi} = 0 \tag{9.128}$$

$$P_y = \Sigma\gamma_i m_i u_{yi} = 0 \tag{9.129}$$

$$P_z = \Sigma\gamma_i m_i u_{zi} = 0 \tag{9.130}$$

where

$$\gamma_i = 1/\surd(1 - u_i^2/c^2)$$

and

$$u_i^2 = u_{xi}^2 + u_{yi}^2 + u_{zi}^2$$

For the present problem equation (9.127) is

$$\gamma_1 m_A + \gamma_2 m_B = E/c^2 = \gamma_3 m_A + \gamma_4 m_B \tag{9.131}$$

and equation (9.128) is

$$\gamma_1 m_A u_1 + \gamma_2 m_B u_2 = 0 = \gamma_3 m_A u_3 + \gamma_4 m_B u_4 \tag{9.132}$$

Now

$$\gamma_1 = \frac{1}{\surd(1 - u_1^2/c^2)}$$

so

$$\frac{u_1}{c} = \surd(1 - \gamma_1^{-2})$$

and

$$\frac{\gamma_1 u_1}{c} = \surd(\gamma_1^2 - 1)$$

By inspection of equations (9.131) and (9.132) there is a trivial solution of $u_3 = u_1$ and $u_4 = u_2$ but there is also the solution

$$u_3 = -u_1 \quad \text{and} \quad u_4 = -u_2$$

This is the same form of answer that would have been expected for a low-speed impact with kinetic energy conserved. It also shows that the speed of approach $(u_1 - u_2)$ is equal to the speed of recession $(u_4 - u_3)$. Another way of looking at this situation is to regard it as a reversible process. That is, if the sequence of events is reversed in time then the process has exactly the same appearance.

Let us now consider the case when the two bodies coalesce; in this case $u_4 = u_3$, so $\gamma_4 = \gamma_3$. Equation (9.132) immediately gives

$$0 = \gamma_3 u_3 (m_A + m_B)$$

which means that $u_3 = 0$ so that $\gamma_3 = 1$.

Equation (9.131) now yields

$$\gamma_1 m_A + \gamma_2 m_B = \frac{E}{c^2} = m_A + m_B$$

which, because $\gamma > 1$, will not balance unless we accept the premise that the total mass on the right hand side of the equation is greater than that on the left. In low-speed mechanics the energy equation would have been balanced by including thermal energy but here we see that this energy manifests itself as an increase in the rest mass. If we imagine this process in reverse then a body at rest disintegrates into two particles having kinetic energy at the expense of the rest mass of the system. In conventional engineering situations this does not occur, but in atomic and nuclear physics it does. As is well known it forms the basis for the operation of nuclear power stations.

The connection between the loss in rest mass (m) and the release of energy should not be confused with the change in apparent mass (γm) which results from Lorentz transformations. Up to the last example the rest mass has been taken to be constant.

9.11 The relativistic Lagrangian

We have defined relativistic kinetic energy in equation (9.111). Equation (9.121) gives the relationship between the time rate of change of energy and the scalar product of force and velocity. This three-vector equation has the same form in classical and relativistic mechanics, that is

$$(f)^T(u) = \frac{dE}{dt} = \frac{dT}{dt}$$

From equation (9.112) we have that

$$\gamma(f)^T = \gamma \frac{d}{dt}(p)^T$$

or

$$(f)^T = \frac{d}{dt}(p)^T$$

Therefore

$$dT = d(p)^T(u)$$

Let us define the co-kinetic energy as

$$T^* = (p)^{\mathrm{T}}(u) - T \tag{9.133}$$

so that

$$\begin{aligned} \mathrm{d}T^* &= \mathrm{d}(p)^{\mathrm{T}}(u) + (p)^{\mathrm{T}}\mathrm{d}(u) - \mathrm{d}T \\ &= (p)^{\mathrm{T}}\mathrm{d}(u) \end{aligned} \tag{9.134}$$

Expanding equation (9.134) gives

$$\sum_i \frac{\partial T^*}{\partial u_i}\mathrm{d}u_i = \sum_i p_i \mathrm{d}u_i$$

Comparing coefficients of $\mathrm{d}u_i$ gives

$$\frac{\partial T^*}{\partial u_i} = p_i \tag{9.135}$$

This suggests that the correct form for the Lagrangian is

$$\mathfrak{L} = T^* - V \tag{9.136}$$

(*see* Chapter 3) where

$$T^* = (p)^{\mathrm{T}}(u) - T = (p)^{\mathrm{T}}(u) - [\gamma m(u) - mc^2] \tag{9.137}$$

Differentiation of this expression with respect to u_i will confirm the above result. Note that for low speeds, where momentum depends linearly on velocity, $T^* = T$.

There is an interesting link with the principle of least action and with Hamilton's principle. A particle moving freely in space with no external forces travels in a straight line. The trace of displacement against time is also straight. Figure 9.7 is a plot of λt against time where λ is a factor with the dimensions of speed. The length of the line between times t_1 and t_2, sometimes called a worldline, is

$$s = \int_{t_1}^{t_2} \sqrt{(\mathrm{d}s^2 + \lambda^2\mathrm{d}t^2)}$$

Let us now define $S = ms$ and seek to minimize this function. Thus we set the variation of S equal to zero,

$$\delta S = \delta \int_{t_1}^{t_2} m\sqrt{(\mathrm{d}s^2 + \lambda^2\mathrm{d}t^2)} = 0$$

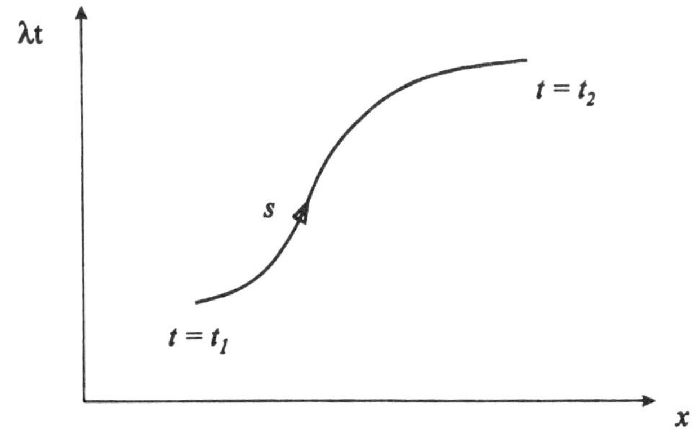

Fig. 9.7

$$= \delta \int_{t_1}^{t_2} m\sqrt{(u^2 + \lambda^2)}\mathrm{d}t = 0$$

$$= \delta \int_{t_1}^{t_2} \frac{mu\delta u}{\sqrt{(u^2 + \lambda^2)}} = 0 \qquad (9.138)$$

If we now let $\lambda^2 = -c^2$ then

$$\delta S = \int_{t_1}^{t_2} \frac{mu\,\delta u}{\mathrm{j}c\sqrt{(1 - u^2/c^2)}} = 0 \qquad (9.139)$$

which, from equation (9.105), reduces to

$$\delta S = \frac{1}{\mathrm{j}c}\int_{t_1}^{t_2} (p\delta u)\,\mathrm{d}t = 0 \qquad (9.140)$$

So, summing for a group of particles and using equation (9.134) we obtain

$$\delta S = \frac{1}{\mathrm{j}c}\int_{t_1}^{t_2} \delta T^* = 0$$

or simply

$$\delta\int_{t_1}^{t_2} T^* = 0 \qquad (9.141)$$

This suggests that Hamilton's principle should read

$$\delta\int_{t_1}^{t_2} (T^* - V) = 0 \qquad (9.142)$$

Again the potential energy, V, poses difficulties for long-range forces.

The above is not a rigorous proof of the relativistic Lagrange equations or Hamilton's principle but an attempt to show that there is a steady transition from classical mechanics to relativistic mechanics. The link is so strong that many authors regard special relativity as being legitimately described as classical.

9.12 Conclusion

In this chapter we started with the statement that the laws of physics take the same form in all inertial frames of reference and in particular that the speed of light shall be invariant. From this we developed the Lorentz transformation and introduced the concept of the four-vector. This results in space and time being inextricably fused into a space–time continuum where simultaneity was also relative.

We then proceeded to develop the four-vector equivalents to the classical velocity and momentum three-vectors. This led us to the first remarkable result that conservation of four-momentum encompassed the conservation of energy. The second remarkable result is the equivalence of mass and energy.

The development has been purely mathematical without attempting to 'explain' the apparent inconsistencies which arise. It is sufficient to state that during most of the twentieth century many experiments have been carried out to test the validity of the predictions and so far no flaw has been detected. The literature abounds with simplifications and pictorial representations of the paradoxes but the reader should beware of attempts to put this topic in a popular science guise. Some are very helpful but others can be misleading.

We have used the position four-vector in the form

$$(E) = (ct \ x \ y \ z)^{\mathrm{T}}$$

but some put time as the last factor, with and without the c.

The product of (E) and its conjugate has here been defined to be

$$(E)^{\mathrm{T}}(\tilde{E}) = (ct)^2 - x^2 - y^2 - z^2$$

Others define it as

$$(E)^{\mathrm{T}}(\tilde{E}) = x^2 + y^2 + z^2 - (ct)^2$$

which is equivalent to replacing the metric matrix $[\eta]$ by its negative.

In texts where the subject is studied in more depth indicial notation is commonly used, in which case the terms covariant and contravariant are used where we have used a standard vector form and its conjugate.

A further variation is to write (E) as

$$(E) = (\mathrm{j}ct \ x \ y \ z)$$

The use of $\mathrm{j} = \sqrt{-1}$ means that there is no need for the metric matrix and this simplifies some equations.

There is no clear winner in the choice of form for (E); it is very much a matter of personal preference.

To conclude this chapter mention should be made of the general theory of relativity. This is far more complex than the special theory and was the real crowning glory of Einstein's work.

The essence of the general theory is known as the principle of equivalence. One form of the theory is

In a small freely falling laboratory the laws of physics are the same as for an inertial frame.

The implication is that locally the effects in a gravitational field of strength $-g$ are the same as those in a frame with an acceleration of g. Thus inertia force, which we hitherto regarded as a fictitious force, is now indistinguishable from a gravitational force. The force of gravity can hence be removed by the proper choice of the frame of reference. The ramifications of this theory are complex and have little bearing on present-day calculations in engineering dynamics. However, some study of the subject as presented in the Bibliography will be rewarding.

One interesting example of the consequences of the principle of equivalence can be found by considering a frame which has a uniform acceleration g. In classical mechanics the path of particles will be parabolic as seen from the accelerating frame and, on a local scale, would be indistinguishable from paths produced in a frame within a gravitational field of strength $-g$. It is now proposed that light will be similarly affected.

In Fig. 9.8 sketches of paths in space are shown for a frame which is accelerating in the y direction and in which particles are being projected in the x direction with increasing initial velocities. These paths will be different. If we now plot a graph of y against time then only one curve is produced irrespective of the initial velocity in the x direction. For small changes in y the curvature will be $d^2y/dt^2 = -g$. This is assumed to be true for any speed, even that of light. Plots of x against time are straight lines with different slopes but the same curvature, in this case zero. This is a crude introduction to the notion of curvature in space–time.

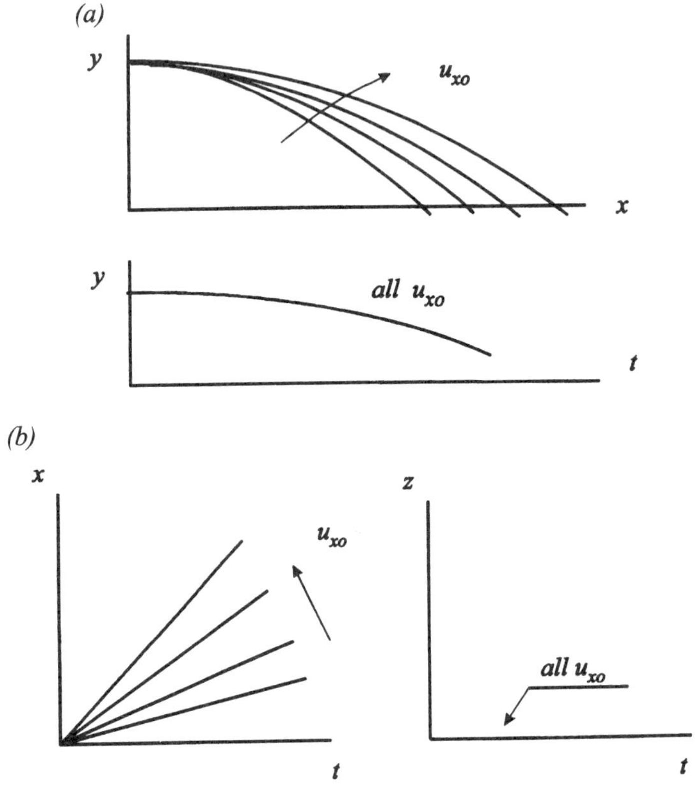

Fig. 9.8 (a) and **(b)**

Problems

1. A small Earth satellite is modelled as a thin spherical shell of mass 20 kg and 1 m in diameter. It is directionally stabilized by a gyro consisting of a thin 4 kg solid disc 200 mm in diameter and mounted on an axle of negligible mass whose frictionless bearings are located on a diameter of the shell.

   The shell is initially not rotating while the gyro is rotating at 3600 rev/min. The satellite is then struck by a small particle which imparts an angular velocity of 0.004 rad/s about an axis perpendicular to the gyro axis.

   Determine the subsequent small perturbation angle of the axis of the gyro and shell.

   Answer: 1.77 mrad at 0.36 Hz

2. A particle is dropped down a vertical chimney situated on the equator. What is the acceleration of the particle normal to the vertical axis after it has fallen 23 m?

   Answer: 2.5 mm/s^2

3. An aircraft is travelling due south along a horizontal path at a constant speed of 700 km/h when it observes a second aircraft that is travelling at a constant speed in a horizontal plane. Tracking equipment on the first aircraft detects the second aircraft and records that the separation is 8 km with a bearing of 45° east and an elevation of 60°. The rate of change of the bearing is 0.0 5rad/s and that of the elevation is 0.002 rad/s.

   Deduce the absolute speed and bearing of the second aircraft.

   Answer: 483 km/h, 107° east of north

4. A spacecraft is on a lunar mission. Set up the equations of motion for free motion under the influence of the gravitational fields of the Earth and the Moon. Use a co-ordinate system centred on the centre of mass of the Earth–Moon system with the x axis directed towards the centre of the Moon. The y axis lies in the plane of motion of the Earth, Moon and spacecraft.

Answer:

$$\ddot{x} = -F_1X_1/R_1 - F_2X_2/R_2 + F_3x/R_3 + 2\Omega\dot{y}$$
$$\ddot{y} = -F_1y/R_1 - F_2y/R_2 + F_3y/R_3 - 2\Omega\dot{x}$$

where

$F_1 = Gm_{\text{Earth}}/R_1^2$, $F_2 = Gm_{\text{Moon}}/R_2^2$ and $F_3 = \Omega^2 R_3$
R_1 = distance of spacecraft from the Earth
R_2 = distance of spacecraft from the Moon
$R_3 = (x_2 + y_2)^{1/2}$
Ω = angular velocity at Earth–Moon axis
G = the universal gravitational constant

5. A gyroscope wheel is mounted in a cage which is carried by light gimbals. The cage consists of three mutually perpendicular hoops so that the moment of inertia of the cage has the same value about any axis through its centre. The *xyz* axes are attached to the cage and the wheel axis coincides with the *z* axis.

   Initially the wheel is spinning at 300 rev/min and the cage is stationary. An impulse is applied to the cage which imparts an angular velocity of 0.1 rad/s about the *x* axis to the cage plus wheel.

   Determine the frequency and amplitude of the small oscillation of the *z* axis.

   The relevant moments of inertia are:

   For the cage 3 kg m^2
   For the wheel about its spin axis (the *z* axis) 1.2 kg m^2
   For the wheel about its *x* or *y* axis 0.6 kg m^2.

   Answer: 16.67 Hz, 0.054°

6. An object is dropped from the top of a tower height *H*. Show that, relative to a plumb line, the object hits the ground to the east of the line by a distance given approximately by

$$\frac{\omega \cos(\gamma)\ g}{3}\left(\frac{2H}{g}\right)^{2/3}$$

   where ω is the angular velocity of the Earth, γ is the angle of latitude and g is the apparent value of the gravitational field strength.

7. An aircraft has a single gas turbine engine the rotor of which rotates at 10 000 rev/min clockwise when viewed from the front. The moment of inertia of the rotor about its spin axis is 15 kg m^2. The engine is mounted on trunnions which allow it to pitch about an axis through the centre of mass. A link is provided between the upper engine casing and the fuselage forwards of the trunnions in order to prevent relative pitching. The moment arm of the force in the link about the centre of mass is 0.5 m.

   Determine the magnitude and sense of the load in the link when the aircraft is making a steady turn to the left at a rate of 3°/s and is banked at 30°.

   Answer: 1424 N, compression

8. Derive an expression for the torque on the shaft of a two-bladed propeller due to gyroscopic action. Consider the propeller blade to be a thin rod.

Answer:

$$\text{Torque} = \frac{mr^2}{2\sqrt{3}}\,\dot{\phi}^2 \sin(2\alpha)$$

where $\dot{\phi}$ is the precession in the plane of rotation and α is the angle of the propeller blade measured from the precession axis.

9. A satellite is launched and attains a velocity of 30 400 km/h relative to the centre of the Earth at a distance of 320 km from the surface. It has been guided into a path which is parallel to the Earth's surface at burnout.

   (a) What is the form of the trajectory?
   (b) What is the furthest distance from the Earth's surface?
   (c) What is the duration of one orbit?

Answer: elliptic, 3600 km, 130 min

10. A motor and gear wheel is modelled as two solid wheels, M and G1, joined by a light shaft S1. The gear wheel G1 meshes with another gear wheel G2 which drives a rotor R via a light shaft S2.
    The moments of inertia of the wheels M, G1, G2 and R are I_M, I_{G1}, I_{G2} and I_R respectively and the torsional stiffnesses of the two shafts are k_{S1} and k_{S2}.
    Derive the equations of motion for the angular motion of the system.

11. Show that the torsional oscillations of a shaft having a circular cross-section are described by the solutions of the wave equation

$$\frac{G}{\rho}\frac{\partial^2\theta}{\partial x^2} = \frac{\partial^2\theta}{\partial t^2}$$

where θ is the rotation of a cross-section. G, ρ, x and t have their usual meanings.
A steel shaft, 20 mm in diameter and 0.5 m long, is fixed at one end. A torque (T) of amplitude 50 N m and varying sinusoidally with a frequency of 2 kHz is applied at the free end. What is the amplitude of vibration at a distance x from the free end? (G = 80 GN/m^2 and ρ = 7750 kg/m^3.)
Hint: For the steady-state response assume a sinusoidal standing wave solution of the form $\theta = X(x)e^{j\omega t}$.

Answer:

$$\theta = \frac{Tc}{IG\omega}\,[\cos(\omega x/c) - \cot(\omega L/c)\sin(\omega x/c)]$$

where $c = \sqrt{(G/\rho)}$, L = length and I = polar second moment of area.

12. A mechanical bandpass filter is constructed from a series of blocks, mass m, separated by axial springs each of stiffness s_1. Also each mass is connected to a rigid foundation by a spring, each having a stiffness of s_2.
    Considering the system as an infinitely long periodic structure show that the dispersion equation is

$$\omega^2 = \frac{s_2}{m} + 4\frac{s_1}{m}\sin^2(k/2)$$

where ω is the circular frequency and k is the wavenumber.

If the passband is 100 Hz to 1000 Hz estimate the maximum speed at which energy can propagate and the associated frequency.

Hint: Assume that the axial displacement

$$u_n = (\text{amplitude})\, e^{j(\omega t - kn)}$$

where n is the block number.

Answer: 2827 block/s at 323 Hz

13. Describe the types of waves that can propagate in a semi-infinite homogeneous, isentropic, linearly elastic solid. Reference should be made to the following points:

    (a) waves in the interior,
    (b) waves on the free surface, and
    (c) reflection and refraction at an interface.

    Sketch the phase velocity/wavenumber curves for waves in an infinite slender bar of constant cross-section.

14. A long uniform rod, with a cross-sectional area A, has a short collar, mass M, fixed a point distant from either end. At one end a compressive pulse is generated which is of constant strain magnitude, $|\varepsilon_0|$, for a short duration τ.

    Sketch the form of the strain pulse transmitted past the collar and pulse reflected from the collar. Derive expressions for the maximum tensile and compressive strains. The bar is long enough so that waves reflected from the ends arrive after the peak values of strain have occurred.

    Answer:

    Maximum transmitted compressive strain = $|\varepsilon_0|\,(1 - e^{-2\mu c\tau})$
    Maximum reflected tensile strain = $|\varepsilon_0|$

    where $c = \sqrt{(E/\rho)}$ and $\mu = EA/(Mc^2)$.

15. A long, straight uniform rod (1) is attached to a short uniform rod (2) of a different material. The free end of rod (1) is subjected to a constant axial velocity for a period τ.

    Show that the maximum force imparted to rod (1) is

    $$VZ_2\left[1 - \left(\frac{Z_2 - Z_1}{Z_2 + Z_1}\right)^n\right]$$

    where $Z = EA/c$ for each respective bar and n is the number of reflections occurring at the interface between the two bars during the time τ.

16. A semi-infinite medium having low impedance (ρc) is bounded by a rigid plate forming the $z = 0$ plane. The surface is lubricated so that the shear stress between the plate and the medium is zero.

Show that the dilatational wave is reflected at the surface without the generation of a transverse wave.

Comment on the reflections generated by an incident transverse wave.

Assume that the potential function for the incident dilatational wave has the form

$$ø = D\, e^{j[wt - k_1 \mathrm{Sin}(ax) + k_1 \mathrm{Cos}(az)]}$$

Note that

$$u_z = \frac{\partial ø}{\partial z} + \frac{\partial \psi}{\partial x}$$

and

$$\sigma_{xz} = 2G \frac{\partial^2 ø}{\partial x \partial z} - \frac{1}{2}\left(\frac{\partial^2 \psi}{\partial z^2} - \frac{\partial^2 \psi}{\partial x^2}\right)$$

17. An electric motor is connected to the input shaft of a gearbox via an elastic shaft with a torsional stiffness of 3 MN m/rad. The gearbox has a 3:1 reduction, the input pinion has negligible inertia and the output gear wheels each have a moment of inertia of 4 kg m². Each output shaft drives identical rotors of moment of inertia 40 kg m² through identical shafts each of torsional stiffness 0.5 MN m/rad.

    If one of the rotors is fixed set up the equations of motion for torsional vibration of the system using the twist in each shaft as the three generalized co-ordinates.

Answer:

$$\begin{bmatrix} 102 & 40 & 18 \\ 40 & 40 & 0 \\ 18 & 0 & 6 \end{bmatrix} \begin{bmatrix} \ddot{\theta}_1 \\ \ddot{\theta}_2 \\ \ddot{\theta}_3 \end{bmatrix} + 10^5 \begin{bmatrix} 5 & 0 & 0 \\ 0 & 5 & 0 \\ 0 & 0 & 30 \end{bmatrix} \begin{bmatrix} \theta_1 \\ \theta_2 \\ \theta_3 \end{bmatrix} = \begin{bmatrix} 0 \\ 0 \\ 0 \end{bmatrix}$$

18. Show that the differential equation for transverse waves in a uniform bar which has a constant tensile force T applied is

$$T \frac{\partial u^2}{\partial x^2} - EI \frac{\partial^4 u}{\partial x^4} = \rho A \frac{\partial^2 u}{\partial t^2}$$

where u is the lateral displacement, EI is the flexural rigidity and ρA is the mass per unit length. Rotary inertia and shear distortion have been neglected.

19. A uniform long steel rod has a rubber block attached at its left hand end. The block is assumed to behave as an ideal massless spring of stiffness s. The end of the rubber block is displaced axially such that the displacement rises linearly in time T to an amplitude h and reduces linearly to zero in a further interval of T.

    Derive expressions for the strain in the steel bar for the region $0 < (ct - x) < 3cT$.

Answer:

For $0 < (ct - x) < cT$

$$\varepsilon = \frac{-h}{cT}(1 - e^{-\mu z})$$

for $ct < (ct - x) < 2ct$

$$\varepsilon = \frac{-h}{cT}[-1 + e^{-\mu z}(-1 + 2e^{\mu cT})]$$

and for $2cT < (ct - x)$

$$\varepsilon = \frac{-h}{cT}[e^{-\mu z}(-e^{2\mu cT} + 2e^{\mu cT} - 1)]$$

where $\mu = s/(EA)$ and $z = (ct - x)$.

20. Use Hamilton's principle to derive the equation given in problem 18.
Derive expressions for phase velocity and group velocity.

Answer:

$$c_p = \left(\frac{T}{\rho A} + \frac{EI}{\rho A}k^2\right)^{1/2}$$

$$c_g = \left(\frac{T}{\rho A} + 2\frac{EI}{\rho A}k^2\right)\frac{1}{c_p}$$

21. Two uniform bars of equal square cross-sections ($b \times b$) are welded together to form a 'T'. The structure is given a sinusoidal input at the joint in the direction of the vertical part of the 'T'. As a result an axial wave is generated in the vertical part and symmetrical bending waves are generated in the side arms.
Assuming the simple wave equation for axial waves and Euler's equation for bending waves obtain an expression for the point impedance at the input point. Point impedance is defined to be the complex ratio of force/velocity at that point.

Answer:

$$Z = b^2\sqrt{(E\rho)}\left[1 + (1 + j)\left(\frac{b\omega}{c_0\sqrt{12}}\right)^{1/2}\right]$$

22. Construct the rotation matrix for a rotation of 30° about the OZ axis, followed by a rotation of 60° about the OX axis, followed by a rotation of 90° about the OY axis.

23. Determine the transformation matrix, T, for a rotation of α about the OX axis, followed by a translation of b along the OZ axis, followed by a rotation of ø about the OV axis.

24. A Stanford-type robot is shown in its home position in Fig. P24. The constants are d_1 = 200 mm and d_2 = 100 mm.
The arm is now moved to $\theta_1 = 90°$, $\theta_2 = -120°$ and $d_3 = 220$ mm.

(a) Draw up the table for θ_i, α_i, a_i, d_i.

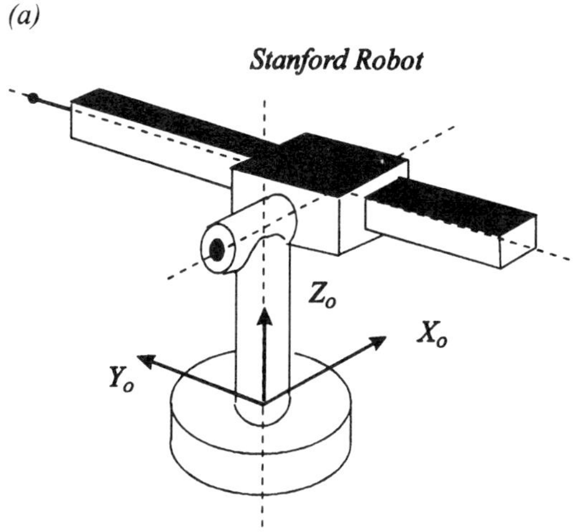

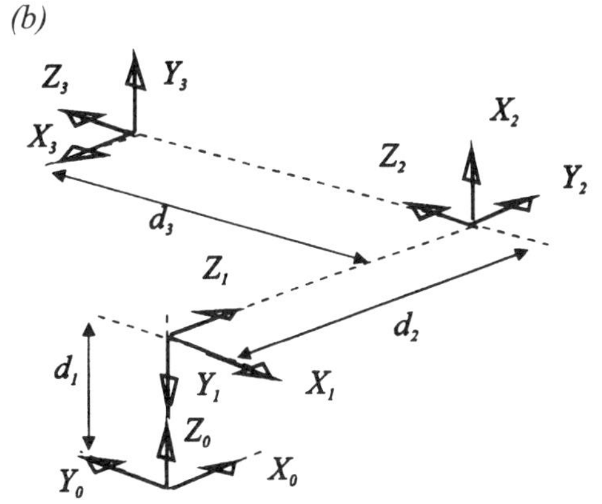

Fig. P24(a) and **(b)**

(b) Calculate the ${}_0A_3$ matrix.
(c) Find the co-ordinates for the origin of the $(XYZ)_3$ set of axes in terms of $X_0Y_0Z_0$.

25. Figure P25 shows a Minimover robot in an extended position with the arm in the X_0Z_0 plane. The co-ordinate system shown satisfies the Denavit–Hartenberg representation.

(a) With reference to the data sheet given at the end of Chapter 8 complete the table of values of θ, α, *a* and *d* for all five joints and links.

Joint/link	θ	α	*a* (mm)	*d* (mm)
1	0	−90°	0	
2		0		0
3		0	175	0
4	+20°		0	0
5	±90°	0	0	100

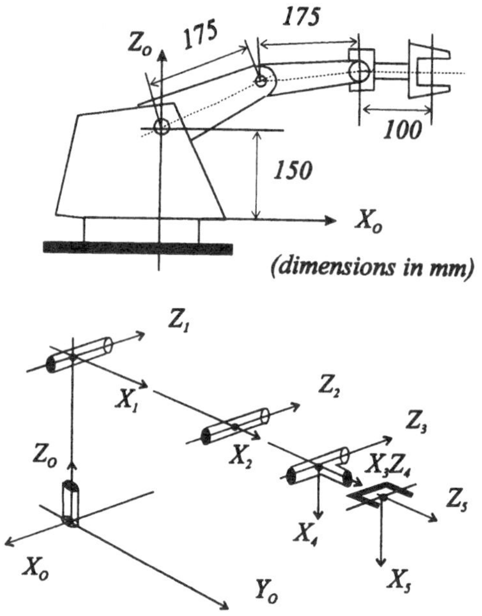

Fig. P25

(b) Determine the transformation matrix ${}_3A_4$.

(c) If the overall transformation matrix ${}_0T_5$ is constructed it is found that three of its elements are

$$T_{14} = -100C_1S_{23}S_4 + 100C_1C_{23}C_4 + 175C_1(C_{23} + C_2)$$

$$T_{24} = -100S_1S_{23}S_4 + 100S_1C_{23}C_4 + 175S_1(C_{23} + C_2)$$

$$T_{34} = 100C_{23}S_4 + 100S_{23}C_4 - 175(S_{23} + S_2)$$

where $C_1 = \cos\theta_1$, $S_1 = \sin\theta_1$ and $C_{23} = \cos(\theta_2 + \theta_3)$ etc.

Show that these elements, in general, give the co-ordinates of the origin of the axes attached to the end effector. Evaluate these co-ordinates.

26. Figure P26 shows a robot of the Stanford type which is moving such that the gripper and arm remain in a horizontal plane. A dynamic model is depicted in the figure. I_1 is the mass moment of inertia of the whole assembly about the Z_0 axis excluding the arm AB. The mass of the arm is represented by two concentrated masses m_2 and m_3.

Using Lagrange's equations, or otherwise, derive general expressions for:

(a) The torque required from the motor causing the rotation, θ, about the Z_0 axis.

(b) The thrust required from the unit producing the extension d.

(c) During the main part of the movement the co-ordinates θ and d are controlled so that their derivatives are constant, the values being

$$\frac{d\theta}{dt} = 0.4 \text{ rad/s} \qquad \frac{dd}{dt} = 0.8 \text{ m/s}$$

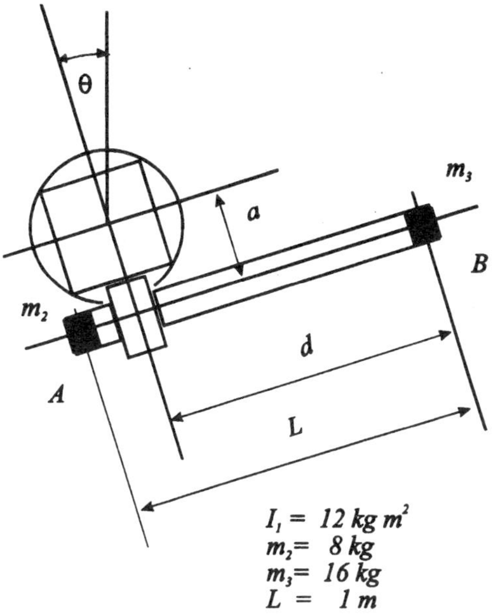

Fig. P26

At the instant when $d = 0.6$ m evaluate the torque and thrust as defined in (a) and (b).

Answer: (c) 7.168 N m, 4.096 N

27. Figure P27 shows an exploded view of a Puma-type robot. The links are numbered and a co-ordinate system is given which satisfies the requirements of the Denavit–Hartenberg representation. Draw up a table for the six links and the corresponding constants α, *a* and *d* and the variables θ.

The three non-zero dimensions are *p*, *q* and r and the variable angles are *s*, *t*, *u*, *v*, *w* and ø. Insert the correct values for all angles.

Answer:

Joint/link	*a*	*d*	α	θ
1	0	0	−90°	90°
2	*p*	0	0	0
3	0	0	90°	90°
4	0	*q*	−90°	0
5	0	0	90°	0
6	0	*r*	0	0

28. A car has a mass of 1300 kg and a wheelbase of 2.5 m. The centre of mass for the unladen car is 1.2 m behind the front axle. The lateral force coefficient for all four tyres is 50 000 N/rad. Determine the static margin.

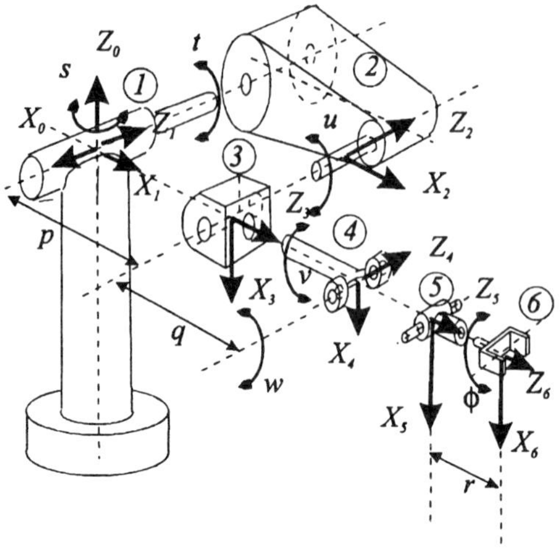

Fig. P27

The vehicle is then overloaded with a 300 kg load in the boot the centre of mass of the load is 0.1 m behind the rear axle. Determine the new static margin and the critical speed.

Answer: 1%, −9%, 48 mph

29. A car has a mass of 1300 kg and a wheelbase of 2.5 m. The centre of mass is midway between the axles and the tyre side force coefficient for all tyres is 50 000 N/rad. The radius of gyration about a vertical axis through the centre of mass is 1.6 m.

The car is travelling along a straight road at 30 mph when it is hit by a sudden gust of wind. Working from first principles show that yawing motion decays exponentially. Evaluate the time to half the initial amplitude.

If the weight distribution is 60% on the front axle and all other data are as above what is the periodic time for oscillations following a disturbance?

Answer: 0.20 s, 2.67 s

30. An aircraft has the following data:

Structural

Mass 20 000 kg

Radius of gyration about the y axis through the centre of mass 8 m

Wing area 50 m^2

Wing aspect ratio 8

Tail arm 9 m

Tailplane area 10 m^2
Position of centre of mass behind the aerodynamic centre 1 m

Aerodynamic

Gradient of lift coefficient – incidence curve 4.5
Gradient of the tailplane lift coefficient – incidence curve 3.5
Gradient of the fuselage pitching moment coefficient – incidence curve 0.5.

The aircraft is in level flight at 240 m/s at an altitude where the density of the air is 0.615 kg/m^3.

(a) Calculate the stick-fixed static margin.
(b) The lift coefficient.
(c) The tail volume ratio.
(d) The lift/drag ratio.

Answer: 0.049, 0.22, 0.72, 10.1

31. The aircraft described in problem 30 receives a small disturbance which sets up a pitching oscillation. Determine the periods of the damped motion.

Answer: 7.79 s and 167 s

32. A rocket with payload has a take-off mass of 6000 kg of which 4800 kg is fuel. The fuel has a specific impulse of 2900 Ns/kg and the rocket motor thrust is 70.63 kN. The rocket is fired vertically from the surface of the Earth and during its flight the aerodynamic forces are negligible and gravity is assumed constant.

Let μ = (mass at burn out/initial mass) and R = (thrust of rocket (assumed constant)/take-off weight). With I = specific impulse obtain expressions for the velocity and height at burn out in terms of μ, R, I and g.

For the data given determine the values of velocity and height at burn out.

Answer:

$$v_b = I[\ln(1/\mu) - (1 - \mu)/R]$$

$$h_b = \frac{I^2}{gR}\left[\mu \ln(\mu) - \mu + 1\right] - \frac{I^2}{2gR^2}(1 - \mu)^2$$

v_b = 2558 m/s, h_b = 183 km and t_b = 3.58 minutes

Appendix 1
Vectors, Tensors and Matrices

Cartesian co-ordinates in three dimensions

In our study of dynamics we have come across three types of physical quantity. The first type is a scalar and requires only a single number for its definition; this is a scalar. The second requires three numbers and is a vector. The third form needs nine numbers for a complete definition. All three can be considered to be tensors of different rank or order.

A tensor of the zeroth rank is the scalar. A tensor of the first rank is a vector and may be written in several ways. A three-dimensional Cartesian vector is

$$\boldsymbol{V} = x\boldsymbol{i} + y\boldsymbol{j} + z\boldsymbol{k} \tag{A1.1}$$

where $\boldsymbol{i}, \boldsymbol{j}$ and $\boldsymbol{k}$ are the respective unit vectors.

In matrix form we have

$$\boldsymbol{V} = (\boldsymbol{i}\,\boldsymbol{j}\,\boldsymbol{k})\begin{bmatrix} x \\ y \\ z \end{bmatrix} = (e)^{\mathrm{T}}(V) \tag{A1.2}$$

It is common practice to refer to a vector simply by its components (V) where it is understood that all vectors in an equation are referred to the same basis (e).

It is convenient to replace $(x\;y\;z)$ with $(x_1\;\;x_2\;\;x_3)$ so that we may write

$$V = x_i \quad i \text{ from 1 to 3} \tag{A1.3}$$

This tensor is said to be of rank 1 because only one index is needed.

A dyad is defined by the following expression

$$\boldsymbol{AB}\cdot\boldsymbol{C} = \boldsymbol{A}(\boldsymbol{B}\cdot\boldsymbol{C}) = \boldsymbol{E} \tag{A1.4}$$

where $\boldsymbol{AB}$ is the dyad and $\boldsymbol{A}$, $\boldsymbol{B}$, $\boldsymbol{C}$ and $\boldsymbol{E}$ are vectors. In three dimensions a dyad may be written

$$\mathbf{D} = \boldsymbol{AB} = (e)^{\mathrm{T}}(A)\,(B)^{\mathrm{T}}(e) = (\boldsymbol{i}\,\boldsymbol{j}\,\boldsymbol{k})\begin{bmatrix} A_1 \\ A_2 \\ A_3 \end{bmatrix}(B_1\;B_2\;B_3)\begin{bmatrix} \boldsymbol{i} \\ \boldsymbol{j} \\ \boldsymbol{k} \end{bmatrix} \tag{A1.5}$$

or

$$\mathbf{D} = (e)^{\mathrm{T}} \begin{bmatrix} A_1B_1 & A_1B_2 & A_1B_3 \\ A_2B_1 & A_2B_2 & A_2B_3 \\ A_3B_1 & A_3B_2 & A_3B_3 \end{bmatrix} (e) \tag{A1.6}$$

The square matrix is the matrix representation of the dyad and can be written

$$D_{ij} = A_iB_j \tag{A1.7}$$

Thus the dyad is a tensor of rank 2 as it requires two indices to define its elements. The sum of two or more dyads is termed a dyadic.

The majority of rank 2 tensors encountered in physics are either symmetric or anti-symmetric. For a symmetric tensor $D_{ij} = D_{ji}$, and thus there are only six independent elements. For an anti-symmetric tensor, $D_{ij} = -D_{ji}$ and, because this implies that $D_{ii} = 0$, there are only three independent elements; this is similar to a vector.

The process of outer multiplication of two tensors is defined typically by

$$A_{ijk}B_{lm} = C_{ijklm} \tag{A1.8}$$

where C is a tensor of rank 5.

If both tensors A and B are of rank 2 then the element

$$C_{ijkl} = A_{ij}B_{kl} \tag{A1.9}$$

Thus, if the indices range from 1 to 3 then C will have 3^4 elements.

We now make $j = k$ and sum over all values of j (or k) to obtain

$$C_{il} = \sum_{j=1}^{j=3} A_{ij}B_{jl} \tag{A1.10}$$

Further, we could omit the summation sign if it is assumed that summation is over the repeated index. This is known as Einstein's summation convention. Thus in compact form

$$C_{il} = A_{ij}B_{jl} \tag{A1.11}$$

The process of making two suffices the same is known as contraction, and outer multiplication followed by a contraction is called inner multiplication.

In the case of two rank 2 tensors the process is identical to that of matrix multiplication of two square matrices.

If we consider two tensors of the first rank (vectors) then outer multiplication is

$$C_{ij} = A_iB_j \tag{A1.12}$$

and these can be thought of as the components of a square matrix. In matrix notation,

$$[C] = (A)\,(B)^{\mathrm{T}} \tag{A1.13}$$

If we now perform a contraction

$$C = A_iB_i = \left(\sum_{i=1}^{i=3} A_iB_i \right) \tag{A1.14}$$

we have inner multiplication, which in matrix notation is

$$C = (A)^{\mathrm{T}}(B) \tag{A1.15}$$

and this is the scalar product.

Alternatively, because $(\mathrm{e}) \cdot (\mathrm{e})^{\mathrm{T}} = [\mathrm{I}]$, the identity matrix, we may write

$$\boldsymbol{C} = \boldsymbol{A \cdot B} = (A)^{\mathrm{T}}(e)\cdot(e)^{\mathrm{T}}(B) = (A)^{\mathrm{T}}(B) \tag{A1.16}$$

The vector product of two vectors is written

$$\boldsymbol{C} = \boldsymbol{A} \times \boldsymbol{B} \tag{A1.17}$$

and is defined as

$$\boldsymbol{C} = AB \sin \alpha\, \boldsymbol{e} \tag{A1.18}$$

where α is the smallest angle between $\boldsymbol{A}$ and $\boldsymbol{B}$ and $\boldsymbol{e}$ is a unit vector normal to both $\boldsymbol{A}$ and $\boldsymbol{B}$ in a sense given by the right hand rule. In matrix notation it can be demonstrated that

$$\begin{aligned} \boldsymbol{C} = \; & (-A_3B_2 + A_2B_3)\, \boldsymbol{i} \\ & + (A_3B_1 - A_1B_3)\, \boldsymbol{j} \\ & + (-A_2B_1 + A_1B_2)\, \boldsymbol{k} \end{aligned}$$

or

$$\boldsymbol{C} = (\mathrm{e})^{\mathrm{T}}(C) = (\boldsymbol{ijk}) \begin{bmatrix} 0 & -A_3 & A_2 \\ A_3 & 0 & -A_1 \\ -A_2 & A_1 & 0 \end{bmatrix} \begin{bmatrix} B_1 \\ B_2 \\ B_3 \end{bmatrix} \tag{A1.19}$$

The square matrix, in this book, is denoted by $[A]^{\mathrm{x}}$ so that equation (A1.19) may be written

$$\boldsymbol{C} = (\mathrm{e})^{\mathrm{T}}[A]^{\mathrm{x}}(B) \tag{A1.20}$$

or, since $(\mathrm{e})\cdot(\mathrm{e})^{\mathrm{T}} = [1]$, the unit matrix,

$$\begin{aligned} \boldsymbol{C} &= (\mathrm{e})^{\mathrm{T}}[\mathrm{A}]^{\mathrm{x}}(\mathrm{e})\cdot(\mathrm{e})^{\mathrm{T}}(B) \\ &= \mathsf{A}^{\mathrm{x}}\cdot\boldsymbol{B} \end{aligned} \tag{A1.21}$$

where $\mathsf{A}^{\mathrm{x}} = (\mathrm{e})^{\mathrm{T}}[A]^{\mathrm{x}}(\mathrm{e})$ is a tensor operator of rank 2.

In tensor notation it can be shown that the vector product is given by

$$C_i = \varepsilon_{ijk} A_j B_k \tag{A1.22}$$

where ε_{ijk} is the alternating tensor, defined as

$$\begin{aligned} \varepsilon_{ijk} &= +1 \quad \text{if } ijk \text{ is a cyclic permutation of } (1\ 2\ 3) \\ &= -1 \quad \text{if } ijk \text{ is an anti-cyclic permutation of } (1\ 2\ 3) \\ &= 0 \quad\;\; \text{otherwise} \end{aligned} \tag{A1.23}$$

Equation (A1.22) may be written

$$C_i = (\varepsilon_{ijk} A_j) B_k \tag{A1.24}$$

Now let us define the tensor

$$T_{ik} = \varepsilon_{ijk} A_j \tag{A1.25}$$

If we change the order of i and k then, because of the definition of the alternating tensor, $T_{ik} = -T_{ki}$; therefore T is anti-symmetric.

The elements are then

$$T_{12} = \varepsilon_{112}A_1 + \varepsilon_{122}A_2 + \varepsilon_{132}A_3 = -A_3 = -T_{21}$$

$$T_{13} = \varepsilon_{113}A_1 + \varepsilon_{123}A_2 + \varepsilon_{133}A_3 = +A_2 = -T_{31}$$

$$T_{23} = \varepsilon_{213}A_1 + \varepsilon_{223}A_2 + \varepsilon_{233}A_3 = -A_1 = -T_{32}$$

and the diagonal terms are all zero. These three equations may be written in matrix form as

$$T_{ik} = \begin{bmatrix} 0 & -A_3 & +A_2 \\ +A_3 & 0 & -A_1 \\ -A_2 & +A_1 & 0 \end{bmatrix} \tag{A1.26}$$

which is the expected result.

In summary the vector product of two vectors $\boldsymbol{A}$ and $\boldsymbol{B}$ may be written

$$\boldsymbol{C} = \boldsymbol{A} \times \boldsymbol{B},$$

$$(e)^{\mathrm{T}}(C) = (e)^{\mathrm{T}}[A]^{\mathrm{x}}(e)\cdot(e)^{\mathrm{T}}(B)$$

or

$$(C) = [A]^{\mathrm{x}}(B)$$

and

$$C_i = \mathrm{e}_{ijk}A_jB_k \quad \text{(summing over } j \text{ and } k)$$

$$= T_{ik}B_k \quad \text{(summing over } k)$$

Transformation of co-ordinates

We shall consider the transformation of three-dimensional Cartesian co-ordinates due to a rotation of the axes about the origin. In fact, mathematical texts define tensors by the way in which they transform. For example, a second-order tensor A is defined as a multi-directional quantity which transforms from one set of co-ordinate axes to another according to the rule

$$A'_{mn} = l_{mi}l_{nj}A_{ij}$$

The original set of coordinates will be designated x_1, x_2, x_3 and the associated unit vectors will be $\boldsymbol{e}_1$, $\boldsymbol{e}_2$, $\boldsymbol{e}_3$. In these terms a position vector $\boldsymbol{V}$ will be

$$\boldsymbol{V} = x_1\boldsymbol{e}_1 + x_2\boldsymbol{e}_2 + x_3\boldsymbol{e}_3 = x_i\boldsymbol{e}_i \tag{A1.27}$$

Using a primed set of coordinates the same vector will be

$$\boldsymbol{V} = x'_1\boldsymbol{e}'_1 + x'_2\boldsymbol{e}'_2 + x'_3\boldsymbol{e}'_3 = x'_i\boldsymbol{e}'_i \tag{A1.28}$$

The primed unit vectors are related to the original unit vectors by

$$\boldsymbol{e}'_1 = l\boldsymbol{e}_1 + m\boldsymbol{e}_2 + n\boldsymbol{e}_3 \tag{A1.29}$$

where l, m and n are the direction cosines between the primed unit vector in the x'_1 direction and those in the original set. We shall now adopt the following notation

$$\begin{aligned} e_1' &= a_{11}e_1 + a_{12}e_2 + a_{13}e_3 \\ &= a_{ij}e_j \end{aligned} \tag{A1.30}$$

with similar expressions for the other two unit vectors. Using the summation convention,

$$e_i' = a_{ij}e_j \tag{A1.31}$$

In matrix form

$$\begin{bmatrix} e_1' \\ e_2' \\ e_3' \end{bmatrix} = \begin{bmatrix} a_{11} & a_{12} & a_{13} \\ a_{21} & a_{22} & a_{23} \\ a_{31} & a_{32} & a_{33} \end{bmatrix} \begin{bmatrix} e_1 \\ e_2 \\ e_3 \end{bmatrix} \tag{A1.32}$$

and the inverse transform, b_{ij}, is such that

$$\begin{bmatrix} e_1 \\ e_2 \\ e_3 \end{bmatrix} = \begin{bmatrix} b_{11} & b_{12} & b_{13} \\ b_{21} & b_{22} & b_{23} \\ b_{31} & b_{32} & b_{33} \end{bmatrix} \begin{bmatrix} e_1' \\ e_2' \\ e_3' \end{bmatrix} \tag{A1.33}$$

It is seen that a_{13} is the direction cosine of the angle between e_1' and e_3 whilst b_{31} is the direction cosine of the angle between e_3 and e_1'; thus $a_{13} = b_{31}$. Therefore b_{ij} is the transpose of a_{ij}, that is $b_{ij} = a_{ji}$.

The transformation tensor a_{ij} is such that its inverse is its transpose, in matrix form $[A][A]^T = [1]$. Such a transformation is said to be orthogonal.

Now

$$V = e_i x_i = e_j' x_j' \tag{A1.34}$$

so premultiplying both sides by e_j' gives

$$e_j' \cdot e_i x_i = x_j' \tag{A1.35}$$

or

$$x_j' = a_{ji} x_i \tag{A1.36}$$

It should be noted that

$$x_i' = a_{ij} x_j$$

is equivalent to the previous equation as only the arrangement of indices is significant.

In matrix notation

$$(V) = (e)^T(x) = (e')^T(x') \tag{A1.37}$$

but $(e') = [a](e)$, and therefore

$$(e)^T(x) = (e)^T[a]^T(x')$$

Premultiplying each side by (e) gives

$$(x) = [a]^T(x')$$

and inverting we obtain

$$(x') = [a](x) \tag{A1.38}$$

The square of the magnitude of a vector is

$$V^2 = (x)^{\mathrm{T}}(x) = (x')^{\mathrm{T}}(x')$$
$$= (x)^{\mathrm{T}}[a]^{\mathrm{T}}[a](x) \tag{A1.39}$$

and because (x) is arbitrary it follows that

$$[a]^{\mathrm{T}}[a] = [1] = [b][a] \tag{A1.40}$$

where

$$[b] = [a]^{\mathrm{T}} = [a]^{-1}$$

In tensor notation this equation is

$$b_{ij}a_{jl} = a_{ji}a_{jl} = \delta_{il} \tag{A1.41}$$

where δ_{il} is the Kronecker delta defined to be 1 when $i = l$ and 0 otherwise.

Because $a_{ji}a_{jl} = a_{jl}a_{ji}$, equation (A1.41) yields six relationships between the nine elements a_{ij}, and this implies that only three independent constants are required to define the transformation. These three constants are not arbitrary if they are to relate to proper rotations; for example, they must all lie between -1 and $+1$. Another condition which has to be met is that the triple scalar product of the unit vectors must be unity as this represents the volume of a unit cube. So

$$\boldsymbol{e}_1 \cdot (\boldsymbol{e}_2 \times \boldsymbol{e}_3) = \boldsymbol{e}_1' \cdot (\boldsymbol{e}_2' \times \boldsymbol{e}_3') = 1 \tag{A1.42}$$

since

$$\boldsymbol{e}_1' = a_{11}\boldsymbol{e}_1 + a_{12}\boldsymbol{e}_2 + a_{13}\boldsymbol{e}_3 \quad \text{etc.}$$

We can use the well-known determinant form for the triple product and write

$$\begin{vmatrix} a_{11} & a_{12} & a_{13} \\ a_{21} & a_{22} & a_{23} \\ a_{31} & a_{32} & a_{33} \end{vmatrix} = 1 \tag{A1.43}$$

or

$$\mathrm{Det}\,[a] = 1$$

The above argument only holds if the original set of axes and the transformed set are both right handed (or both left handed). If the handedness is changed by, for example, the direction of the z' axis being reversed then the bottom row of the determinant would all be of opposite sign, so the value of the determinant would be -1. It is interesting to note that no way of formally defining a left- or right-handed system has been devised; it is only the difference that is recognized.

In general vectors which require the use of the right hand rule to define their sense transform differently when changing from right- to left-handed systems. Such vectors are called axial vectors or pseudo vectors in contrast to polar vectors.

Examples of polar vectors are position, displacement, velocity, acceleration and force. Examples of axial vectors are angular velocity and moment of force. It can be demonstrated that the vector product of a polar vector and an axial vector is a polar vector. Another interesting point is that the vector of a 3×3 anti-symmetric tensor is an axial vector. This point does not affect any of the arguments in this book because we are always dealing with right-handed systems and pure rotation does not change the handedness of the axes. However, if the reader wishes to delve deeper into relativistic mechanics this distinction is of some importance.

Diagonalization of a second-order tensor

We shall consider a 3 × 3 second-order symmetric Cartesian tensor which may represent moment of inertia, stress or strain. Let this tensor be $T = T_{ij}$ and the matrix of its elements be $[T]$. The transformation tensor is $A = A_{ij}$ and its matrix is $[A]$. The transformed tensor is

$$[T'] = [A]^{T}[T][A] \tag{A1.44}$$

Let us now assume that the transformed matrix is diagonal so

$$[T'] = \begin{bmatrix} \lambda_1 & 0 & 0 \\ 0 & \lambda_2 & 0 \\ 0 & 0 & \lambda_3 \end{bmatrix} \tag{A1.45}$$

If this dyad acts on a vector (C) the result is

$$\begin{aligned} C_1' &= \lambda_1 C_1 \\ C_2' &= \lambda_2 C_2 \\ C_3' &= \lambda_3 C_3 \end{aligned} \tag{A1.46}$$

Thus if the vector is wholly in the x' direction the vector $T'x'$ would still be in the x' direction, but multiplied by λ_1.

Therefore the vectors $C_1'\boldsymbol{i}'$, $C_2'\boldsymbol{j}'$ and $C_3'\boldsymbol{k}'$ form a unique set of orthogonal axes which are known as the principal axes. From the point of view of the original set of axes if a vector lies along any one of the principal axes then its direction will remain unaltered. Such a vector is called an eigenvector. In symbol form

$$T_{ij}C_j = \lambda C_i \tag{A1.47}$$

or

$$[T]\,(C) = \lambda(C) \tag{A1.48}$$

Rearranging equation (A1.48) gives

$$\{[T] - \lambda[1]\}(C) = (0)$$

where [1] is the unit matrix. In detail

$$\begin{bmatrix} (T_{11} - \lambda) & T_{12} & T_{13} \\ T_{21} & (T_{22} - \lambda) & T_{23} \\ T_{31} & T_{32} & (T_{33} - \lambda) \end{bmatrix} \begin{bmatrix} C_1 \\ C_2 \\ C_3 \end{bmatrix} = \begin{bmatrix} 0 \\ 0 \\ 0 \end{bmatrix} \tag{A1.49}$$

This expands to three homogeneous equations which have the trivial solution of $(C) = (0)$. The theory of linear equations states that for a non-trivial solution the determinant of the square matrix has to be zero. That is,

$$\begin{bmatrix} (T_{11} - \lambda) & T_{12} & T_{13} \\ T_{21} & (T_{22} - \lambda) & T_{23} \\ T_{31} & T_{32} & (T_{33} - \lambda) \end{bmatrix} = 0 \qquad \text{(A1.50)}$$

This leads to a cubic in λ thus yielding the three roots which are known as the eigenvalues. Associated with each eigenvalue is an eigenvector, all of which can be shown to be mutually orthogonal. The eigenvectors only define a direction because their magnitudes are arbitrary.

Let us consider a special case for which $T_{12} = T_{21} = 0$ and $T_{13} = T_{31} = 0$. In this case for a vector $(C) = (1\ 0\ 0)^T$ the product $[T](C)$ yields a vector $(T_{11}\ 0\ 0)^T$, which is in the same direction as (C). Therefore the x_1 direction is a principal axis and the x_2, x_3 plane is a plane of symmetry. Equation (A1.50) now becomes

$$(T_{11} - \lambda)[(T_{22} - \lambda)(T - \lambda) - T_{23}^2] = 0 \qquad \text{(A1.51)}$$

In general a symmetric tensor when referred to its principal co-ordinates takes the form

$$[T] = \begin{bmatrix} \lambda_1 & 0 & 0 \\ 0 & \lambda_2 & 0 \\ 0 & 0 & \lambda_3 \end{bmatrix} \qquad \text{(A1.52)}$$

and when it operates on an arbitrary vector (C) the result is

$$[T](C) = \begin{bmatrix} \lambda_1 C_1 \\ \lambda_2 C_2 \\ \lambda_3 C_3 \end{bmatrix} \qquad \text{(A1.53)}$$

Let us now consider the case of degeneracy with $\lambda_3 = \lambda_2$. It is easily seen that if (C) lies in the x_2x_3 plane, that is $(C) = (0\ C_2\ C_3)^T$, then

$$[T](C) = \lambda_2 \begin{bmatrix} 0 \\ C_2 \\ C_3 \end{bmatrix} \qquad \text{(A1.54)}$$

from which we see that the vector remains in the x_2x_3 plane and is in the same direction. This also implies that the directions of the x_2 and x_3 axes can lie anywhere in the plane normal to the x_1 axis. This would be true if the x_1 axis is an axis of symmetry.

If the eigenvalues are triply degenerate, that is they are all equal, then any arbitrary vector will have its direction unaltered, from which it follows that all axes are principal axes.

The orthogonality of the eigenvectors is readily proved by reference to equation (A1.48). Each eigenvector will satisfy this equation with the appropriate eigenvalue thus

$$[T](C)_1 = \lambda_1(C)_1 \qquad \text{(A1.55)}$$

and

$$[T](C)_2 = \lambda_2(C)_2 \qquad \text{(A1.56)}$$

We premultiply equation (A1.55) by $(C)_2^T$ and equation (A1.56) by $(C)_1^T$ to obtain the scalars

$$(C)_2^T[T](C)_1 = \lambda_1(C)_2^T(C)_1 \qquad \text{(A1.57)}$$

and

$$(C)_1^T[T](C)_2 = \lambda_2(C)_1^T(C)_2 \tag{A1.58}$$

Transposing both sides of the last equation, remembering that $[T]$ is symmetrical, gives

$$(C)_2^T[T](C)_1 = \lambda_2(C)_2^T(C)_1 \tag{A1.59}$$

and subtracting equation (A1.59) from (A1.57) gives

$$0 = (\lambda_1 - \lambda_2)(C)_2^T(C)_1 \tag{A1.60}$$

so when $\lambda_1 \neq \lambda_2$ we have that $(C)_2^T(C)_1 = 0$; that is, the vectors are orthogonal.

Appendix 2

ANALYTICAL DYNAMICS

Introduction

The term analytical dynamics is usually confined to the discussion of systems of particles moving under the action of ideal workless constraints. The most important methods are Lagrange's equations which are dealt with in Chapter 2 and Hamilton's principle which was discussed in Chapter 3. Both methods start by formulating the kinetic and potential energies of the system. In the Lagrange method the Lagrangian (kinetic energy less the potential energy) is operated on directly to produce a set of second-order differential equations of motion. Hamilton's principle seeks to find a stationary value of a time integral of the Lagrangian. Either method can be used to generate the other and both may be derived from the principle of virtual work and D'Alembert's principle.

Virtual work and D'Alembert's principle are regarded as the fundamentals of analytical dynamics but there are many variations on this theme, two of which we have just mentioned. The main attraction of these two methods is that the Lagrangian is a function of position, velocity and time and does not involve acceleration. Another feature is that in certain circumstances (cyclic or ignorable co-ordinates) integrals of the equations are readily deduced. For some constrained systems, particularly those with non-holonomic constraints, the solution requires the use of Lagrange multipliers which may require some manipulation. In this case other methods may be advantageous. Even if this is not the case the methods are of interest in their own right and help to develop a deeper understanding of dynamics.

Constraints and virtual work

Constraints are usually expressed as some form of kinematic relationship between co-ordinates and time. In the case of holonomic constraints the equations are of the form

$$f_j(q_i t) = 0 \qquad \text{(A2.1)}$$

$1 \leq i \leq m$ and $1 \leq j \leq r$.

For non-holonomic constraints where the relationships between the differentials cannot be integrated we have

$$a_{ji}\mathrm{d}q_i + c_j \mathrm{d}t = 0 \qquad \text{(A2.2)}$$

Differentiating equation (A2.1) we obtain

$$\frac{\partial f_j}{\partial q_i}\,\mathrm{d}q_i + \frac{\partial f_j}{\partial t}\,\mathrm{d}t = 0 \tag{A2.3}$$

which has the same form as equation (A2.2).

In the above equations we have assumed that there are m generalized co-ordinates and r equations of constraint. We have made use of the summation convention.

For constraint equations of the form of (A2.1) it is theoretically possible to reduce the number of co-ordinates required to specify the system from m to $n = m - r$, where n is the number of degrees of freedom of the system.

Dividing equation (A2.2) through by $\mathrm{d}t$ gives

$$a_{ji}\dot{q}_i + c_j = 0 \tag{A2.4}$$

and this may be differentiated with respect to time to give

$$a_{ji}\ddot{q}_i + \dot{a}_{ji}\dot{q}_i + \dot{c}_j = 0$$

or

$$a_{ji}\ddot{q}_i = b_j \tag{A2.5}$$

where $b_j = -(\dot{a}_{ji}\dot{q}_i + \dot{c}_j)$. Note that a, $\dot{a}$, b and c may, in general, all be functions of $\dot{q}$, q and t.

By definition a virtual displacement is any possible displacement which satisfies the constraints at a given instant of time (i.e. time is held fixed). Therefore from equation (A2.3) a virtual displacement δq_i will be any vector such that

$$a_{ij}\delta q_i = 0 \tag{A2.6}$$

There is no reason why we should not replace the virtual displacements δq_i by virtual velocities v_i provided that the velocities are consistent with the constraints. The principle of virtual work can then be called the principle of virtual velocities or even virtual power.

D'Alembert argued that the motion due to the impressed forces, less the motion which the masses would have acquired had they been free, would be produced by a set of forces which are in equilibrium. Motion here is taken to be momentum but the argument is equally valid if we use the change of momentum or the mass acceleration vectors. This difference in motion is just that due to the forces of constraint so we may say that the constraint forces have zero resultant. If we now restrict the constraints to ideal constraints (i.e. frictionless or workless) then the virtual work done by the constraint forces will be zero. In mathematical terms the sum of the impressed force plus the constraint force gives

$$F_i^{\mathrm{i}} + F_i^{\mathrm{c}} = m_i\ddot{r}_i \tag{A2.7}$$

and the impressed force alone gives

$$F_i^{\mathrm{i}} = m_i a_i \tag{A2.8}$$

Therefore the constraint force is

$$F_i^{\mathrm{c}} = m_i(\ddot{r}_i - a_i) \tag{A2.9}$$

Now the principle of virtual work states that

$$\sum F_i^{\mathrm{c}}\cdot\delta r_i = 0 \tag{A2.10}$$

or

$$\sum m_i(\ddot{r}_i - a_i)\cdot\delta r_i = 0 \tag{A2.11}$$

or

$$\sum (m_i\ddot{r}_i - F_i^{\mathrm{i}})\cdot\delta r_i = 0 \tag{A2.12}$$

Gauss's principle

A very interesting principle, also known as the principle of least constraint, was introduced by Gauss in 1829. Gauss himself stated that there is no new principle in the (classical) science of equilibrium or motion which cannot be deduced from the principle of virtual velocities and D'Alembert's principle. However, he considered that his principle allowed the laws of nature to be seen from a different and advantageous point of view.

Referring to Fig. A2.1 we see that point a is the position of particle i having mass m_i and velocity v_i. Point c is the position of the particle at a time Δt later. Point b is the position that the particle would have achieved under the action of the impressed forces only. Gauss asserted that the function

$$G = \sum m_i \overrightarrow{bc}_i^2 \tag{A2.13}$$

will always be a minimum.

For the small time interval Δt we can write

$$\overrightarrow{ab}_i = \boldsymbol{v}_i\Delta t + \frac{1}{2}\Delta t^2 \frac{\boldsymbol{F}_i^{\mathrm{i}}}{m_i} \tag{A2.14}$$

and

$$\overrightarrow{ac}_i = \boldsymbol{v}_i\Delta t + \frac{1}{2}\Delta t^2 \left(\frac{\boldsymbol{F}_i^{\mathrm{i}}}{m_i} + \frac{\boldsymbol{F}_i^{\mathrm{c}}}{m_i}\right) \tag{A2.15}$$

Therefore

$$\overrightarrow{bc}_i = \overrightarrow{ac}_i - \overrightarrow{ab}_i = \frac{1}{2}\Delta t^2 \frac{\boldsymbol{F}_i^{\mathrm{c}}}{m_i} \tag{A2.16}$$

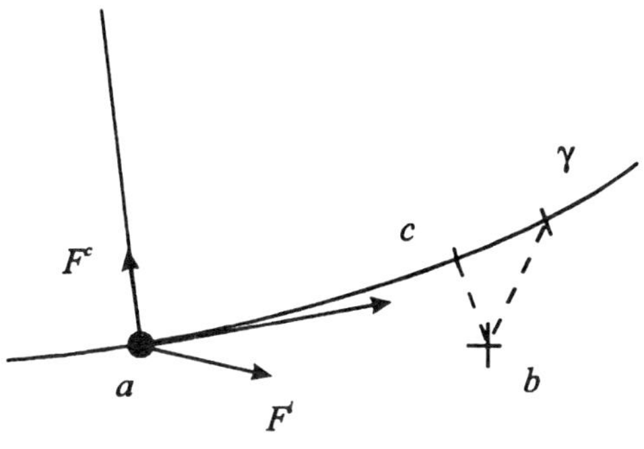

Fig. A2.1

so that

$$m_i\overrightarrow{bc}_i \;\alpha\; \boldsymbol{F}_i^{\mathrm{c}} \tag{A2.17}$$

Now let γ be another point on the path so it is clear that $\overrightarrow{c\gamma}$ is a possible displacement consistent with the constraints. The new Gaussian function will be

$$(G + \Delta G) = \sum m_i(\overrightarrow{bc}_i + \overrightarrow{c\gamma}_i)^2 \tag{A2.18}$$

$$= \sum m_i\overrightarrow{bc}_i^2 + \sum m_i\overrightarrow{c\gamma}_i^2 + 2\sum m_i\overrightarrow{bc}_i \cdot \overrightarrow{c\gamma}_i \tag{A2.19}$$

Because $m\overrightarrow{bc}$ is proportional to the force of constraint and $\overrightarrow{c\gamma}$ is a virtual displacement the principle of virtual work dictates that the third term on the right will be zero. The first term on the right is simply G so we have that

$$\Delta G = \sum m_i\overrightarrow{c\gamma}_i^2 \geqslant 0 \tag{A2.20}$$

Therefore Gauss concluded that, since the sum cannot be negative, then $(G + \Delta G) \geqslant G$, so that G must always be a minimum.

The Gaussian could also be written in the form

$$G = \sum m_i(F_i^{\mathrm{c}}/m_i)^2 = \sum m_i(\ddot{\boldsymbol{r}}_i - \boldsymbol{a}_i)^2 \tag{A2.21}$$

from which it is apparent that the true set of constraint vectors or the true set of acceleration vectors are those which minimize G.

It must be emphasized that the constraint forces are workless and as such act in a direction which is normal to the true path.

Gibbs–Appell equations

The Gibbs–Appell formulation is also based on acceleration and starts with the definition of the Gibbs function S for a system of n particles. This is

$$S = \sum_{i=1}^{i=3n} \frac{1}{2} m_i\ddot{x}_i^2 \tag{A2.22}$$

Clearly

$$\frac{\partial S}{\partial \ddot{x}_i} = m_i\ddot{x}_i = F_i \tag{A2.23}$$

If the displacements are expressible in terms of m generalized co-ordinates in the form

$$x_i = x_i(q_1 \ldots q_m t) \tag{A2.24}$$

then, as in the treatment of Lagrange's equations,

$$\mathrm{d}x_i = \frac{\partial x_i}{\partial q_j}\mathrm{d}q_j + \frac{\partial x_i}{\partial t}\mathrm{d}t \tag{A2.25}$$

and

$$\dot{x}_i = \frac{\partial x_i}{\partial q_j}\dot{q}_j + \frac{\partial x_i}{\partial t} \tag{A2.26}$$

We shall consider the differentials of the generalized co-ordinates to be the sum of two groups: the first group dq_i (i from 1 to $m-r$), and the second group $d\gamma_j$ (j from $m-r+1$ to m). The difference between the two groups is that dq can be integrated to give q whereas $d\gamma$ cannot be integrated. The velocities are expressed as $\dot{q}_i$ and $\dot{\gamma}_j$. The latter group is formed from quasi-velocities, so called because they satisfy the constraints but are not necessarily associated with any identifiable displacement. Quasi-velocities can be chosen in much the same way as generalized co-ordinates are chosen, that is they must satisfy the constraints. The number of quasi-velocities must be no smaller than the number of non-holonomic constraint equations but all velocities can be considered to be quasi and it is common practice to do so.

The quasi-velocities can be expressed in terms of the generalized velocities as

$$\dot{\gamma}_i = u_{ij}\dot{q}_j + g_i \tag{A2.27}$$

and for linear equations inversion gives

$$\dot{q}_i = v_{ij}\dot{\gamma}_j + h_i \tag{A2.28}$$

Substituting this expression in equation (A2.5) leads to a constraint equation of the form

$$A_{ij}\dot{\gamma}_j + B_i = 0 \tag{A2.29}$$

and differentiating with respect to time gives

$$A_{ij}\ddot{\gamma}_j + (\dot{A}_{ij}\dot{\gamma}_j + \dot{B}_i) = 0 \tag{A2.30}$$

Now from equation (A2.28) we have

$$dq_i = v_{ij}d\gamma_j + h_i dt \tag{A2.31}$$

Therefore a virtual displacement, for which time is held constant, is

$$\delta q_i = v_{ij}\delta\gamma_j \tag{A2.32}$$

Similarly the constraint equation (A2.6) for virtual displacements becomes

$$A_{ij}\delta\gamma_j = 0 \tag{A2.33}$$

If the Gibbs function is expressed in terms of the generalized co-ordinates then the usual generalized force Q is

$$Q_i = \frac{\partial S}{\partial \ddot{q}_i} \tag{A2.34}$$

The total virtual work done by the generalized forces is

$$\delta W = Q_i\delta q_i = Q_i(v_{ij}\delta\gamma_j) = \Gamma_j\delta\gamma_j \tag{A2.35}$$

where Γ is the quasi-generalized force and is related to the usual generalized force by

$$\Gamma_j = Q_i v_{ij} \tag{A2.36}$$

Other methods

A further development has been proposed which is also based on acceleration. One form of Gauss's principle, equation (A2.21), gives

$$G = m_i(\ddot{r}_i - a_i)^2 \tag{A2.37}$$

so that

$$(G + \Delta G) = m_i(\ddot{r}_i - a_i + \delta\ddot{r}_i)^2 \tag{A2.38}$$

Thus

$$\Delta G = m_i(\delta\ddot{r}_i)^2 + 2m_i(\ddot{r}_i - a_i)\cdot\delta\ddot{r}_i \tag{A2.39}$$

Now from the constraint equation (A2.5) and changing to upper case to avoid confusion with other terms

$$A_{ij}\ddot{r}_i = B_i \tag{A2.40}$$

so if $\delta\ddot{r}_i$ satisfies the constraints then

$$A_{ij}(\ddot{r}_i + \delta\ddot{r}_i) = B_i \tag{A2.41}$$

from which it follows that

$$A_{ij}\delta\ddot{r}_i = 0 \tag{A2.42}$$

Since any set of values for $\delta\ddot{r}_i$ which satisfy the constraints may be used it follows that in equation (A2.39)

$$m_i(\ddot{r}_i - a_i)\cdot\delta\ddot{r}_i = 0 \tag{A2.43}$$

The new method is expressed by a 'fundamental equation' which in matrix form is

$$(\ddot{r}) = (a) + [M]^{-1/2}\{[A][M]^{-1/2}\}^{+}\{(B) - [A](a)\} \tag{A2.44}$$

where the superscript + signifies the pseudo-inverse or Moore–Penrose inverse; $(\ddot{r})$ is the actual column vector of accelerations and (a) is the column vector of the unconstrained system.

The pseudo-inverse of any matrix $[A]$, square or non-square, is such that it satisfies the following conditions

$$[A][A]^{+}[A] = [A]$$

$$[A]^{+}[A][A]^{+} = [A]^{+}$$

and $[A][A]^{+}$ and $[A]^{+}[A]$ are both symmetric.

We now write equations (A2.43), (A2.40) and (A2.42) in matrix form. Thus

$$(\delta\ddot{r})^{\mathrm{T}}[M]\{(\ddot{r}) - (a)\} = 0 \tag{A2.45}$$

$$[A](\ddot{r}) = (B) \tag{A2.46}$$

$$[A](\delta\ddot{r}) = (0) \tag{A2.47}$$

It is known that $\{[X][Y]\}^{+} = [Y]^{+}[X]^{+}$, and the pseudo-inverse of a column matrix (y) is easily shown to be

$$(y)^{+} = \frac{(y)^{\mathrm{T}}}{(y)^{\mathrm{T}}(y)} \tag{A2.48}$$

Therefore inverting equation (A2.47) gives

$$(\delta\ddot{r})^{\mathrm{T}}[A]^{+} = (0) \tag{A2.49}$$

In order to show that the 'fundamental equation' satisfies both the constraint equation and the principle of virtual work it is convenient to express the acceleration vectors in a weighted form. The following definitions will be used

$$(\underline{\ddot{r}}) = [M]^{1/2}(\ddot{r}) \tag{A2.50}$$

$$(\underline{a}) = [M]^{1/2}(a) \tag{A2.51}$$

$$[\underline{A}] = [A][M]^{-1/2} \tag{A2.52}$$

Premultiplying equation (A2.45) by $[M]^{1/2}$ and using the above weighted terms we arrive at

$$(\underline{\ddot{r}}) = (\underline{a}) + [\underline{A}]^{+}\{(B) - [\underline{A}](\underline{a})\} \tag{A2.53}$$

We shall now demonstrate that this equation satisfies the constraint equation. Equation (A2.46) can be written as

$$[A][M]^{-1/2}[M]^{1/2}(\ddot{r}) = (B)$$

or

$$[\underline{A}](\underline{\ddot{r}}) = (B) \tag{A2.54}$$

Thus substitution into (A2.53) followed by a slight rearrangement gives

$$\{(\underline{\ddot{r}}) - (\underline{a})\} = [\underline{A}]^{+}\{[\underline{A}](\underline{\ddot{r}}) - [\underline{A}](\underline{a})\} \tag{A2.55}$$

Premultiplying both sides by $[\underline{A}]$ leads to

$$[\underline{A}]\{(\underline{\ddot{r}}) - (\underline{a})\} = [\underline{A}][\underline{A}]^{+}[\underline{A}]\{(\underline{\ddot{r}}) - (\underline{a})\} \tag{A2.56}$$

The pseudo-inverse is defined such that

$$[\underline{A}][\underline{A}]^{+}[\underline{A}] = [\underline{A}] \tag{A2.57}$$

Therefore the constraint equation is satisfied.

To show that the principle of virtual work is satisfied we premultiply equation (A2.55) by $(\delta\underline{\ddot{r}})^{\mathrm{T}}$ and rearrange it to give

$$(\delta\underline{\ddot{r}})^{\mathrm{T}}\{(\underline{\ddot{r}}) - (\underline{a})\} = (\delta\underline{\ddot{r}})^{\mathrm{T}}[\underline{A}]^{+}\{(B) - [\underline{A}](\underline{a})\} \tag{A2.58}$$

The left hand side is zero because of equation (A2.45) and the right hand side is zero because of equation (A2.49). Thus the equality is proved

The 'fundamental equation' therefore satisfies the constraints and basic equations of analytical dynamics. Any advantage that this method may have is that the constraint equation is not affected by whether the constraints are holonomic or not. The disadvantage is that the unconstrained accelerations have first to be determined. For systems involving only particles, such that the mass matrix is diagonal, the unconstrained accelerations are readily found and some advantage may be obtained.

Appendix 3
Curvilinear co-ordinate systems

In the chapter dealing with continuous media all the equations were presented using Cartesian co-ordinates, but in many instances it is advantageous to use curvilinear co-ordinates such as cylindrical or spherical. We shall develop a general approach and then apply it to the two systems mentioned above.

In Cartesian co-ordinates the differential vector operator (equation (7.4)) is

$$\nabla = (\boldsymbol{i}\ \boldsymbol{j}\ \boldsymbol{k})(\partial/\partial x\ \ \partial/\partial y\ \ \partial/\partial z)^{\mathrm{T}}$$
$$= (e)^{\mathrm{T}}(\nabla)$$

Equation (7.12) defines rotation

$$\boldsymbol{\Omega} = \frac{1}{2}(e)^{\mathrm{T}}(\nabla)^{\times}(u) = \frac{1}{2}\ \nabla\times\boldsymbol{u}$$
$$= \frac{1}{2}\ \text{curl}\ \boldsymbol{u}$$

Equation (7.32) defines dilatation

$$\Delta = (\nabla)^{\mathrm{T}}(u) = \boldsymbol{\nabla}\cdot\boldsymbol{u}$$
$$= \text{divergence}\ \boldsymbol{u}$$

The equations of motion can be written as a single vector equation

$$\rho\frac{\partial^2\boldsymbol{u}}{\partial t^2} = (\lambda + 2\mu)\text{grad}\,\Delta - 2\mu\,\text{curl}\,\boldsymbol{\Omega} \tag{A3.1}$$

or

$$\rho\frac{\partial^2\boldsymbol{u}}{\partial t^2} = (\lambda + 2\mu)\boldsymbol{\nabla}\Delta - 2\mu\,\boldsymbol{\nabla}\times\boldsymbol{\Omega}$$

Premultiplying both sides by $\boldsymbol{\nabla}$, that is taking the divergence of both sides, gives

$$\rho\frac{\partial^2\Delta}{\partial t^2} = (\lambda + 2\mu)\nabla^2\Delta$$

since $\boldsymbol{\nabla}\cdot\boldsymbol{\nabla}\times\boldsymbol{\Omega} = 0$

Premultiplying both sides by $\nabla\times$, that is taking the curl of both sides, yields

$$\rho\frac{\partial^2(2\boldsymbol{\Omega})}{\partial t^2} = -2\mu\nabla\times(\nabla\times\boldsymbol{\Omega})$$

$$= -2\mu[(\nabla\cdot\boldsymbol{\Omega}) - (\nabla\cdot\nabla)\boldsymbol{\Omega}]$$

but $\nabla\cdot\boldsymbol{\Omega} = \nabla\cdot(\nabla\times\boldsymbol{u})/2 = 0$ and therefore

$$\rho\frac{\partial^2\boldsymbol{\Omega}}{\partial t^2} = \mu\nabla^2\boldsymbol{\Omega}$$

Curvilinear co-ordinates

Curvilinear co-ordinates will be described by the magnitudes q_1, q_2 and q_3 with the corresponding unit vectors $\boldsymbol{e}_1$, $\boldsymbol{e}_2$ and $\boldsymbol{e}_3$. A small change in the position vector is

$$\begin{aligned} \mathrm{d}\boldsymbol{r} &= e_1h_1\mathrm{d}q_1 + e_2h_2\mathrm{d}q_2 + e_3h_3\mathrm{d}q_3 \\ &= (e)^{\mathrm{T}}[h](\mathrm{d}q) \\ &= e_ih_i\mathrm{d}q_i \end{aligned} \qquad \text{(A3.2)}$$

where

$$[h] = \begin{bmatrix} h_1 & 0 & 0 \\ 0 & h_2 & 0 \\ 0 & 0 & h_3 \end{bmatrix}$$

The scale factors, $[h]$, are defined so that $h_i\mathrm{d}q_i$ is the elemental length. For Cartesian co-ordinates the scale factors are each unity.

From Fig. A3.1 we see that the areas of the faces of the elemental volume are

$$\mathrm{d}A_1 = h_2\mathrm{d}q_2h_3\mathrm{d}q_3 \quad \text{etc.}$$

so

$$(\mathrm{d}A) = (\mathrm{d}q_2\mathrm{d}q_3h_2h_3 \;\; \mathrm{d}q_3\mathrm{d}q_1h_3h_1 \;\; \mathrm{d}q_1\mathrm{d}q_2h_1h_2)^{\mathrm{T}} \qquad \text{(A3.3)}$$

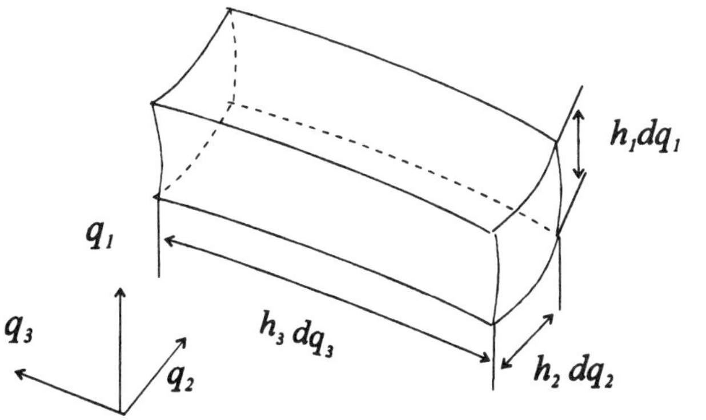

Fig. A3.1

The volume is

$$dV = dq_1 dq_2 dq_3 h_1 h_2 h_3 \tag{A3.4}$$

Divergence

The divergence of a vector function $\boldsymbol{F}$ can be found by use of the divergence theorem

$$\int_{\text{vol}} (\text{div}\,\boldsymbol{F}) dV = \int_{\text{surface}} \boldsymbol{F} \cdot d\boldsymbol{A} \tag{A3.5}$$

The integral on the right hand side is known as the flux through the surface. If the vector represents the velocity of a fluid then the flux would be the volumetric flow rate of fluid leaving the volume enclosed by the surface. If we now make the volume tend to zero

$$\text{div}\,\boldsymbol{F} = \lim_{V \to 0} \frac{1}{V} \int_{\text{surface}} \boldsymbol{F} \cdot d\boldsymbol{A} \tag{A3.6}$$

For the elemental volume shown in Fig. A3.2

$$\text{div}\,\boldsymbol{F} = \frac{1}{dV} \left(\sum_i \frac{\partial (F_i A_i)}{\partial q_i} dq_i \right)$$

$$= \frac{1}{h_1 h_2 h_3} \left(\frac{\partial}{\partial q_1} (h_2 h_3 F_1) + \frac{\partial}{\partial q_2} (h_3 h_1 F_2) + \frac{\partial}{\partial q_3} (h_1 h_2 F_3) \right) \tag{A3.7}$$

Curl

The component in the $\boldsymbol{n}$ direction of the curl of a vector function can be defined by the following integral, which is known as Stoke's theorem

$$(\text{curl}\,\boldsymbol{F}) \cdot \boldsymbol{n} = \lim_{A_n \to 0} \frac{1}{A_n} \oint_{\text{line}} \boldsymbol{F} \cdot d\boldsymbol{s} \tag{A3.8}$$

The integral is known as the circulation due to its interpretation if the vector $\boldsymbol{F}$ is again the velocity of a fluid.

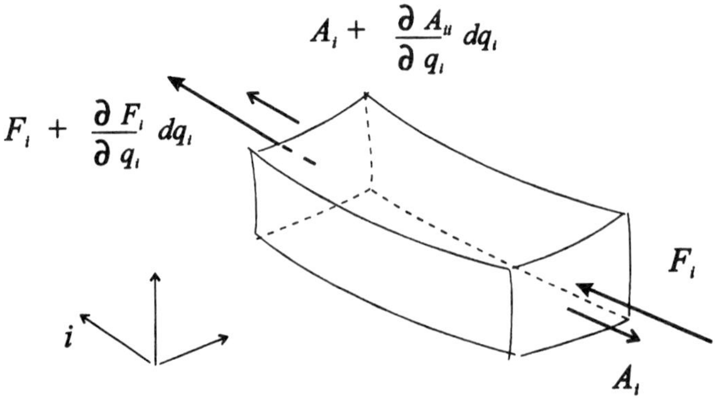

Fig. A3.2

If we take $\boldsymbol{n}$ to be $\boldsymbol{e}_1$ then, from Fig. A3.3, for the elemental area

$$\oint \boldsymbol{F}\cdot\mathrm{d}\boldsymbol{s} = F_2h_2\mathrm{d}q_2 - \left(F_2 + \frac{\partial F_2}{\partial q_3}\mathrm{d}q_3\right)\left(h_2\mathrm{d}q_2 + \frac{\partial h_2}{\partial q_3}\mathrm{d}q_3\mathrm{d}q_2\right)$$

$$- F_3h_3\mathrm{d}q_3 + \left(F_3 + \frac{\partial F_3}{\partial q_2}\mathrm{d}q_2\right)\left(h_3\mathrm{d}q_3 + \frac{\partial h_3}{\partial q_2}\mathrm{d}q_2\mathrm{d}q_3\right)$$

$$= \left(-\frac{\partial(F_2h_2)}{\partial q_3} + \frac{\partial(F_3h_3)}{\partial q_2}\right)\mathrm{d}q_2\mathrm{d}q_3$$

Thus, since $\mathrm{d}A_1 = h_2\mathrm{d}q_2h_3\mathrm{d}q_3$ we have

$$(\mathrm{curl}\,\boldsymbol{F})\cdot\boldsymbol{e}_1 = \frac{1}{h_2h_3}\left(\frac{\partial(F_3h_3)}{\partial q_2} - \frac{\partial(F_2h_2)}{\partial q_3}\right) \tag{A3.9}$$

with similar expressions for the other two components.

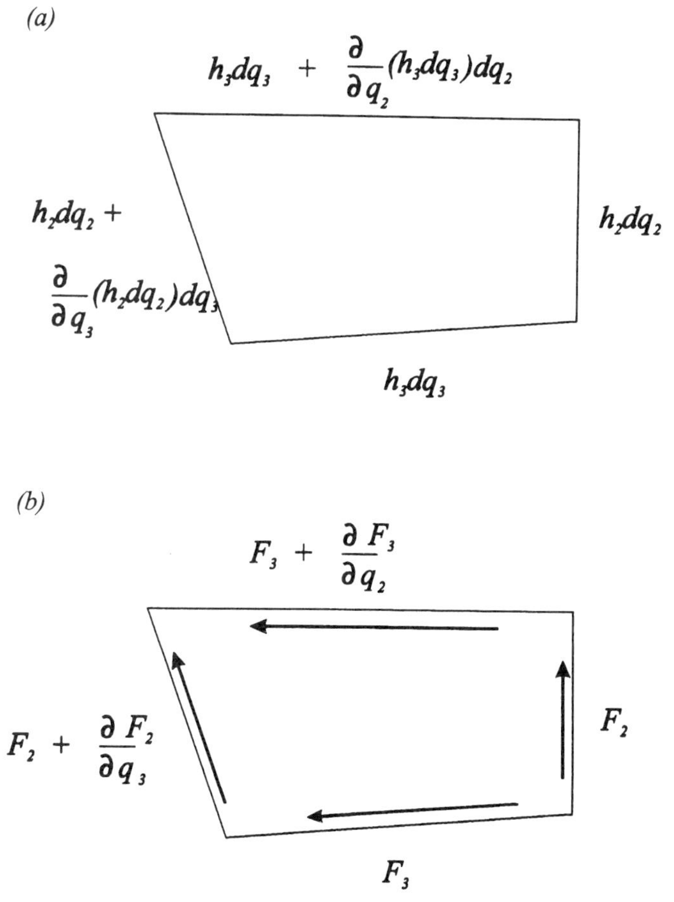

Fig. A3.3(a) and **(b)**

Gradient

The gradient of a scalar function f is defined so that its component in any given direction, $\boldsymbol{n}$, is the directional derivative in that direction so

$$\mathrm{grad}(f)\cdot\boldsymbol{n} = \frac{\partial f}{\partial s} \tag{A3.10}$$

where s is the distance along the curve parallel to $\boldsymbol{n}$. Now

$$\mathrm{d}\boldsymbol{r} = h_1 \mathrm{d}q_1 \boldsymbol{e}_1 + h_2 \mathrm{d}q_2 \boldsymbol{e}_2 + h_3 \mathrm{d}q_3 \boldsymbol{e}_3$$

so if we choose $\boldsymbol{n} = \boldsymbol{e}_1$

$$(\mathrm{grad}(f))\cdot\boldsymbol{e}_1 = \lim_{\Delta_s \to 0} \frac{\Delta f}{\Delta s} = \lim_{\Delta q_1 \to 0} \frac{\Delta f}{h_1 \Delta q_1} = \frac{1}{h_1}\frac{\partial f}{\partial q_1}$$

Thus

$$\mathrm{grad}(f) = \frac{1}{h_1}\frac{\partial f}{\partial q_1}\boldsymbol{e}_1 + \frac{1}{h_2}\frac{\partial f}{\partial q_2}\boldsymbol{e}_2 + \frac{1}{h_3}\frac{\partial f}{\partial q_3}\boldsymbol{e}_3 \tag{A3.11}$$

Cylindrical co-ordinates

From Fig. A3.4 we have

$$\boldsymbol{r} = r\boldsymbol{e}_r + z\boldsymbol{k}$$

Therefore

$$\mathrm{d}\boldsymbol{r} = \mathrm{d}r\boldsymbol{e}_r + r\,\mathrm{d}\boldsymbol{e}_r + \mathrm{d}z\,\boldsymbol{k}$$

but $\mathrm{d}\boldsymbol{e}_r = \mathrm{d}\theta\,\boldsymbol{e}_\theta$, so

$$\mathrm{d}\boldsymbol{r} = 1\,\mathrm{d}r\,\boldsymbol{e}_r + r\,\mathrm{d}\theta\,\boldsymbol{e}_\theta + 1\,\mathrm{d}z\,\boldsymbol{k}$$

Thus

$$h_1 = 1 \quad h_2 = r \quad \text{and} \quad h_3 = 1$$

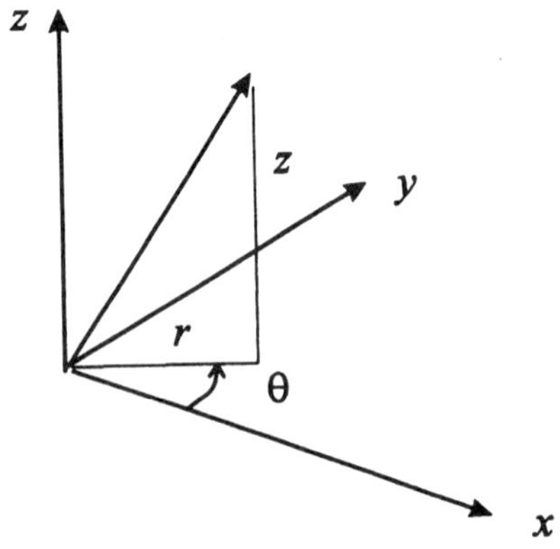

Fig. A3.4 Cylindrical co-ordinates

Spherical co-ordinates

Figure A3.5(a) shows one definition of the spherical co-ordinates

$$\boldsymbol{r} = r\,\boldsymbol{e}_r$$

and

$$\mathrm{d}\boldsymbol{r} = \mathrm{d}r\,\boldsymbol{e}_r + r\,\mathrm{d}\boldsymbol{e}_r$$

but

$$\mathrm{d}\boldsymbol{e}_r = r\cos\phi\,\mathrm{d}\theta\,\boldsymbol{e}_\theta + r\,\mathrm{d}\phi\,\boldsymbol{e}_\phi$$

so

$$\mathrm{d}\boldsymbol{r} = 1\,\mathrm{d}r\,\boldsymbol{e}_r + r\cos\phi\,\mathrm{d}\theta\,\boldsymbol{e}_\theta + r\,\mathrm{d}\phi\,\boldsymbol{e}_\phi$$

Thus the factors for the co-ordinates r, θ and ϕ are

$$h_1 = 1 \quad h_2 = r\cos\phi \quad \text{and} \quad h_3 = r$$

Figure A3.5(b) shows an alternative definition which leads to the following expressions for the factors for co-ordinates r, θ and ϕ

$$h_1 = 1 \quad h_2 = r \quad \text{and} \quad h_3 = r\sin\theta$$

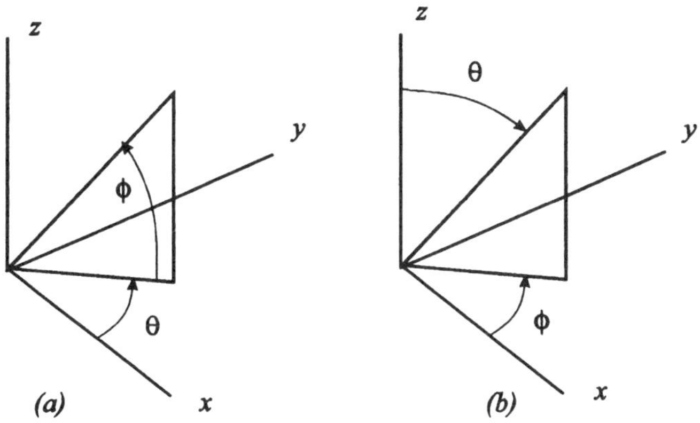

Fig. A3.5 Spherical co-ordinates

Expressions for div, grad and curl in cylindrical and spherical co-ordinates

Direct substitution of the scale factors into equations (A3.7), (A3.9) and (A3.11) will generate the required vector formulae.

Let

$$[\tilde{L}] = [h_1\ h_2\ h_3]_{\text{diag}}$$

$$[\tilde{A}] = [h_2h_3\ h_3h_1\ h_1h_2]_{\text{diag}}$$

and

$$\tilde{V} = h_1h_2h_3$$

which are the scale factors which relate to length, area and volume respectively. In these terms the general expressions are

$$\text{div}(F) = \tilde{V}^{-1}(\nabla)^{T}\{[\tilde{A}](F)\}$$

$$\text{curl}(F) = [\tilde{A}]^{-1}[\nabla]^{x}\{[\tilde{L}](F)\}$$

$$\text{grad}(f) = [\tilde{L}]^{-1}(\nabla)(F)$$

where

$$(\nabla) = \left(\frac{\partial}{\partial q_1} \ \frac{\partial}{\partial q_2} \ \frac{\partial}{\partial q_3}\right)^{T}$$

For spherical co-ordinates corresponding to Fig. A3.5(a)

$$\text{div}\ (F) = \frac{1}{r^2\cos\phi}\left(\frac{\partial}{\partial r}(r^2\cos\phi F_r) + \frac{\partial}{\partial\theta}(r\ F_\theta) + \frac{\partial}{\partial\theta}(F_\phi)\right)$$

$$= \frac{1}{r^2}\frac{\partial(r^2F_r)}{\partial r} + \frac{1}{r\cos\phi}\frac{\partial F_\theta}{\partial\theta} + \frac{1}{r\cos\phi}\frac{\partial(\cos\phi\ F_\phi)}{\partial\phi}$$

$$\text{curl}\ (F) = \frac{1}{r\cos\phi}\left(-\frac{\partial}{\partial\phi}(\cos\phi F_\theta) + \frac{\partial}{\partial\theta}(F_\theta)\right)e_r$$

$$+ \frac{1}{r}\left(\frac{\partial F_r}{\partial\phi} - \frac{\partial(rF_\theta)}{\partial r}\right)e_\theta$$

$$+ \frac{1}{r\cos\phi}\left(-\frac{\partial F_r}{\partial\theta} + \cos\phi\frac{\partial(rF_\theta)}{\partial r}\right)e_\phi$$

$$\text{grad}\ (f) = \frac{\partial f}{\partial r}e_r + \frac{1}{r\cos\phi}\frac{\partial f}{\partial\theta}e_\theta + \frac{1}{r}\frac{\partial f}{\partial\phi}e_\phi$$

For spherical co-ordinates corresponding to Fig. A3.5(b)

$$\text{div}\ (F) = \frac{1}{r^2}\frac{\partial(r^2F_r)}{\partial r} + \frac{1}{r\sin\theta}\frac{\partial(\sin\theta\ F_\theta)}{\partial\theta} + \frac{1}{r\sin\theta}\frac{\partial F_\phi}{\partial\phi}$$

$$\text{curl}\ (F) = \frac{1}{r\sin\theta}\left(\frac{\partial}{\partial\theta}(\sin\theta\ F_\phi) - \frac{\partial}{\partial\phi}(F_\theta)\right)e_r$$

$$+ \frac{1}{r\sin\theta}\left(\frac{\partial F_r}{\partial\phi} - \sin\theta\frac{\partial(rF_\phi)}{\partial r}\right)e_\theta$$

$$+ \frac{1}{r}\left(\frac{\partial(rF_\theta)}{\partial r} - \frac{\partial(F_r)}{\partial\phi}\right)e_\phi$$

$$\text{grad}\ (f) = \frac{\partial f}{\partial r}e_r + \frac{1}{r}\frac{\partial f}{\partial\theta}e_\theta + \frac{1}{r\sin\theta}\frac{\partial f}{\partial\phi}e_\phi$$

In cylindrical co-ordinates

$$\operatorname{div}(F) = \frac{1}{r}\frac{\partial}{\partial r}(rF_r) + \frac{1}{r}\frac{\partial}{\partial \theta}(F_\theta) + \frac{\partial}{\partial z}(F_z)$$

$$\operatorname{curl}(F) = \left(\frac{1}{r}\frac{\partial}{\partial \theta}F_z - \frac{\partial}{\partial z}F_\theta\right)\boldsymbol{e}_r$$

$$+ \left(\frac{\partial}{\partial z}F_r - \frac{\partial}{\partial r}F_z\right)\boldsymbol{e}_\theta$$

$$+ \left(\frac{1}{r}\frac{\partial}{\partial r}(rF_\theta) - \frac{1}{r}\frac{\partial}{\partial \theta}F_r\right)\boldsymbol{e}_z$$

$$\operatorname{grad}(f) = \frac{\partial f}{\partial r}\boldsymbol{e}_r + \frac{1}{r}\frac{\partial f}{\partial \theta}\boldsymbol{e}_\theta + \frac{\partial f}{\partial z}\boldsymbol{e}_z$$

Strain

In cylindrical co-ordinates

$$e_{rr} = \frac{\partial u_r}{\partial r} \qquad e_{\theta\theta} = \frac{1}{r}\frac{\partial u_\theta}{\partial \theta} \qquad e_{zz} = \frac{\partial u_z}{\partial z}$$

$$e_{\theta z} = \frac{1}{r}\frac{\partial u_z}{\partial \theta} + \frac{\partial u_\theta}{\partial z} \qquad e_{zr} = \frac{\partial u_r}{\partial z} + \frac{\partial u_z}{\partial r}$$

$$e_{r\theta} = \frac{\partial u_\theta}{\partial r} - \frac{u_\theta}{r} + \frac{1}{r}\frac{\partial u_r}{\partial \theta}$$

For spherical co-ordinates corresponding to Fig. A3.5(b)

$$e_{rr} = \frac{\partial u_r}{\partial r} \qquad e_{\theta\theta} = \frac{1}{r}\frac{\partial u_\theta}{\partial \theta} + \frac{u_r}{r}$$

$$e_{\phi\phi} = \frac{1}{r\sin\theta}\frac{\partial u_\phi}{\partial \phi} + \frac{u_\theta}{r}\cot\theta + \frac{u_r}{r}$$

$$e_{\phi\theta} = \frac{1}{r}\frac{\partial u_\phi}{\partial \theta} - \frac{u_\phi}{r}\cot\theta + \frac{1}{r\sin\theta}\frac{\partial u_\theta}{\partial \phi}$$

$$e_{\phi r} = \frac{1}{r\sin\theta}\frac{\partial u_r}{\partial \phi} + \frac{\partial u_\phi}{\partial r} - \frac{u_\phi}{r}$$

$$e_{r\theta} = \frac{\partial u_\theta}{\partial r} - \frac{u_\theta}{r} + \frac{1}{r}\frac{\partial u_r}{\partial \theta}$$

Stress

For any orthogonal co-ordinate system the stresses in an isotropic linear solid are related to the strains by

$$\sigma_{ii} = \lambda(e_{ii} + e_{jj} + e_{kk}) + 2\mu e_{ii}$$

$$\sigma_{jj} = \lambda(e_{ii} + e_{jj} + e_{kk}) + 2\mu e_{jj}$$

$$\sigma_{kk} = \lambda(e_{ii} + e_{jj} + e_{kk}) + 2\mu e_{kk}$$

and

$$\sigma_{ij} = \mu e_{ij}$$

$$\sigma_{jk} = \mu e_{jk}$$

$$\sigma_{ki} = \mu e_{ki}$$

where λ and μ are the Lamé constants.

In Cartesian co-ordinates $i = x, j = y$ and $k = z$. For cylindrical co-ordinates $i = r, j = \theta$ and $k = z$. In spherical co-ordinates $i = r, j = \theta$ and $k = ø$.

Bibliography

Arnold and Maunder, L., 1961: *Gyrodynamics*. Academic Press.

Clough, R.W. and Penzien, J., 1975: *Dynamics of structures*. McGraw-Hill.

Dugas, R., 1986: *A history of mechanics*. Dover.

Einstein, A., 1967: *The meaning of relativity*, 6th edn. Chapman and Hall (or Princeton University Press, 5th edn, 1955).

Ellis, J.R., 1969: *Vehicle dynamics*. London Business Books.

Fu, K.S., Gonzalez, R.C. and Lee, C.S.G., 1987: *Robotics*. McGraw-Hill.

Ginsberg, J.H. 1995: *Advanced engineering dynamics*, 2nd edn. Cambridge University Press.

Goldsmith, W. 1960: *Impact*. Edward Arnold.

Goldstein, H. 1980: *Classical mechanics*, 2nd edn. Addison-Wesley.

Harrison, H.R. and Nettleton, T., 1994: *Principles of engineering mechanics*, 2nd edn. Edward Arnold.

Houghton, E.L. and Carruthers, N.B. 1982: *Aerodynamics*, 3rd edn. Edward Arnold.

Hunter, S.C. 1983: *Mechanics of continuous media*. Simon and Schuster.

Johnson, W. 1972: *Impact strength of materials*. Edward Arnold.

Kolsky, H. 1953: *Stress waves in solids*. Oxford University Press.

Longair, M.S. 1984: *Theoretical concepts in physics*. Cambridge University Press.

McCloy, D. and Harris, M. 1986: *Robotics: An introduction*. Open University Press.

Meirovitch, L. 1967: *Analytical methods in vibrations*. McGraw-Hill.

Meirovitch, L. 1970: *Methods of analytical dynamics*. McGraw-Hill.

Rao, S.R., 1986: *Mechanical vibrations*. Addison-Wesley.

Redwood, M., 1960: *Mechanical waveguides*. Pergamon.

Routh, E.J., 1891/1892: *Dynamics of a system of rigid bodies*, Part 1/Part 2. Macmillan.

Snowdon, 1964: *Shock and vibration in damped mechanical systems*. Wiley.

Symon, K.R., 1971: *Mechanics*. 3rd edn. Addison-Wesley.

Thomson, W.T. 1986: *Introduction to space dynamics*. Dover.

Udwadia, F.E., and Kalaba, R.E. 1996: *Analytical dynamics, a new approach*. Cambridge University Press.

Index